THE NATURE OF SC

Science & Technology Education Library

VOLUME 5

SERIES EDITOR

Ken Tobin, *University of Pennsylvania, Philadelphia, USA*

EDITORIAL BOARD

Beverley Bell, *University of Waikato, Hamilton, New Zealand*
Reinders Duit, *University of Kiel, Germany*
Kathlene Fisher, *San diego State University, California, USA*
Barry Fraser, *Curtin University of Technology, Perth, Australia*
Chao-Ti Hsiung, *National Taipei Teachers College, Australia*
Doris Jorde, *University of Oslo, Norway*
Michael Khan, *Centre for Education Policy Development, Braamfontein, South Africa*
Vince Lunetta, *Pennsylvania State University, University Park, Pennsylvania, USA*
Pinchas Tamir, *Hebrew University, Jerusalem, Israel*

SCOPE

The book series *Science & Technology Education Library* provides a publication forum for scholarship in science and technology education. It aims to publish innovative books which are at the forefront of the field. Monographs as well as collections of papers will be published.

The titles published in this series are listed at the end of this volume.

The Nature of Science in Science Education

Rationales and Strategies

edited by

WILLIAM F. McCOMAS

*University of Southern California,
Los Angeles, California, U.S.A.*

KLUWER ACADEMIC PUBLISHERS
DORDRECHT / BOSTON / LONDON

A C.I.P. Ctalogue record for this book is available from the Library of Congress.

ISBN 0-7923-6168-7 (PB)

Published by Kluwer Academic Publishers,
P.O. Box 17, 3300 AA Dordrecht, The Netherlands.

Sold and distributed in North, Central and South America
by Kluwer Academic Publishers,
101 Philip Drive, Norwell, MA 02061, U.S.A.

In all other countries, sold and distributed
by Kluwer Academic Publishers,
P.O. Box 322, 3300 AH Dordrecht, The Netherlands.

Printed on acid-free paper

All Rights Reserved
© 2000 Kluwer Academic Publishers
No part of the material protected by this copyright notice may be reproduced or
utilized in any form or by any means, electronic or mechanical,
including photocopying, recording or by any information storage and
retrieval system, without written permission from the copyright owner.

Printed in the Netherlands.

TABLE OF CONTENTS

Acknowledgments ix

Foreword and Introduction xi
Michael Matthews

SECTION I: RATIONALES FOR THE NATURE OF SCIENCE IN SCIENCE INSTRUCTION

1. The Role And Character of The Nature of Science in Science Education 3
   William F. McComas, Michael P. Clough & Hiya Almazroa

2. The Nature of Science in International Science Education Standards Documents 41
   William F. McComas & Joanne K. Olson

3. The Principal Elements of the Nature of Science: Dispelling the Myths 53
   William F. McComas

SECTION II: COMMUNICATING THE NATURE OF SCIENCE PLANS, APPROACHES AND STRATEGIES

4. The Card Exchange: Introducing the Philosophy of Science 73
   William W. Cobern & Cathleen C. Loving

5. Avoiding De-Natured Science: Activities that Promote Understandings of the Nature of Science 83
   Norman Lederman & Fouad Abd-El-Khalick

TABLE OF CONTENTS

6. Confronting Students' Conceptions of the Nature of Science 127
    with Cooperative Controversy
    Penny L. Hammerich

7. Nature of Science Activities Using the Scientific Theory Profile: 137
    From the Hawking-Gould Dichotomy to a Philosophy Checklist
    Cathleen C. Loving

8. Learning By Designing: A Case of 151
    Heuristic Theory Development in Science Teaching
    Fred Jansen & Peter Voogt

9. Using Historical Case Studies in Biology to Explore the 163
    Nature of Science: A Professional Development Program
    for High School Teachers
    Karen R. Dawkins & Allan A. Glatthorn

10. A History of Science Approach to the Nature of Science: 177
    Learning Science by Rediscovering It
    Nahum Kipnis

11. Integrating the Nature of Science with Student Teaching: 197
    Rationales and Strategies
    Michael P. Clough

PART III: COMMUNICATING THE NATURE OF SCIENCE COURSES AND COURSE ELEMENTS

12. A Thematic Introduction to the Nature of Science: The Rationale 211
    and Content of a Course for Science Educators
    William F. McComas

13. The Nature of Science: 223
    Achieving Science Literacy by Doing Science
    John O. Matson & Sharon Parsons

14. Elementary Science Methods: Developing and Measuring 231
    Student Views about the Nature of Science
    Yvonne Meichtry

TABLE OF CONTENTS vii

15.	Nature of Science: Implications for Education: An Undergraduate Course for Prospective Teachers *Karen Sullenger & Steve Turner*	243
16.	The Use of Real and Imaginary Cases in Communicating the Nature of Science: A Course Outline *David Boersema*	255
17.	Teaching the Nature of Science as an Element of Science, Technology and Society *Barbara Spector, Paschal Strong & Thomas LaPorta*	267
18.	Of Starting Points and Destinations: Teacher Education and the Nature of Science *Michael L. Bentley & Steve C. Fleury*	277
19.	A Programme for Developing Understanding of the Nature of Science in Teacher Education *Mick Nott & Jerry Wellington*	293
20.	The Nature of Science as a Foundation for Teaching Science: Evolution as a Case Study *Craig E. Nelson, Martin K. Nickels & Jean Beard*	315

SECTION IV: ASSESSING NATURE OF SCIENCE UNDERSTANDING

21.	Assessing Understanding of the Nature of Science: A Historical Perspective *Norman Lederman, Philip Wade & Randy L. Bell*	331
Notes on Contributors		351
Index		361

ACKNOWLEDGMENTS

The editor wishes to acknowledge

The seminal contributions of Michael Martin and James Robinson who began the conversation about the relevance of the nature of science to science education;

Richard Duschl, Derek Hodson, Michael Matthews and Norman Lederman who, through their continuing scholarship, stimulate and inform the study of the nature of science in science education;

René Stofflett, University of Illinois at Urbana-Champaign and J. Steve Oliver, of the University of Georgia, the external reviewers, whose suggestions greatly improved the manuscript;

Michael Clough for insightful conversation and countless useful suggestions;

George Cossman of the University of Iowa, who awakened the editor's interest in this important field;

Beverly Franco who helped prepare the manuscript for publication and struggled with me through the countless revisions;

Joanne Olson, who helped greatly with the index and in polishing the manuscript;

Ken Tobin, for his support of this project as the Science & Technology Education Library series editor;

The publications board of the Association for the Education of Teachers in Science who have supported the development of the book from the outset;

Joy Carp and Irene van den Reydt of Kluwer Academic Publishers who have responded with patience to my countless questions; and

The many authors who have developed the plans and strategies contained in this book and then entrusted me with their innumerable excellent ideas. To you, most sincerely, I offer my profound thanks.

MICHAEL R. MATTHEWS

FOREWORD AND INTRODUCTION

It is a pleasure to contribute the foreword to this book. The editor has done science education a service by overseeing and coordinating the publication of perhaps the first anthology whose contributions address applied, pedagogical and curricular aspects of teaching school students and trainee teachers about the nature of science. Although there has been a long tradition of theoretical writing on the subject, of establishing the cultural, educational and scientific benefits of teaching about the nature of science – just how this is to be put into practice has been less attended to.

The theoretical tradition includes Ernst Mach's arguments in the late nineteenth century (Mach 1886/1943), John Dewey's work at the turn of the century (Dewey 1910), F. W. Westaway's publications in the nineteen-twenties (Westaway 1919, 1929), Joseph Schwab's writings in the nineteen-forties and nineteen-fifties (Schwab 1945, 1958), the books of Leo Klopfer and James Robinson in the nineteen-sixties (Klopfer 1969; Robinson 1968), the publications of Jim Rutherford and Michael Martin in the nineteen-seventies (Rutherford 1972; Martin 1972), and more recently the work of Derek Hodson, Rick Duschl, Norman Lederman, Joan Solomon, and a number of others, including myself, many of whom are associated with the International History, Philosophy and Science Teaching Group and whose work has appeared in the journal *Science & Education* (Hodson 1986, 1988; Duschl 1985, 1990; Lederman 1986, 1992; Solomon 1989, 1991; Matthews 1992, 1994, 1998).

This theoretical tradition has not been without some advice for teachers about how to address matters concerning the nature of science, but this advice has generally been in the margins. Which may be one reason why the theoretical tradition itself has been marginalised within the science education community: teachers reasonably enough require not only the destination, but some roadmap showing how to get there. Pleasingly, contributors to this anthology directly address these neglected and practical classroom matters.

Questions about the nature of science have long been of concern to science teachers and curriculum developers. It has been hoped that science education would enhance society and have a beneficial impact on the quality of culture and public life in virtue of students appreciating something of the nature of science, internalising something of the scientific spirit, and developing a scientific frame of mind that might carry over into other spheres of life. John Dewey expressed well this Enlightenment hope for science education when he said:

> Our predilection for premature acceptance and assertion, our aversion to suspended judgment, are signs that we tend naturally to cut short the process of testing. We are satisfied with superficial and immediate short-visioned applications. . . . Science represents the safeguard of the race against these natural propensities and the evils which flow from them. . . . It is artificial (an acquired art), not spontaneous; learned, not native. To this fact is due the unique, the invaluable place of science in education. (Dewey 1916, p. 189)

Some, with a more narrow focus, maintain that if we want students to learn and become competent in science, then they must be taught something about the nature of science. For instance, Frederick Reif in a recent publication, has said:

> All too often introductory physics courses 'cover' numerous topics, but the knowledge actually acquired by students is often nominal rather than functional. If students are to acquire basic physics knowledge. . . It is necessary to understand better the requisite thought processes and to teach these more explicitly. . . . if one wants to improve significantly students' learning of physics. . . . It is also necessary to modify students' naive notions about the nature of science. (Reif 1995, p. 281)

Curriculum documents in the US (especially Project 2061 and the National Science Standards), Canada (especially Science for Every Student), and Europe (particularly those from UK, Denmark, and Spain) are giving increased prominence to students understanding the nature of science. William McComas and Joanne Olson, in their contribution to this anthology, discuss some of these international curricula (but not the European material, discussions of which can be found in Solomon (1991) and Nielsen and Thomson (1991)).

There has been an imposing amount of research conducted over many decades concerning the curricular, pedagogical and educational aspects of the nature of science. Much of this research tradition is reviewed in Lederman (1992). Students' understanding of the nature of science have been studied (Mackay 1971; Rubba, Horner & Smith 1981; Griffith & Barman 1995; Griffith & Barry 1993; Solomon, Scott & Duveen 1996). Teachers' epistemologies and beliefs about the nature of science have been documented (Kimball 1967; Carey & Strauss 1968; Rowell & Cawthron 1982; Koulaidis & Ogborn 1989, 1995; Gallagher 1991; Lederman 1986; Lakin & Wellington 1994). The impact, or otherwise, of teachers' epistemology on their classroom practice has been debated (Lederman & Zeidler 1987; Duschl & Wright 1989; Brickhouse 1989; Mellado 1997; Tobin & McRobbie 1997). The influence of students' epistemologies on their learning of science has been investigated (Hammer 1995). The history of the linkage between curricular definitions of scientific literacy and knowledge of the nature of science has been documented (Meichtry 1993). The classroom processes whereby teachers' beliefs about the nature of science interact with curriculum and influence students' epistemologies have been researched (Tobin, Tippins & Hook 1994; Mellado 1997). The justification, effectiveness, and practicality of nature of science courses in teacher education programmes have been canvassed (Scheffler 1970; Loving 1991; Clemison 1990; Eichinger, Abell & Dagher 1997). And the epistemological assumptions, or commitments, underlying various 'nature of science' tests have been identified and criticized by Lederman, Wade and Bell in this volume and by others

(Lucas, 1975; Doran, Guerin & Cavalieri, 1974; Rubba & Anderson, 1978).
Nature of science questions are becoming, at the end of the century, perhaps more pressing than they were during the period from Mach to Martin. During those one hundred years there was a degree of cultural and philosophical unanimity about the nature and purpose of science. Of course there was some dispute about the topic – inductivists versus falsificationists, positivists versus realists, Kuhnians versus Popperians, empiricists versus rationalists, Bernalian state-interventionists versus Polanyian free-enterprisers, etc. – but these disputes were basically domestic ones. There was general agreement that science was a good thing, that it was a cognitive enterprise abiding by intellectual standards, that it valued objectivity, that it sought to find truths about the world, and that it gave us the best possible understanding of nature and reality. Robert Merton's characterization of science as: open minded, universalist, disinterested and communal (Merton 1942) summed up professional and lay opinion on the matter. Anyone who has been reading anything of the 'Science Wars' that have been raging in professional journals and the popular press, will know that times have changed concerning this hitherto cultural agreement about science. (If you have missed the war and want to catch up, see Gross and Levitt 1994; Gross, Levitt and Lewis 1996.)

A well known Canadian science educator, Glen Aikenhead, has recently stated, 'the social studies of science' reveal science as:

> . . . mechanistic, materialist, reductionist, empirical, rational, decontextualized, mathematically idealized, communal, ideological, masculine, elitist, competitive, exploitive, impersonal, and violent. (Aikenhead 1997, p. 220)

The contrast with Dewey, Merton, and the orthodox tradition, is palpable. Times certainly have changed. And science teachers, and science educators (the teachers of the teachers), are caught up in the war. Teachers, as usually happens when cultural and social wars erupt, suffer collateral damage.

The science teacher has to master not only his or her subject matter, and the techniques of making it interesting and intelligible to students, but also has to get some grip on orthodox, or we might say modernist, understandings of the nature of science, and now the legitimacy, or otherwise, of postmodernist challenges to the orthodox understanding. This is a hard call, but it cannot be avoided. The business of science teaching cannot be conducted as usual; or if it is so conducted, then it must be done with recognition of, and with answers to, the challenges posed by postmodernist critics. To do otherwise is professionally irresponsible.

That these epistemological debates concerning the nature of science are not educationally idle has been dramatically shown in decisions by a string of major USA cities to adopt the *Portland Baseline Essays* in their science programmes (Martel 1991). The science component of these essays lists a number of commitments that are flatly at odds with a Western-scientific view of the world (see critical discussion in Good and Demastes 1995, and Montellano 1996). And the claim is made that the

Western view is just one of a number of equally valid scientific views, and thus the purportedly African science contained in the *Portland Baseline Essays* should be included in the curriculum. For black children, and others in these cities, the value of their education hinges on the correctness or otherwise of teachers' and administrators' decisions regarding the historical and epistemological claims made in the *Baseline Essays*. Of the *Essays,* one critic has said:

> People who are genuinely concerned with improving science education in the schools and with increasing the number of minority scientists should vigorously oppose the inclusion of material into the curriculum that makes unsupported claims, introduces religion under the guise of science, and claims that the paranormal exists. The critical need to increase the supply of minority scientists requires that they be taught science at its best rather than a parody. (Montellano 1996, p. 569)

The ability to distinguish good science from parodies and pseudoscience depends on a grasp of the nature of science. More generally, the widespread debates occurring over multi cultural science also, in part, hinge upon answers to the question: What is the nature of science? That is, there is an epistemological question at issue in the debate, which unfortunately often gets confused with ethical and political questions (Matthews 1994, chap. 9; Siegel 1997). The Creation Science debates and trials in the USA in large part hinged upon answers to the question: What is the nature of science? This was explicitly the case in the 1981 Little Rock trial over the constitutionality of Arkansas's Act 590 requiring equal time for the teaching of creation science and evolutionary science (see contributions to Ruse 1988).

Similarly, nature of science questions are at the heart of many feminist critiques of contemporary science and of feminist proposals for the reform of science curricula. Feminism has long ceased to be merely a political movement; it also encompasses a variety of epistemological positions, some of which are in open conflict with mainstream or orthodox understanding of science (see, for instance, contributions to Nelson and Nelson (1996)). Sandra Harding's 'standpoint' feminist epistemology is an example of one such position that has had some influence in science education circles (Harding 1986, 1991). But it also needs to be recognized that there are many critics of feminist epistemology who maintain that feminists have got both the history and the philosophy of science wrong, and further that they draw wrong conclusions from the parts that they get right (Koertge 1981, 1996; Pinnick 1994). For instance, Noretta Koertge, a feminist and one of the first philosophers of science to write on science education (Koertge 1969) has said, in one piece critical of feminist philosophy of science, that:

> If it really could be shown that patriarchal thinking not only played a crucial role in the Scientific Revolution but is also necessary for carrying out scientific inquiry as we now know it, that would constitute the strongest argument for patriarchy that I can think of. (Koertge 1981, p. 354)

Epistemological questions are also prompted by much environmentalist, 'deep ecology', and 'new science' writing to which students are increasingly exposed. It is a rare science teacher who has not in the recent past been asked: 'Do you believe in the Gaia hypothesis?' or, 'Does the Big Bang prove God's existence?' or, 'Do whales have rights?

Further there has been increased stress on metacognitive awareness and epistemological development as important outcomes of science instruction (White & Gunstone 1989). In the words of Rosalind Driver and her co-workers at Leeds University, 'Science learning, viewed from a constructivist perspective, involves epistemological as well as conceptual development' (Driver *et al.* 1994, p. 219). Any specification of 'epistemological development' brings to the fore questions about the methodology and nature of science. Just as one cannot specify 'moral development' without some normative account of morality, so too one cannot identify 'epistemological development' without some normative account of epistemology.

Perhaps more than anything else, what has pushed nature of science considerations to the forefront of contemporary science education, is the prominence that constructivism has gained in the science (and mathematics and literature) education community. Peter Fensham, a well-placed observer has said: 'The most conspicuous psychological influence on curriculum thinking in science since 1980 has been the constructivist view of learning' (Fensham 1992, p. 801). Russell Yeany, another keen observer, has said: 'A unification of thinking, research, curriculum development, and teacher education appears to now be occurring under the theme of constructivism' (Yeany 1991, p. 1). And Ken Tobin, a past president of the National Association for Research in Science Teaching, has remarked that: 'constructivism has become increasingly popular . . . in the past ten years . . . it represents a paradigm change in science education' (Tobin 1993, p. ix).

I have argued in a number of places (Matthews 1993, 1994, chap. 7, 1997) that constructivism is at its core, as it was with Piaget, an epistemological doctrine and it is standardly coupled with commitments to certain views (postpositivist, postmodernist, anti-realist, instrumentalist) about the nature of science. The issues here are complex, (see Volume Six, Numbers One and Two of the journal *Science & Education*), but for the present purpose it suffices to say that science teachers and science educators who engage with, or adopt, constructivist philosophy, must address the central issue of the nature of science. As Peter Slezak has written:

There could scarcely be a more fundamental contribution to science education than the one offered by constructivist sociological theories, since they purport to overturn 'the very idea' of science as a distinctive intellectual enterprise with its special values. (Slezak 1994, p. 291)

The foregoing discussion continues the *theoretical* analysis of the place of nature of science considerations in science teaching. What needs to be done, and what is done in this anthology, is move on to the *practical* matter of how to teach about the nature of science, how to engage students in the debate or conversation about the nature of science. Karen Sullenger and Steve Turner, in their contribution, have nicely articulated the matter in saying:

> ... we wanted to find a way to encourage preservice teachers to clarify and articulate their own, tacit notions of science and teaching science by challenging them with alternative conceptions of science held within the history of science, science education, and science communities. ... We wanted to involve them as participants in these debates. Further, we wanted prospective teachers of science to construct more complex understandings of science, particularly the understanding of science they would us as a basis for making pedagogical decisions.

It is in everyone's interest that students, and trainee teachers, develop 'a more complex understanding of science'. William McComas's contribution on 'Dispelling the Myths of Science' is a nice beginning to this task. We all know science is a complex matter: that metaphysics is an indispensable part of science, that theories are under-determined by empirical evidence, that science makes enormous use of metaphorical reasoning, that observation is colored by theoretical preconceptions, that political and personal factors enter into scientific decision making, and so on. Science effects culture, and is in turn effected by it. Any half-decent study of Newton's mechanics illustrates all of these matters.

But recognition of this complexity simply does not mean that 'anything goes', or 'all ideas and world views are equal', or 'science is just a social construction'. The recognition that science is a complex human construction by no means implies the relativism and instrumentalism frequently associated with constructivist philosophy. The choice is not, as it is often put, simplistic understanding of science, or constructivist understanding of science. If, in the course of studying science, children are routinely made aware of its intellectual complexity, then they might be more immune to snake-oil merchants, and others, who gain adherents by promising to initiate followers into complex and 'mind-blowing' ideas. If children have thought carefully enough about simple old Newtonian 'action-at-a-distance', then their interest in complexity and metaphysics is being productively encouraged.

One important consideration in teaching about the nature of science is that pointed out by Karen Dawkins and Allan Glatthorn in their contribution, namely:

> Since it takes historians and philosophers of science decades of study to derive generalizations about the nature of science from the historical evidence, it is unrealistic to expect teachers or students to develop profound understandings from a necessarily limited exposure to the world of science.

It is unrealistic to expect students, or trainee teachers, to become competent historians, sociologists or philosophers of science. We should have limited aims: a more complex understanding of science, not a total, or even a very complex, understanding. There are numerous low-level questions that students can be engaged by: What is a scientific explanation? What is a controlled experiment? What is a crucial experiment? How do models function in science? How much confirmation does a hypothesis require before it is established? Are there ways of evaluating the worth of competing research programmes? Did Newton's religious belief affect his science? Was Darwin's 'damaged book' analogy a competent reply to critics who pointed to all the evidence that contradicted his evolutionary theory? Was Planck culpable for remaining in Nazi Germany and continuing his scientific research during the war? There is no need to overwhelm students with 'cutting-edge' questions. They have to crawl before they can walk, and walk before they can run. This is no more than commonsensical pedagogical practice: simple pendulums are dealt with before compound pendulums, addition and subtraction before multiplication and division, Euclidean geometry before non-Euclidean geometry, proportional reasoning is mastered before algebra, differentiation before integration, and so on. It may be that students do not get beyond the simple questions, but teachers should feel sanguine about that. Rather, students genuinely struggle to grasp the simple questions rather than just repeat popular nostrums, or their teacher's prejudice about the complex questions. Voicing opinions about postmodernism, when the basics of modernism have not been understood, is akin to the rattling of an empty vessel.

To this end, the contributions in Part II of the anthology, are ideal. Following Mach last century, and James Conant and Leo Klopfer in the nineteen-fifties, I personally believe that historical case studies are ideal vehicles for achieving this limited, but expandable, goal. A good number of such case studies are included in this anthology.

These considerations lead onto another question which, unfortunately, is often not asked. This question is, "What is the purpose in teaching about the nature of science, or including nature of science considerations in the teacher-education curriculum?" The question seems so straight-forward as to not be worth asking. But there is an unfortunate tendency in current writing that suggests that this question needs to be attended to.

The tendency is one which says: 'The reason we want teachers to know more about the constructivist / positivist / realist / feminist / Marxist / multiculturalist / universalist nature of science is so that their students will better know or embrace constructivist / positivist / realist / feminist / Marxist / multiculturalist / universalist views about the nature of science'. That is, there is a tendency that says that the educational purpose of promoting any particular view about the nature of science among teachers is so they can, in turn, promote it among their students. This view needs to be questioned. At one level it is profoundly anti-educational.

Naturally we like people to believe what we believe. Be it in politics, morals, religion, economics, environmental matters, or whatever else. Teachers have their share of this tendency. The problem for teachers arises when this natural tendency nullifies their role as educators.

John Locke expressed this Socratic point in his 1689 *Essay Concerning Human Understanding*, where he said:

> The floating of other men's opinions in our brains makes us not one jot more knowing, though they happen to be true. What in them was science is in us but opiniatertry, whilst we give up our assent only to reverend names, and do not, as they did, employ our own reason to understand those truths which gave them reputation. (Locke 1690/1924, p. 40)

And then proceeded, memorably, to say:

> Such borrowed wealth, like fairy money, though it be gold in the hand from which he received it, will be but leaves and dust when it comes to use. (ibid)

If Locke's argument holds against *truths* it certainly holds against *contested* or *false* opinions. Bringing epistemology and philosophy into focus in science education, and putting the nature of science into curriculum documents, will be to no great avail if it merely becomes the occasion for students repeating the opinions of their teachers. If HPS in science teaching becomes a catechism – like *dimat* in the former Soviet Union – then it defeats its potentially major educative purpose.

Bertrand Russell, an out-spoken and fierce partisan of numerous philosophical and social causes, recognized the anti-educational effect of indoctrination. In his essay *On Education*, written during the First Great War, he pointed out that:

> If the children themselves were considered, education would not aim at making them belong to this party or that, but at enabling them to choose intelligently between the parties; it would aim at making them able to think, not at making them think what their teachers think ... we should educate them so as to give them the knowledge and the mental habits required for forming independent opinions... (Egner & Denonn, 1961, p. 401)

Israel Scheffler (1973) and Harvey Siegel (1988) have elaborated this basic Platonic insight in terms of reason-giving being the *raison d'être* of education. Education, as distinct from indoctrination, conditioning, brain-washing etc., is marked by students having a concern for reasons and for the justification of beliefs. And these in turn depend upon free and informed inquiry. As Plato so long ago insisted, education is not just the having of correct beliefs, it is the having of adequate reasons for these beliefs. Without informed and adequate reasons, beliefs are just Locke's 'fairy money'.

Science programmes in any scheme of liberal education will include discussion of the nature of science – its history, methodology, philosophy, social and cultural impacts, and its relation to other forms of knowledge. And all but the most impoverished technical programmes will also include aspects of the nature of science – scientific method, hypothesis generation and testing, and so on. The art of the teacher is to judge the sophistication of his or her students, and present aspects of the nature of science that are intelligible to them without being overwhelming.

This anthology goes a long way towards making available to teachers intelligible, but not overwhelming, nature of science material that can be utilised in classrooms.

University of New South Wales, Sydney, Australia

REFERENCES

Aikenhead, G.S. (1997). 'Towards a first nations cross-cultural science and technology curriculum', *Science Education* (81), 217-238.

Brickhouse, N.W. (1989). 'The teaching of the philosophy of science in secondary classrooms: Case studies of teachers personal theories', *International Journal of Science Education* (11), 437-450.

Carey, R.L., & Strauss, N.G. (1968). 'An analysis of the understanding of the nature of science by secondary school science teachers', *Science Education* (58), 358-363.

Cleminson, A. (1990). 'Establishing an epistemological base for science teaching in the light of contemporary notions of the nature of science and of how children learn science', *Journal of Research in Science Teaching* (27), 429-445.

Dewey, J. (1910). 'Science as subject-matter and as method', *Science* (31), 121-127, Reproduced in *Science & Education*, 1995, (4), 391-398.

Dewey, J. (1916). *Democracy and education*, New York, Macmillan Company.

Doran, L., Guerin, O. & Cavalieri, J. (1974). 'An analysis of several instruments measuring "Nature of science" Objectives', *Science Education* (58), 321-329.

Driver, R. et al. (1994). 'Working from children's ideas: Planning and teaching a chemistry topic from a constructivist perspective', in P. Fensham, R. Gunstone & R. White (eds.), *The content of science: A constructivist approach to its teaching and learning*, London, Falmer Press, 201-220.

Duschl, R.A. (1985). 'Science education and philosophy of science twenty-five, years of mutually exclusive development', *School Science and Mathematics* (87), 541-555.

Duschl, R.A. (1990). *Restructuring science education: The importance of theories and their development*, New York, Teachers College Press.

Duschl, R.A. & Wright, E. (1989). 'A case study of high school teachers' Decision making models for planning and teaching science', *Journal of Research in Science Teaching* (26), 467-501.

Egner, R.E. & Denonn, L.E. (eds.), (1961). *The basic writings of Bertrand Russell*, London, George Allen & Unwin.

Eichinger, D.C., Abell, S.K., & Dagher, Z.R. (1997). 'Developing a graduate level science education course on the nature of science', *Science & Education* (6), 417-429.

Fensham, P.J. (1992). 'Science and technology'. In P.W. Jackson (ed.), *Handbook of Research on Curriculum*, New York, Macmillan, pp. 789-829.

Gallagher, J. (1991). 'Prospective and practicing secondary school science teachers' knowledge and beliefs about the philosophy of science of science', *Science Education* (75), 121-134.

Good, R.G. & Demastes, S. (1995). 'The diminished role of nature in postmodern views of science and science education', in F. Finley et al. (eds.), *Proceedings of the Third International History, Philosophy, and Science Teaching Conference*, Minneapolis, MN, University of Minnesota, 480-487.

Griffiths, A.K. & Barman, C.R. (1995). 'High school students' views about the nature of science: Results from three countries', *School Science and Mathematics* (95), 248-255.

Griffiths, A.K. & Barry, M. (1991). 'Secondary school students' understanding of the nature of science', *Research in Science Education* (21), 141-150.

Gross, P.R. & Levitt, N. (1994). *Higher superstition: The academic left and its quarrels with science*, Baltimore, MD, Johns Hopkins University Press.

Gross, P.R., Levitt, N. & Lewis, M.W. (eds.), (1996). *The flight from science and reason*, New York, New York Academy of Sciences, (distributed by Johns Hopkins University Press, Baltimore, MD).

Hammer, D. (1995). 'Epistemological considerations in teaching introductory physics', *Science Education* (79)), 393-413.

Harding, S.G. (1986). *The science question in feminism*, Ithaca, NY, Cornell University Press.

Harding, S.G. (1991). *Whose science? Whose knowledge?*, Ithaca, NY, Cornell University Press.

Hodson, D. (1986). Philosophy of science and the science curriculum', *Journal of Philosophy of Education* (20), 241-251. Reprinted in M.R. Matthews (ed.), *History, Philosophy and Science Teaching: Selected Readings*, Toronto, OISE Press, 1991.

Hodson, D. (1988). 'Toward a philosophically more valid science curriculum', *Science Education* (72), 19-40.

Kimball, M. (1967). 'Understanding the nature of science: A comparison of scientists and science teachers', *Journal of Research in Science Teaching* (5), 110-120.

Klopfer, L.E. (1969). *Case histories and science education*, San Francisco, CA, Wadsworth Publishing Company.

Koertge, N. (1969). 'Towards an integration of content and method in the science curriculum', *Curriculum Theory Network* (4), 26-43. Reprinted in *Science & Education* (5), 391-402, (1996).

Koertge, N. (1981). 'Methodology, ideology and feminist critiques of science', in P. D. Asquith & R. N. Giere (eds.), *Proceedings of the Philosophy of Science Association 1980*, Ann Arbor, MI, Edwards Bros, 346-359.

Koertge, N. (1996). 'Feminist epistemology: Stalking an un-dead horse', in P.R. Gross, N. Levitt & M.W. Lewis (eds.), *The flight from science and reason*, New York, New York Academy of Sciences, 413-419.

Koulaidis, V. & Ogborn, J. (1989). 'Philosophy of science: An empirical study of teachers' views', *International Journal of Science Education* (11), 173-184.

Koulaidis, V. & Ogborn, J. (1995). 'Science teachers' philosophical assumptions: How well do we understand them?', *International Journal of Science Education* (17), 273-283.

Lakin, S. & Wellington, J. (1994). 'Who will teach the "nature of science"?: Teachers' views of science and their implications for science education', *International Journal of Science Education* (16), 175-190.

Lederman, N.G. (1986). 'Students' and teachers' understanding of the nature of science: A reassessment', *School Science and Mathematics* (86), 91-99.

Lederman, N.G. (1992). 'Students' and teachers' conceptions of the nature of science: A review of the research', *Journal of Research in Science Teaching* (29), 331-359.

Lederman, N.G. & Zeidler, D.L. (1987). 'Science teachers conceptions of the nature of science: Do they really influence teaching behaviour?', Science Education (71), 721-734.

Locke, J. (1690/1924). *An essay concerning human understanding*, abridged and edited by A.S. Pringle-Pattison, Oxford, Clarendon Press.

Loving, C.G. (1991). 'The scientific theory profile: A philosophy of science model for science teachers', *Journal of Research in Science Teaching* (28), 823-838.

Lucas, A.M. (1975). 'Hidden assumptions in measures of "Knowledge about science and scientists"', *Science Education* (59), 481-485.

Mach, E. (1886/1943). 'On instruction in the classics and the sciences'. In his *Popular Scientific Lectures*, LaSalle, Open Court Publishing Company..

Mackay, L.D. (1971). 'Development of understanding about the nature of science', Journal of Research in Science Teaching (8), 57-66.

Martel, E. (1991). 'How valid are the Portland baseline essays?', *Educational Leadership* Dec/Jan, 20-23.

Martin, M. (1972). *Concepts of science education*, New York, Scott, Foresman & Co, (reprinted, University Press of America, 1985)

Matthews, M.R. (1992). 'History, philosophy and science teaching: The present rapprochement', *Science & Education* (1), 11-48.

Matthews, M.R. (1993). 'Constructivism and science education: Some epistemological problems', *Journal of Science Education and Technology* (2), 359-370.

Matthews, M.R. (1994). *Science teaching: The role of history and philosophy of science*, New York, Routledge.

Matthews, M.R. (1997). 'Introductory comments on philosophy and constructivism in science education', *Science & Education* (6), 5-14.

Matthews, M.R. (1998). *Time for science education: How teaching about the history and philosophy of pendulum motion can contribute to science literacy*, New York, Plenum Publishers.

Meichtry, Y.J. (1993). 'The impact of science curricula on student views about the nature of science', *Journal of Research in Science Teaching* (30), 429-444.

Mellado, V. (1997). 'Preservice teachers' classroom practice and their conceptions of the nature of science', *Science & Education* (4), 331-354.

Merton, R.K. (1942). 'The normative structure of science'. In his *The sociology of science: Theoretical and empirical investigations* (N.W. Storer, ed.), Chicago, IL, University of Chicago Press, 1973, pp.267-280.

Meyling, H. (1997). 'How to change students' conceptions of the epistemology of science', *Science & Education* (6), 397-416.

Montellano, B.R.O. de (1996). 'Afrocentric pseudoscience: The miseducation of African Americans', in P.R. Gross, N. Levitt & M.W. Lewis (eds.), *The Flight from Science and Reason*, New York, New York Academy of Science, pp. 561-573.

Nelson, L.H. & J. (eds.) (1996). *Feminism, science, and the philosophy of science*, Dordrecht, Kluwer Academic Publishers. ISBN, 0-7923-4162-7

Nielsen H. & Thomsen, P. (1990). 'History and philosophy of science in the Danish curriculum', *International Journal of Science Education* (12), 308-316.

Pinnick, C. (1994). 'Feminist epistemology: Implications for philosophy of science', *Philosophy of Science* 61, 646-657.

Reif, F. (1995). 'Understanding and teaching important scientific thought processes', *Journal of Science Education and Technology* (4), 261-282.

Robinson, J.T. (1968). *The nature of science and science teaching*, Belmont, CA, Wadsworth.

Rowell, J.A. & Cawthron, E.R. (1982). 'Images of science: An empirical study', *European Journal of Science Education* 4, 79-94.

Rubba, P., Horner, J., & Smith, J. (1981). 'A study of two misconceptions about the nature of science among junior high school students', *School Science and Mathematics* 81, 221-226.

Rubba, P.A. & Andersen, H. (1978). 'Development of an instrument to assess secondary school students' understanding of the nature of scientific knowledge', *Science Education* 62, 449-458.

Ruse, M. (ed.) (1988). *But is it science? The philosophical question in the creation/evolution controversy*, Albany, NY, Prometheus Books.

Rutherford, F.J. (1972). 'A humanistic approach to science teaching', *National Association of Secondary School Principals Bulletin* 56, 53-63.

Scheffler, I. (1970). 'Philosophy and the curriculum'. In his *Reason and Teaching*, London, Routledge, 1973, pp.31-44. Reprinted in *Science & Education* 1, 385-394.

Scheffler, I. (1973). *Reason and teaching*, Indianapolis, IN, Bobbs-Merrill.

Schwab, J.J. (1945). 'The nature of scientific knowledge as related to liberal education', *Journal of General Education* 3, 245-266. Reprinted in I. Westbury & N.J. Wilkof (eds.), *Joseph J. Schwab: Science, Curriculum, and Liberal Education*, Chicago, IL, University of Chicago Press, 1978.

Schwab, J.J. (1958). 'The teaching of science as inquiry', *Bulletin of Atomic Scientists* 14, 374-379. Reprinted in J.J. Schwab & P.F. Brandwein (eds.), *The Teaching of Science*, Cambridge, MA, Harvard University Press.

Siegel, H. (1988). *Educating reason: Rationality, critical thinking, and education*, London, Routledge.

Siegel, H. (1997). 'Science education: Multi cultural and universal', *Interchange* (28), 97-108.

Slezak, P. (1994). 'Sociology of science and science education: Part I', *Science & Education* (3), 265-294.

Solomon, J. (1989). 'Teaching the history of science: Is nothing sacred?'. in M. Shortland & A. Warick (eds.), *Teaching the History of Science*, Oxford, Basil Blackwell, 42-53.

Solomon, J.(1991). 'Teaching about the nature of science in the British national curriculum', *Science Education* (75), 95-104.

Solomon, J., Scott, L., & Duveen, J. (1996). 'Large-scale exploration of pupils' Understanding of the nature of science', *Science Education* (80), 493-508.

Suchting, W.A. (1995). 'The nature of scientific thought', *Science & Education* (4), 1-22.

Tobin, K. & McRobbie, C.J. (1997). 'Beliefs about the nature of science and the enacted science curriculum', *Science & Education* (6), 355-371.

Tobin, K. (ed.) (1993). *The practice of constructivism in science and mathematics education*, Washington, D.C., AAAS Press.

Tobin, K., Tippins, D.J., & Hook, K.S. (1994). 'Referents for changing a science curriculum: A case study of one teacher's change in beliefs', *Science & Education* (3), 245-264.

Westaway, F.W. (1919). *Scientific method: Its philosophy and practise*, London.

Westaway, F.W. (1929). *Science teaching*, London, Blackie and Son.

Westbury, I. & Wilkof, N.J. (eds.), (1978). *Joseph J. Schwab: Science, curriculum, and liberal education*, Chicago, IL, University of Chicago Press.

White, R.T. & Gunstone, R.F. (1989). 'Metalearning and conceptual change', *International Journal of Science Education* (1), 577-586.

Yeany, R. H. (1991). 'A unifying theme in science education?', *NARST News* (33), 1-3.

SECTION I

RATIONALES FOR INCLUDING NATURE OF SCIENCE IN SCIENCE INSTRUCTION

1. THE ROLE AND CHARACTER OF THE NATURE OF SCIENCE IN SCIENCE EDUCATION

Science has a pervasive but often subtle, impact on virtually every aspect of modern life—both from the technology that flows from it and the profound philosophical implications arising from its ideas. However, despite this enormous effect, few individuals even have an elementary understanding how the scientific enterprise operates. This lack of understanding is potentially harmful, particularly in societies where citizens have a voice in science funding decisions, evaluating policy matters and weighing scientific evidence provided in legal proceedings. At the foundation of many illogical decisions and unreasonable positions are misunderstandings of the character of science.

For almost twenty-five years the National Science Board has surveyed the American public as part of its *Science and Engineering Indicators* study to determine the state of interest in and awareness of fundamental issues in the sciences and technology. The results of this study included the conclusion that the "level of interest in science and technology in the U.S. has remained relatively stable over the past 16 years with approximately 40 percent of Americans expressing a high level of interest in science . . ." (National Science Board, 1996). This finding may be encouraging, but further probing indicates that although Americans may be interested in science they have no clear idea how science operates.

One element of the survey examined individuals' views about how science is conducted. The study designers formulated a series of questions aimed at classifying respondents' positions on a four level hierarchy of nature of science understanding. Those at the highest level (Level I) understand that science is concerned with the development and testing of theory. Those responding who lack this degree of sophistication, but still have an awareness that experiments require a control group would be classified as Level II. Individuals at Level III do not have the comprehension of those in the higher two groups but still see scientific findings based on a foundation of careful and rigorous comparison with precise measurements. Those lacking any understanding of the nature of science were classified as Level IV.

These findings are sobering. Two percent of the two thousand adult respondents were at Level I, 21 percent were at Level II, 13 percent were at Level III and 64 percent were at Level IV. This finding is sobering. Even as measured by the basic nature of science elements contained in this study, more than 60 percent of the American public effectively had no knowledge of how science works. Somewhat encouraging was the finding that individuals having higher levels of education and more science and mathematics education were likely to be in Levels I and II.

The explanation for the general public's poor understanding of how science functions is astoundingly simple. The crux of the matter is that at all levels science teaching and textbooks emphasize the factual recall of science content to the near total exclusion of the knowledge-generation process. Science teachers rarely have opportunities to learn how science functions in their own studies and, not surprisingly, fail to emphasize that aspect of science to their students. Lakin and Wellington, (1994) reported that teachers in their study, having never reflected on issues relating to the nature of science, tended to undervalue such ideas in their teaching. Furthermore, educators who would like to incorporate something of the pageant of science in their science lesson must consult the same textbooks that frequently misrepresent or even omit discussion of the way in which science knowledge is produced. This book, *The Nature of Science in Science Education: Rationales and Strategies,* was designed to addresses these problematic issues by illustrating both why and how to communicate the excitement of the scientific endeavor to others.

WHAT IS THE NATURE OF SCIENCE?

The phrase "history and philosophy of science" (HPS) has been used to describe the interplay of disciplines that inform science education about the character of science itself. However, a more encompassing phrase to describe the scientific enterprise for science education is the "nature of science" (NOS). The nature of science is a fertile hybrid arena which blends aspects of various social studies of science including the history, sociology, and philosophy of science combined with research from the cognitive sciences such as psychology into a rich description of what science is, how it works, how scientists operate as a social group and how society itself both directs and reacts to scientific endeavors. The intersection of the various social studies of science is where the richest view of science is revealed for those who have but a single opportunity to take in the scenery.

The nature of science is not particularly concerned with the natural world in the way that science itself is, at least not directly. The scientific community consists of individuals who devote their careers to better understanding the natural world. This community of experts determines which ideas best account for natural phenomena. Those who study the nature of science come from many diverse fields and investigate science and scientists by asking questions such as "What, if anything, demarcates science from other human endeavors?", "Are science ideas discovered or invented?", and "How is consensus reached in the scientific community?" In brief, the nature of science epitomizes Einstein's, (1933) advice that if you want to know how scientists work, "don't listen to their words, fix your attention on their deeds" (p. 270). Through multiple lenses, the nature of science describes how science functions.

For science educators the phrase "nature of science," is used to describe the intersection of issues addressed by the philosophy, history, sociology, and psychology of science as they apply to and potentially impact science teaching and learning. As such, the nature of science is a fundamental domain for guiding science educators in accurately portraying science to students.

This book has been developed to help science educators cultivate the multiple kinds of knowledge necessary for effective teaching. Shulman, (1986) suggests that teachers' knowledge can be divided into three broad categories—pedagogical, curricular, and subject matter—and defines subject matter knowledge as a discipline's facts, principals *and* structure. He writes (p. 9):

Teachers must not only be capable of defining for students the accepted truths in a domain. They must also be able to explain why a particular proposition is deemed warranted, why it is worth knowing and how it relates to other propositions, both within the discipline and without, both in theory and in practice.

Hollon, Roth and Anderson, (1991) add that ". . . science teachers must develop knowledge that enables them to make two types of decisions—curricular decisions and instructional decisions" (p. 149). The challenge therefore, is for science teachers to translate an understanding of the knowledge generation process into meaningful classroom experiences and appropriate classroom discourse. Early in the book we define what constitutes the nature of science appropriate for guiding science instruction and in the second section a variety of experts have provided a range of strategies that effectively communicate this important knowledge both to teachers and students.

CONSENSUS VIEWS REGARDING THE NATURE OF SCIENCE

Before embarking on the development of any course or unit of study designed to assist teachers or students in the acquisition of a nature of science understanding, one must have some notion of what knowledge is worth possessing for incorporation into curricula and classroom discourse. In spite of significant progress toward characterizing science, much disagreement remains. Almost thirty years ago, Herron, (1969) claimed that no sound and precise description exists concerning the nature and structure of science, and more recent voices echo that sentiment. As an example, Lauden states that ". . . we have no well-confirmed general picture of how science works, no theory of science worthy of general assent" (in Ginev, p. 64). Welch, (1984) and Duschl, (1994) also cites the lack of consensus regarding the appropriate image of scientific inquiry and the growth of scientific knowledge while Lederman (1992) notes that the nature of science is neither universal nor stable. At the level of the fine details, there will always be an active debate regarding the ultimate nature of science. However, one of the central responsibilities of science teachers is to provide an accurate description of the function, processes and limits of science rather than to engage students in the somewhat arcane arguments that occur among philosophers of science. At the level of description, there is significant consensus regarding the nature of science. During the past three decades, a number of scholars including Robinson, 1968; Martin, 1972; Ennis, 1979; Giddings, 1982; Lederman, 1983; Duschl, 1988, and Matthews, 1994 have provided both explicit and implicit suggestions for the characteristics of science to be included in science instruction.

For example, the nature of science recommendations contained in eight international science education standards documents show significant overlap (Table 1). Without question, the positions in the following table in no way convey the range of complex issues surrounding the statements. Such complexities are fundamental to specialists in the social studies of science, but these recommendations are for K-12 science students—not future philosophers of science. Knowing who the recommendations are for and the degree of sophistication appropriate for that target group is an important consideration when crafting nature of science standards. Moreover, the image of science emerging from the social studies of science is sufficiently robust that science educators can move forward with confidence and provide a more realistic picture of the strengths and limitations of this thing called science. (Smith *et al.*, In press). Where consensus does not exist, science teachers should present a plurality of views. As Matthews, (1997) argues, the purpose of nature of science education is not to indoctrinate, but to address reasons for accepting a particular position. Chapter 2 provides a full elucidation of the variety of nature of science elements and the degree of consensus, but here in Table I the most prevalent issues are illustrated.

TABLE I
A consensus view of the nature of science objectives extracted from eight international science standards documents.

- Scientific knowledge while durable, has a tentative character.
- Scientific knowledge relies heavily, but not entirely, on observation, experimental evidence, rational arguments, and skepticism.
- There is no one way to do science (therefore, there is no universal step-by-step scientific method)
- Science is an attempt to explain natural phenomena
- Laws and theories serve different roles in science, therefore students should

note that theories do not become laws even with additional evidence.
- People from all cultures contribute to science
- New knowledge must be reported clearly and openly
- Scientists require accurate record keeping, peer review and replicability
- Observations are theory-laden
- Scientists are creative
- The history of science reveals both an evolutionary and revolutionary character
- Science is part of social and cultural traditions
- Science and technology impact each other
- Scientific ideas are affected by their social & historical milieu

THE NATURE OF SCIENCE IN SCIENCE EDUCATION: HISTORICAL PERSPECTIVE

Advocacy for students' understanding of science and its nature can be traced back to the early years of this century. Although at that time the phrase "understanding the nature of science" was not clearly stated, some elements and characteristics of science were noted as goals worth pursuing in science teaching. For example, Lederman, (1992) reported that the Central Association of Science and Math Teachers in 1907 strongly emphasized the scientific method and processes of science in science teaching. Hodson, (1991) cites Dewey's 1916 argument that understanding scientific method is more important than the acquisition of scientific knowledge. Jaffe, (1938) in his high school textbook *New World of Chemistry* listed nature of science objectives such as a willingness to swing judgment while experiments are in progress, willingness to abandon a theory in light of new evidence, and knowledge that scientific laws may not be the ultimate truth.

In 1946, James Bryan Conant delivered his famous Terry Lectures at Yale advocating a historical approach to science instruction. He suggested that all students must understand the tactics and strategies of science. It was not until the second half of this century that the construct we now call the nature of science was stated explicitly as a major aim of science teaching by the National Society for the Study of Education, (1960):

There are two major aims of science-teaching; one is knowledge, and the other is enterprise. From science courses, pupils should acquire a useful command of science concepts and principles. Science is more than a collection of isolated and assorted facts . . . A student should learn something about the character of scientific knowledge, how it has been developed, and how it is used. (Hurd, 1960, p. 34).

One of the primary justifications for the inclusion of the nature of science in science education comes from Schwab, (1964) who was both a philosopher and science educator. He correctly observed that science is taught as an "unmitigated rhetoric of conclusions in which the current and temporal constructions of scientific knowledge are conveyed as empirical, literal, and irrevocable truths" (p. 24).

With the advent of the 1960s science curriculum projects, an effort was made by some developers to shift science instruction away from the primary focus concerning, "what do scientists know," to an examination of the question "how do scientists know"—reflecting Schwab's, (1960) emphasis on "what do scientists do?" Klopfer's (1964-66) *History of Science Cases* and *Harvard Project Physics* (Rutherford, Holton and Walton, 1970) and Schwab's, (1963) seminal contributions to the Biological Science Curriculum Studies programs were important attempts to illustrate both the process and products of science in formal curricula.

At the turn of the decade, several important books were published advocating and defining elements of the nature of science necessary for inclusion in school science curricula. Robinson, (1968) in *The Nature of Science and Science Teaching* provided science educators ready access to the philosophy of science for the first time. In his book, Robinson provided an overview of the nature of physical reality, aspects of physical description including probability, certainty and causality, and view of the nature of science in various science disciplines. He concluded with considerations for the interplay between science instruction and the nature of science. In *Concepts of Science Education: A Philosophical Analysis*, Martin, (1972) reiterated many of the arguments put forward by Robinson in supporting NOS in science instruction. In addition, he reviewed many of the important concepts from the philosophy of science including the value of inquiry learning, the nature of explanation, and the character of observation both in science and in science learning. In the section on goals of science education he specifically stated that student should acquire a range of scientific propensities.

Despite these early and now-neglected plans and rationales, science teachers and science curricula seem rigidly bound to a tradition of communicating the facts or end products of science while generally neglecting how this knowledge was constructed. After almost fifty years of interest concerning the nature of science in science curricula, little change has occurred. For example, in 1982, Kilborn suggested that science instruction does not present the essential background for understanding the meaning of science. Ten years later, Gallagher, (1991) observed that science lessons revealed an emphasis on the body and terminology of knowledge in science rather than the nature of science. Fleury and Bentley, (1991) refer to the way that science knowledge comes to exist as the "infrastructure of scientific knowledge" and argue that if knowledge of the infrastructure is faulty then any understanding constructed on it will be fallible.

Most sobering is the assertion of Bentley and Garrison, (1991) that for most science students, a description of the NOS is relegated to a few paragraphs at the beginning of the textbook quickly glossed over in favor of the facts and concepts that cram the remainder of the book and generally fill the course. And the ideas put forth in textbooks and school science concerning the nature of science are almost universally incorrect, simplistic, or incomplete. Duschl, (1994) recently argued that students are learning facts, hypothesis, and theories of science—the "what" of science—but they are not learning where this knowledge originated—the "how" of science.

Recently this disheartening situation is facing more aggressive attack. Discussions concerning a role for nature of science in school science have increased rapidly and few now argue with the proposition that school science experiences should include significant attention to how science works including how knowledge is created and established. In the past decade alone a number of international conferences (University of Florida, 1989; Kingston, Ontario, 1992, University of Minnesota, 1995, and, a North American Regional conference was held in Calgary, Alberta, Canada in June of 1997) have addressed a more extensive role for the social studies of science in science education. Furthermore, the main objective of a number of additional meetings (Pavia in 1983, Munich in 1986 and Paris in 1988) has been to investigate why and how history and philosophy of science may be integrated in school science (Nielsen, 1990). Much of this renewed interest in the NOS and its place in science teaching has come from the leadership of Michael Matthews. His 1994 book *Science Teaching; The Role of History and Philosophy of Science* provides a well-reasoned argument for the inclusion of the nature of science in science instruction.

Incorporating the nature of science in school science has been widely embraced by organizations such as the Association for Science Education, (1981) in Britain and organizations in the United States such as the National Science Teachers Association, (1995), the American Association for the Advancement of Science, (1989, 1993) and the National Research Council, (1996). Many contemporary science educators agree that encouraging students' understanding of the nature of science, its presuppositions, values, aims, and limitations should be a central goal of science teaching. As an example, Morris Shamos, (1995) argues in *The Myth of Scientific Literacy* that while knowledge of science content may not be necessary for obtaining science literacy, understanding the nature of science *is* prerequisite to such literacy.

THE NATURE OF SCIENCE IN SCIENCE EDUCATION: A RATIONALE

The Current State of Students' and Teachers' Knowledge of the Nature of Science

A number of studies exist that document students' misconceptions concerning the nature of science (Clough, 1995; Lederman, 1992; Meyling, 1997; Rowell & Cawthron, 1982; Rubba, Horner & Smith, 1981). Ryan and Aikenhead, (1992) collected the responses of more than two-thousand upper secondary students, concluded that they confused science with technology, and were only superficially aware of the private and public side of science and the effect that values have on scientific knowledge. Moreover, they reported that:

- 46% held the view that science could rest on the assumption of an interfering deity;
- Only 17% were certain of the inventive character of scientific knowledge;
- 19% believed that models are actual copies of reality;
- Only 9% chose the contemporary view that scientists "use any method that might get favorable results"; and
- 64% of students expressed a simplistic hierarchical relationship in which hypotheses become theories and theories become laws, depending on the amount of "proof behind the idea."

While acknowledging that a number of out-of-school factors misrepresent the nature of science, overwhelming evidence exists that school science is at least equally culpable. Almost 50 years ago Anderson, (1950) concluded that teachers were more concerned with imparting scientific facts than helping students understand the processes of science—an indication that something was awry regarding teachers' notions of the nature of science. Miller, (1963), after conducting one of the earliest and frequently-cited studies of teachers' views of the image of science, concluded that many science teachers and their students failed to demonstrated understanding of how science works. Several years later, Schmidt, (1967) replicated Miller's study and made essentially the same conclusions. Elkana, (1970) claimed that teachers' understanding of the philosophy of science trailed developments in the philosophy of science by some twenty to thirty years. In 1978, Cawthron and Rowell concluded that science teachers take a naive-realist position of science, maintain that scientists have particular characteristics, and employ scientific method to account for the achievements of science. Four years later they argued (Rowell and Cawthron, 1982) that many science teachers subscribe to an inductivist-empiricist outlook of science. Brush, (1989) noted that science teachers are not generally aware of the social and cultural construction of scientific thought. DeBoer, (1991) in his review of the history of science education states that the positivist view of the philosophy of science from the last century still informs much classroom practice and pervades most available curriculum materials. Melado, (1997) reports that while the preservice science teachers in his small study showed more multi-faceted NOS views that a positivist empirical label implies, they were insecure and contradictory in their statements and admitted they had never before reflected on the epistemology of science. Tragically, too often science teachers simply do not include nature of science issues in the design of science learning plans (Bell, et al. 1997; King 1991; Lakin and Wellington, 1994), likely because they lack knowledge of them.

The Value of Nature of Science for Teaching and Learning

The significant misconceptions that both students and teachers hold regarding the nature of science, by themselves, represent an important justification for including the social studies of science in science courses and preservice science teacher education programs. Driver et al., (1996) have suggested five additional arguments supporting the inclusion of the nature of science as a goal of science instruction. The arguments include the utilitarian view that "an understanding of the nature of science is necessary if people are to make sense of the science and manage the technological objects and processes they encounter . . ." (p. 16). This is related to the democratic view that people must understand the NOS "to make sense of socio-scientific issues and participate in the decision-making process" (p. 18) and the cultural argument that such understanding is necessary "in order to appreciate science as a major element of contemporary culture" (p. 19). The fourth rationale is moral, to understand the ". . . norms of the scientific community, embodying moral commitments which are of general value," (p. 19). Driver's final justification for including the nature of science in science instruction is that it "supports successful learning of science content" (p. 20).

Matthews, (1997) argues that questions regarding the nature of science are inherent in many education issues such as multicultural science, the evolution/creation public education controversy, feminist critiques of modern science and their suggestions for science program reform, the place of religion in science education, environmentalism and new-age science, and the notion that learning science will result in an understanding of its nature while at the same time causing students to become more scientific in solving life's problems.

Moreover, students in a recent study by Meyling, (1997) showed significant interest in the nature of science. Two-thirds of the physics students who experienced instruction regarding epistemological issues showed interest in more epistemology. In contrast, only one-third of students not experiencing such instruction showed interest. Students in this study approved of NOS discussions and most indicated their epistemological conceptions had changed.

NOS to Enhance the Learning of Science Content

Evidence suggests that knowledge of the nature of science assists students in learning science content. For example, Songer and Linn, (1991) illustrated the importance of students having dynamic rather than static views of science in developing a conceptual understanding of topics such as thermodynamics. In a sample of 153 eighth-grade physical science students with instruction emphasizing hands-on experiments, the authors were able to characterize students' views of science as either static, mixed, or dynamic. The static view of science is the idea that science is a group of facts that are best memorized. The dynamic view of science posits that scientific knowledge is tentative, and the best way to understand this knowledge is by understanding what scientific ideas mean and how they are related. Although the authors did not address the mixed view, they did find that students with dynamic views of science acquired a more integrated understanding of thermodynamics than those with static views.

NOS Knowledge to Enhance Understanding of Science

Understanding how science operates is imperative for evaluating the strengths and limitations of science, as well as the value of different types of scientific knowledge. For instance, science teachers may understand the atomic model, Boyle's law, and evolutionary theory, but may not understand what law, theory, and model mean in the discipline of science. Hence, ridiculous statements like, "evolution is *only* a theory" or "when such-and-such a theory is proven it will become a law" may result. One of the major theses of Michael Martin's, (1972) book *Concepts of Science Education: A Philosophical Analysis* is that philosophy of science study is beneficial to the science educator. Studies in philosophy of science will clarify teachers' thought about the nature of science and help them understand the roles and methods which guide study in the discipline. As Manuel, (1981) writes:

This more philosophical background advocated for teachers would, it is believed, enable them to handle their science teaching in a more informed and versatile manner and to be in a more effective position to help their pupils build up the coherent picture of science-appropriate to age and ability -- which is so often lacking (p. 771)

Furthermore, those who comprehend the durable, yet revisionary, nature of scientific knowledge will not be confused by changing science concepts or the disappearance of particular science ideas learned earlier. Individuals who understand how science works will likely be less cynical about the scientific enterprise (Connelly et al., 1977, p. 7-8). Because science is often wrongly perceived primarily as a body of literal truths, entire fields of knowledge are sometimes questioned when single facts are revised. Perceiving science as a process of improving our understanding of the natural world turns the notion of tentativeness into a strength rather than a weakness.

Surely, lack of knowledge about the history and philosophy of science would hinder teachers' incorporation of philosophical aspects of science in their teaching. In a 1991 study, King showed that the science teachers he investigated attributed the difficulties of incorporating ideas such as discovery and relevance in their science instructions to their ignorance about history and philosophy of science.

NOS to Enhance Interest in Science

A sensitivity to the development of scientific knowledge may also make science itself and science education more interesting. Tobias, (1990) maintains that a number of potential university science students—those she calls the second tier—lament that science classes ignore the historical, philosophical, and sociological foundations of science. Incorporating the nature of science while teaching science content humanizes the sciences and conveys a great adventure rather than memorizing trivial outcomes of the process. The purpose is not to teach students philosophy of science as a pure discipline but to help them be aware of the processes in the development of scientific knowledge (Matthews, 1989). Here we see justification for Driver's, (1996) "cultural argument" for learning about the nature of science.

NOS Knowledge to Enhance Decision Making

The "democratic argument" for the nature of science instruction (Driver et al., 1996, p. 18) may be illustrated in a number of ways, but certainly having accurate views about how science functions is vital for informed decision making. For example, the funding of scientific and technological research is an increasingly important topic with respect to governmental budget decisions. Ample evidence exists (Ryan and Aikenhead, 1992) to suggest that science is often confused as technology leading the public to support science because they wrongly see it as providing society with gadgets, vaccines, and other practical outcomes that improve everyday living. However, basic research is not directly concerned with practical societal outcomes, but rather an understanding of the natural world for its own sake. The public's failure to see the importance of basic research in technological innovations may have significant societal consequences when funding decisions are made (Elmer-Dewitt, 1994).

The public education controversy involving evolution and creationism nicely illustrates the importance of an informed citizenry regarding the nature of science. John Moore, (1983) writes:

It becomes evermore important to understand what is science and what is not. Somehow we have failed to let our students in on that secret. We find as a consequence, that we have a large and effective group of creationists who seek to scuttle the basic concept of the science of biology . . . It is hard to think of a more terrible indictment of the way we have taught science . . .

Far too many secondary science teachers avoid teaching biological evolution. However, Scharmann and Harris, (1992) found that promoting an applied understanding of the nature of science reduced teachers' anxiety toward teaching this fundamental idea. Johnson and Peeples, (1987) found that as students' understanding of the nature of science increases, they are more likely to accept evolutionary theory. Dagher and Boujaoude, (1997) investigated how students with different religious backgrounds accommodated their beliefs with biological evolution and recommended that teachers devote significant attention to values, beliefs and the nature of science. Clough, (1994) made several practical suggestions regarding NOS instruction for reducing students' conflicts with the theory of evolution. If students and teachers simply understood the distinction between science and religion, that alone would ease the occasional tension caused by discussions of evolution.

NOS Knowledge to Enhance Instructional Delivery

Matthews, (1994) has argued for the inclusion of NOS courses in science teacher education programs. The examples he provided demonstrates that a firm grounding in the nature of science is likely to enhance teachers' ability to implement conceptual change models of instruction. Studying the process of historical conceptual development in science may shed some light on individual cognitive development (Wandersee, 1986). For example, many students' ideas parallel that of early scientific ideas, suggesting that "alternative conceptions" may sometimes be a better description than "misconceptions." The persistence of students' naive ideas in science suggests that teachers could use the historical development of scientific concepts to help illuminate the conceptual journey students must make away from their own naive misconceptions. In other words, teachers' interest in NOS could assist in understanding the psychology of students' learning (Matthews, 1994).

The history of science confirms that scientific knowledge is not exclusively determined empirically. The construction of scientific knowledge (Latour, 1987; Latour & Woolger, 1986; Knorr-Cetina, 1981; Kuhn, 1970; Mendelsohn, 1977; Mulkay & Gilbert, 1982; Shapin, 1982) has much in common with conceptual change. This makes the nature of science useful as a disequilibrating agent in changing science teachers' views of learning and teaching. For example, some of the resistance to conceptual change theory among classroom teachers arises from the mistaken notion that knowledge of the natural world is completely objective—existing independently of the searching individual. This view of science gives the impression that learning is a fairly straightforward process of replacing what is known with that which the scientific community has *discovered* is right. However, the history of science may also reveal a fierce battle to *construct* meaning concerning the natural world. This construction, sometimes requiring enormous effort and time, is not a straightforward process. When science is seen in this light, children's misconceptions and difficulties in learning contemporary science ideas are understandable. To assist in this process of inquiry-based or constructivist teaching, Duschl, (1987) states that teachers themselves need to have an adequate understanding of the nature of science.

THE EFFECT OF TEACHERS' KNOWLEDGE OF THE NATURE OF SCIENCE ON STUDENT UNDERSTANDING

A Simplistic Assumption

Teachers represent the most important variable in the classroom learning equation. Even well-designed NOS instructional packages that are at odds with the philosophical orientations of teachers may not be effective. Duschl, (1987) writes that in spite of attempts to "teacher proof" schooling through the enforcement of strict curriculum guidelines and teaching models, teachers will continue to make the most critical decisions in the education of students. Regarding NOS instruction, Hodson, (1988) argues that "the most important factors determining [attitudes toward science] are teaching style (Evans & Baker, 1977; Rubba, Horner & Smith, 1981) and the teacher's own image of science" (Jungwirth, 1971). What this suggests for nature of science instruction is sobering given science teachers' dismal understanding of the nature of science (Hodson, 1988).

Hence, bolstering teachers' understanding of NOS is clearly a prerequisite for effective science teaching with many adherents (Table II).

TABLE II
Researchers' assumptions about the relationship between
teachers' views on science and their teaching behavior

Author	Relevant Statement
Abimbola	Knowledge of the philosophy of science will enable science teachers to know what science they are teaching and this knowledge will affect all their instructional practices (1983, p. 189).
Carey and Stauss	If the teacher's understanding and philosophy of science is not congruent with the current interpretations of the nature of science, . . . then the instructional outcomes will not be representative of science (1970, p. 368).
Gill	The type of science imparted to students depends upon the teachers own views of the nature of science (1977, p.4).
Hurd	It is undoubtedly true [that] a teacher's conception of what science is influences not only what he teaches, but how he teaches (1969, p. 16).

Nunan   . . . I wish to suggest that a teacher's "image of science" which contains [the teacher's notion of the nature of science and its links to society] is a vital factor in directing personal stances towards content and methodological concerns. . . . it is irresponsible of science teachers to suggest that historical, philosophical, or sociological examinations of science have no relevance to them as in the final analysis their teaching preferences convey values based upon studies in these areas. (1977)

Ogunniyi   If science teachers hold inadequate views of science then the instructional outcomes will inevitably be a corruption of what they are supposed to teach (1983, p.193).

Robinson   It is assumed that a teacher's conception of nature of science is an important force in shaping his classroom behavior (1969, p. 99).

Scheffler   . . . whatever [the science teacher] does is likely to be qualified by his [philosophical] reflections on the field of science (1973, p. 36).

A Complex Reality

Only in the last decade has the assumed relationship between teachers' knowledge of NOS and their classroom behaviors been put to the test. In 1982, Dibbs was among the first to investigate the relationship between teachers' practices and students' views regarding the nature of science. Dibbs determined that a questionnaire he developed could be used to delineate four broad philosophical categories among science teachers verificationist (V-type), inductivist (I-type), hypothetico-deductivist (H-type), and no discernable philosophy of science (O-type). Interviews, occurring some months after completion of the questionnaire, were held with selected "extreme" V-type, I-type and H-type teachers to (a) determine whether teachers' views expressed in a more searching interview agreed with their responses to the questionnaire, and (b) probe for possible links between the teachers' stated views on the philosophy of science and their reported teaching styles. Dibbs determined that H-type teachers have a problem solving approach to teaching science, V-type teachers had a factual or informative approach to teaching science, and I-type teachers have an observational approach to the teaching of science. Dibbs, writes:

Teachers of the first three extreme types not only have more clearly formulated views about the philosophy of science, but are influenced by their views when planning their lessons. The ways in which they describe their teaching styles and report using practical work in their lessons shows a strong positive relationship between these and their beliefs concerning the philosophy of science (p. 225).

Information from the "extreme" teachers' teaching style and particularly the manner in which they introduced and used practical work in their lessons was then used in planning teaching practices for the groups of pupils involved in the teaching experiment. Five normal science classes in a natural setting were subjected to biased teaching styles in some of their science lessons for approximately two school terms. According to Dibbs:

Pupils' views about the philosophy of science are influenced by the way in which they are taught science even if their teacher does not attempt to do so explicitly. The teaching style used implies that the teacher holds a certain philosophy, and this implicit philosophy can have effects upon the pupils. Pupils taught in different ways exhibit a different pattern of responses on the specially designed measure of science understanding. Pupils given H type teaching achieve significantly higher scores on the scale measuring agreement with the hypothetico-deductivist philosophy. Pupils who have received I type or V type teaching obtain their highest scores on the inductivist and verificationist scales respectively (p. 226).

Particularly noteworthy is the effect H, V, and I-type teaching appears to have on student vocabulary used to describe scientists and their activities. With the cooperation of an English teacher who taught many of the students in the experiment, students were asked to write a science fiction story. No guidance was provided about the plot of the story, but students were asked to make the scientists in the story behave as authentic scientists would if faced with the situation described in their story. Over half of the students in the experiment wrote stories. Simply counting the number of times certain words or phrases appeared in the stories revealed some interesting differences between students who had received the three different teaching styles. For example (pp. 202-3):

Pupils who had been exposed to the H style of teaching used words such as 'idea', 'think', 'thought', 'problem', 'question', 'test', 'testing', 'check', and 'clues' more often than did those who had been in the I or V groups. Pupils from the I group seemed to favor words such as 'record', 'recording', 'noted', 'sample', 'specimen', 'notice', 'look', 'observations', 'information', 'discovery', 'theory', 'pattern', and 'conclusion' . . . All mentioned experiments frequently but pupils from the V groups more often said that they 'prove' something.

In another study, Smith and Anderson, (1984) observed an elementary science teacher teaching an activity-based unit on plant growth and photosynthesis. They reported that the teacher was surprised to learn that her students could not predict whether or not seeds beginning to grown in the dark would survive. This was unfortunate because the students spent two weeks measuring and observing two planted grass seeds both in the light and dark. The teacher hoped that her students would derive photosynthesis, a theoretical construct, from empirical observations of plants grown in the light and dark. The teacher's expectation for her students betrays her personal views of science in that she believes scientific theories are implied from data through inductive logic.

A study by Lantz and Kass, (1987) is particularly instructive regarding the influence of teachers' beliefs on their classroom practice. Three chemistry teachers who adopted the same curriculum materials implemented them in ways that reflected their views of chemistry knowledge rather than the explicit orientation of the curriculum materials. For example, two teachers who viewed chemistry as a stable body of concepts, principles, and theories, had difficulty finishing the course because they attempted to teach everything as fundamental. The third teacher perceived chemistry as a constantly-developing body of knowledge, and limited the presentation of topics to those deemed essential.

Brickhouse, (1989) also considered the influence of teachers' beliefs about the nature of science on their classroom practice. She conducted extensive interviews and observations of three science teachers and concluded that teachers' nature of science conceptions do influence their decisions about what they teach. For example, the first teacher viewed theories as truths uncovered through rigid experimentation, and, not surprisingly, the intent of instruction in this classroom was for students to learn the "truth." Students' performance in science activities was evaluated solely by the outcome of the activity, not on the process. This first teacher also perceived scientific processes as inductive, and therefore, lab instruction included precise procedures to acquire the "right" answer. Viewing science as an accumulation of knowledge, students were told, "every experiment from this page on proves the rest of the chapter, each and every one of them." The second teacher, on the other hand, thought of theories as tools to solve problems and, therefore, students used theories to explain observations and to resolve problems. The third teacher, viewed science as the accumulation of knowledge, a position clearly reflected in classroom instruction regarding the development of atomic theory. For example, each change in our historical conception of atomic structure was presented as simply the building on prior conceptions, and each scientist's contribution was conveyed to students as simply increased detail to the former model of the atom. In summary, Brickhouse concluded that teachers' science philosophies influence laboratory instruction, the way in which demonstrations are used, and instructional goals.

Not all research supports the notion that a teacher's views concerning the nature of science will influence their classroom practice. Mellado, (1997) performed an in-depth study of four preservice science teachers and concluded that no correspondence existed between their NOS views and classroom practice. In 1987, Lederman and Zeidler reported the results of a study of 18 biology teachers each of whom was characterized by a rich data set. A systematic pair-wise qualitative comparison among the 18 data sets classified differences in the classrooms and in instruction used by the teachers. As a result, 44 classroom variables emerged that distinguished the 18 teachers' behaviors. The processes of deriving the classroom variables was done without knowledge about the teacher's conceptions of the nature of science. Zeidler administered the *Nature of Scientific Knowledge Scale* (NSKS) to the science teachers and categorized teachers into two groups, high and low, depending upon their scores on the NSKS. To examine the assumption that a teacher's views of the nature of science influence their classroom behaviors, they tested the capability of each classroom variable statistically to differentiate between high and low teachers. They concluded that "teachers' classroom behavior doesn't vary as a direct result of his/her conception" (p. 731).

Tobin and McRobbie, (1997) in an in-depth study of one chemistry teacher, his students, and the instruction that took place found that the implemented curriculum was at odds with both the teacher's and students' views of science. The curriculum seemed to be most influenced by the teacher's notions of how students learn, beliefs about how teachers should yield power in a classroom, and how extensively the teacher accepted perceived constraints without struggle. Interestingly, the students' and teacher's goals for the course matched perfectly, hence significant change in instruction, including the nature of science, would most likely have been met with some resistance.

In 1989, Duschl and Wright also addressed this question with their investigation of high school teachers' decision-making models for planning and teaching science. Based on this study, they argued that student development, curriculum guide objectives, and pressures of accountability significantly affected science teachers' decision-making in the selection, implementation, and development of instructional tasks. Significantly missing in this decision-making process was any consideration to the nature or role of scientific theories or structure of the subject matter. In 1992, Lederman used this conclusion as evidence against the presumed relationship between teachers' conception and their teaching behavior. Lederman reported "the nature and role of scientific theories are not integral components in the constellation of influences affecting teachers' education decisions" (p. 347). However, Duschl and Wright pointed out that none of the teachers held contemporary views about the nature of science. They reasoned that "the lack of consideration for an accurate portrayal of the cognitive activities of science can be explained by teacher's lack of knowledge about the nature of science" (p. 493). Hence, if those teachers had been informed about the most fundamental issues in the nature of science, then these new views may have influenced teachers' decision making. They indicated that heavy emphasis is typically placed on memorizing the vocabulary of science with limited weight on matters related to the nature of science, and that this is related to the fact that "teacher's knowledge of science is limited to the body of knowledge of science" (p. 125).

In spite of some views to the contrary, we take the position from this overview of research that science teachers' knowledge and understanding of the nature of science do influence the teachers' classroom behaviors. That is, without explicitly thinking about such things, teachers consider the nature of what they perceive their discipline to be and those views are translated in ways they themselves may find surprising.

Moreover, the influence of a subject matter understanding on teachers' behavior is not limited to science education. For example, in mathematics, Thompson, (1982) found that teachers' conception of the nature and structure of mathematics influence their behavior. Scholars who have examined teachers' thinking such as Olson, (1980), (1981), Elbaz, (1981), Thompson, (1982), and Holton and Anderson, (1987) have shown that a teacher's beliefs, values, and personal philosophies impact a teacher's instruction and curriculum reforming (Tuan, 1991). In particular, a summary of research on teachers' thinking by Shavelson and Stern, (1981) considered teachers' conceptions of their subject matter to be one of the factors that affect teachers' pedagogical decisions and judgments.

Obviously, teachers must consider myriad factors when planning and implementing lessons, and that may explain the difficulty in always finding a direct relationship between teachers' views regarding the nature of science and their classroom practices. Clark, (1988) maintains that research on teachers' preconceptions and classroom instruction "has documented the many heretofore unappreciated ways in which the practice of teaching can be as complex and cognitively demanding as the practice of medicine, law, architecture" (p. 8). Therefore, the complexity of teaching may mitigate or hide the influence of teachers' views concerning the nature of science.

Lederman, (1992), following an extensive review of the literature, writes that "the presumed relationships between teachers' conceptions of science and those of their students as well as that between teachers' conceptions and instructional behaviors were finally directly tested and demonstrated to be too simplistic . . ." (p.347). Simply because a teacher has an understanding of the nature of science does not mean they know how to integrate forcefully such topics within instruction. Apparently, an understanding of the nature of science is a necessary, but insufficient condition, for purposeful teaching to facilitate student understanding of the nature of science. However, as Lederman, (1992) adds, ". . . the most important variables that influence students' beliefs about the nature of science are those specific instructional behaviors, activities, and decisions implemented within the context of a lesson" (p. 351).

COMMUNICATING THE NATURE OF SCIENCE

The Role of Textbooks and Activities

Science teachers' dependence on textbooks is well documented (Weiss, 1993) and the United States has the disconcerting distinction of ranking first in frequency of textbook reading as a means of instruction (Lapointe, Mead, and Phillips, 1989). The situation is much the same as described by Stake and Easley, (1978) who found that more than 90 percent of the science teachers surveyed indicated they used textbooks or other instructional materials 90 to 95 percent of the time. They write

Behind nearly every teacher-learner transaction . . . lay an instructional product waiting to play its dual role as medium and message. They command teacher's and learner's attention. In a way, they virtually dictated the curriculum. The curriculum did not venture beyond the boundaries set by the instructional materials (p. 66).

Thomas Kuhn, in *The Structure of Scientific Revolutions* (1970) makes clear that science textbooks convey an image of what science is and how it works. He writes that "[m]ore than any other single aspect of science, [the textbook] has determined our image of the nature of science and of the role of discovery and invention in its advance" (p. 143).

The significant role in instruction played by textbooks necessitates a look at how they portray the nature of science. Munby, (1976) speculated that the language appearing in curriculum materials may significantly affect students' understanding of the nature of science. In the science textbooks he analyzed, two distinct positions--instrumentalism and realism-- regarding the nature of scientific knowledge were implicit in the language used. Consider the following passage taken from *Physics: Matter, Energy, and the Universe* by Harnwell and Legge, (1967):

If a glass rod and a piece of fur are chosen as the test materials, it is found initially that they show no tendency either to attract or repel one another. If they are rubbed smartly together and quickly separated, it is found that they tend to attract one another. Our description of the process says that the materials have become electrified or charged, and that the force of attraction arises in consequence . . . On the evidence of such simple experiments . . . , early experimenters were led to construct a qualitative description of electrification in terms of the separation of some quality or substance associated with the materials being rubbed together (in Munby, 1976, p.121).

Munby, (1976) states that when students read phrases like ". . . our description of the process says . . ." and " . . . experimenters were led to construct . . . ," they may easily interpret this to mean that science ideas have an inventive character and, thus, may not exactly describe reality. The position that scientific ideas are useful tools to help us understand the natural world fits nicely with an instrumentalist view of science knowledge.

Consider the passage from *Conceptual Physics: a New Introduction to Your Environment* by Hewitt (published in 1971):

Although the innermost electrons in an atom are bound very tightly to the oppositely charged nucleus, the outermost electrons of many atoms are bound very loosely and can be easily dislodged. The force with which the outer electrons are held in the atom varies for different substances. The electrons are held more firmly in rubber than in fur, for example. Hence, when a rubber rod is rubbed by a piece of fur, electrons transfer from the fur to the rubber rod. The rubber therefore has an excess of electrons and is said to be negatively charged. The fur, in turn, has a deficiency of electrons and is said to be positively charged. (In Munby, 1976, p. 121)

The description of electrical phenomena in this passage (e.g., electrons are bound, dislodged, transferred) conveys an entirely different message concerning the ontological status of electrons. Here the role of humans in producing explanations for phenomena is missing and students are left to infer that scientists simply found electrons while doing experiments. Therefore, Munby argued, students may derive different views about the nature of science depending on the way that science knowledge is communicated in textbooks. He further suggested that the societal attitudes toward science can be explained by the way that science has been communicated in schools. "The fact that science is viewed as a source of true, reliable, and dependable knowledge might be a consequence of it being taught that way" (Munby, 1976, p. 123).

In a more recent study, Jacoby and Spargo, (1989) examined a total of sixteen physical science textbooks from Britain, the United States, and South Africa for their portrayal of the nature of science. They concluded that "With the exception of one textbook . . . , all the texts examined revealed a predominance of an inductivist-empiricist approach" (p. 45). Textbook readings will, in ways not anticipated by most science teachers, convey a definite view about the status of scientific knowledge and how that knowledge came to be.

Other curricula decisions also impact students' notions about the nature of science. As an example, the form of laboratory activities conveys much about science processes and the construction of knowledge. Unfortunately, these experiences are often cookbook or verification type laboratory activities which again portray science as a rhetoric of conclusions totally divorced from human influence. Clough and Clark, (1994) have suggested a different approach to laboratory exercises that more actively engage students in science content and accurately portray many significant issues in the nature of science. They advocate placing students in small research teams that are responsible for developing experiments to investigate a particular question posed by the instructor. Students must make important decisions concerning the experimental set-up, collection of relevant data, its interpretation, and judgements regarding the veracity of their work. Ensuing negotiation of meaning conveys a very different picture of science than typical cookbook/verification lab work where students are following recipes to get preordained results. This suggestion is supported by the work of Burbules, (1991) who suggests that classrooms should look like a research laboratory where students participate in science activities as part of a social group. Earlier, Haukoos and Penick, (1983) compared the influence of two classroom climates on students' learning of science process skills and content achievement in college level classes. While maintaining even gains in biology content, the students in the discovery climate classroom achieved significantly higher scores in science process skills as measured by the Welch Science Process Inventory.

The Role of the Teacher

Despite the pervasive and critical role of curricula, evidence is clear and substantial that teachers are the most influential factor in educational change (Duffee and Aikenhead, 1992; Eylon and Linn, 1989; Fullan, 1991; Good and Brophy, 1987; Koballa and Crawley, 1985; Laforgia, 1988; Langer and Applebee, 1987; and Shymansky and Penick, 1981) and that teachers make exemplary programs (Berliner, 1989; Penick, Yager, and Bonnstetter, 1986). For instance, after observing how science teachers assimilated new writing activities into their old ways of thinking, Langer and Applebee (1987, pp. 87, 137) wrote:

> For those who wish to reform education through the introduction of new curricula, the results suggest a different message. We are unlikely to make fundamental changes in instruction simply by changing curricula and activities without attention to the purposes the activities serve for the teacher as well as for the student. It may be much more important to give teachers new frameworks for understanding what to count as learning than it is to give them new activities or curricula. ... [T]o summarize bluntly, given traditional notions of instruction, it may be impossible to implement successfully the approaches we have championed.

Teachers translate the written curriculum into a form ready for classroom application and decide what, how and why to learn. As Eisner, (1985, p. 59) writes, "In the final analysis, what teachers do in the classroom and what students experience define the educational process." In fact, curricula has been claimed to constitute only 5 percent of the variance in students' learning (Welch, 1979), while science teachers' beliefs, knowledge, and practices represent the bulk of what the science instructional experience is for students (Smith, 1980).

One of the dominant activities in the classroom is teacher talk, and, therefore, important implications for student understanding could be derived from an analysis of teachers' verbal behavior. Munby's analysis of language applies equally well to teachers' verbal patterns.

Zeidler and Lederman, (1989) extended Munby's findings by investigating whether or not teachers' presentation of subject matter has an impact on students' formulation of a world view of science. They administered the Nature of Scientific Knowledge Scale (NSKS) to 18 science teachers and their 409 students at the beginning and end of a semester and isolated six variables that "reflected Munby's distinction between Realist and Instrumentalist language." The six variables "represent teachers' conceptions of the nature of science by way of the language teachers use to convey the subject matter." In an earlier paper Zeidler and Lederman, (1987) concluded from this analysis (pp. 6-7) that:

> The results reveal that the variables Testable, Developmental, Arbitrary Constructs, Anthropomorphic Language, Creativity and Subjective were highly significant in distinguishing between instrumental and realist conceptions of the nature of science with respect to teachers' language and subsequent changes in students' orientation. ... it is concluded that the ordinary language teachers use to communicate science content does provide the context in which students formulate their own conceptions of the nature of science.

For example, when the teacher used ordinary language in discussing science constructs, the students tended to have a realist conception of science. A selection in which the teachers' use of language is clear comes from Zeidler and Lederman, (1989); "This portion of the amino acid is called the amino group. It contains a nitrogen atom and two hydrogens. Always and forever . . . Exactly, always and forever" (p. 780).

Alternatively, students tended to develop an instrumentalist view of science when the teacher used precise language in presenting science constructs. For example, . . . the periodic table is just something created by scientists to organize all the elements . . . This brings up again another problem that always exists in classification. Remember I told you that living organisms don't always fit into the neat little classifications that we have made up . . ." (Zeidler and Lederman, 1989, p. 778).

The conclusion is inescapable. The teachers' use of language influences their students' views of the nature of science. Lederman, (1986), Haukoos and Penick, (1983), Yager, (1966), and Dibbs, (1982) have all concluded that the way teachers conduct instruction in the classroom influences the way students think of science. In particular, the way teachers verbally present scientific enterprise has implications for the way in which students will form their views of science (Munby, 1976 and Ziedler and Lederman, 1989).

NATURE OF SCIENCE AND SCIENCE EXPERIENCES

Requiring that preservice teachers take more science courses and experience authentic laboratory research projects are often thought to be the solution to improving prospective science teachers' understanding of the nature of science. A number of studies have investigated whether a relationship exists between knowledge of science content and an understanding of the nature of science. For instance, Craven (1966) investigated the relationships between the total science grade point average earned by prospective teachers and their understanding of science measured by TOUS. The data revealed no significant correlation between the two factors.

Carey and Stauss, (1968) administrated the Wisconsin Inventory Science Process (WISP) test to prospective and experienced science teachers, (1970). The WISP scores were correlated variables including the number of high school science units, college mathematics, biology, physical science and total science units. Further correlations were performed with college mathematics, biology, physical science and total science grade point averages. No significant correlations were found, leading Carey and Stauss to conclude that simply taking science classes or even doing well in such classes does not communicate much in the way of NOS knowledge. This conclusion is substantiated by Olsted, (1969) who showed that no relationship exists between an understanding of the nature of science (as measured by the Test on

Understanding Science) and science content knowledge as measured by the Advanced General Science Test.

Anderson, Harty, and Samuel, (1986) used the Nature of Science Scale (NOSS) in comparing two groups (1969 and 1984) of science teachers' views concerning the nature of science. The two groups differed in the number of semester hours of science completed for certification requirements. Even though both groups demonstrated faulty views of science, the 1984 group with 51 semester hours of science had a significantly higher NOSS score than the 1969 group with 64 semester hours of science.

Rymonda, (1986) analyzed prospective secondary school science teachers' understanding of the nature of scientific knowledge. She identified academic variables that might contribute to the teachers' views of science and compared these to the subjects' scores on the Nature of Scientific Knowledge Scale (NSKS). With the exception of "orientation toward science process," she found no significant correlation between teachers' understanding of the nature of science and science major GPA, overall science GPA, number of science major semester hours, and total number of science semester hours.

Other researchers have investigated the influence of authentic research experiences on participants' views of science. Benhke, (1961) investigated a group of scientists and a group of secondary science teachers' regarding the nature of science, science and society, the scientist and society, and the teaching of science. More than one half of the teachers and 20% of the scientists incorrectly viewed the content of science as fixed and unchangeable. He found that teachers differed most from the scientists on statements which involve an understanding of the scientific enterprise in some depth. Teachers and scientists disagree most frequently on statements related to the goals and limitations of science.

Kimball, (1967) used the NOSS nature of science instrument and administered it to college graduates. He was interested in investigating the differences in nature of science knowledge between those science majors who entered professional science careers and those who became science teachers. He wanted to know if the difference in understanding the nature of science could be attributed to training or experience. Using NOSS, he concluded that no significant difference in understanding the nature of science existed between scientists and science teachers.

Visavateeranon, (1992) studied the effect of research experience on teachers' perceptions of the nature of science. The study involved thirty-one teachers who worked on research projects with university faculty members in the lab or in the field for two to four weeks. The researcher compared teachers' pre- and post-experience perceptions of science using the Nature of Science - Key Features Test, lesson plans, interviews, and journals. Only 18-27% of teachers as measured by various instruments changed their traditional views of science to more contemporary ones. Interestingly, the researcher recommended linking history of science reading materials and organized seminars about the nature of science to the research experience so teachers would better understand the nature of science.

An interesting explanation for the misunderstanding of science held by science teachers and scientists was provided by Pomeroy, (1993) who used a fifty-item survey to investigate scientists' and elementary and secondary science teachers' views about the nature of science. She found that science teachers and scientists expressed traditional views of the nature of science. They perceived science as objective, empirical, and involved with issues of the control of nature. She suggested that the positivistic ideas expressed by scientists and science teachers are due to the "scientists and secondary science teacher's deep initiation into the norms of the scientific community" (p. 269). They see themselves as role models and as such are likely to present normative rather than realistic views of science. Another reason, she suggested, could be that the scientists are working within the "normal science" paradigm (Kuhn, 1970) and so used accepted theories to answer new questions. Therefore, their work may resemble positivistic modes of scientific method.

Despite naive commonsense solutions for improving science teachers' views regarding the nature of science, simply requiring extensive science course work and research experience will not help. The general conclusion from these studies is that neither extensive science course work nor research experience is sufficient for promoting an understanding of significant issues in the social studies of science.

NATURE OF SCIENCE IN SCIENCE TEACHER EDUCATION

A dynamic understanding of science requires significant background in the social studies of science—taught to teachers in a manner so this knowledge is connected to what Lederman, (1992) argues and Clough, (1997) describes are those "specific instructional behaviors, activities, and decisions implemented within the context of a lesson." Obviously, science teacher education programs are in the best position to ensure that these practices are taught to science teachers. Without question, the calls for improving teachers' views of the nature of science through the inclusion of history and philosophy of science (HPS) courses in teacher education programs has a long and consistent history (Abimbola, 1983; Anderson, et al., 1986; Gill, 1977; Harms & Yager, 1981; Kimball, 1967; King, 1991; Loving, 1991; Manuel, 1981; Martin, 1972; Matthews, 1989, 1990, 1994; Nunan, 1977; Robinson, 1969 and Summers, 1982).

However, Summers, (1982) noted that undergraduate science and science teacher education curricula do not emphasize the philosophical background of science content. As recently as 1988, Hodson lamented that "the frequent calls for philosophy of science to become a major component in teacher-training courses have gone largely unheeded, so that there are now several generations of serving teachers with little or no understanding of basic issues in the philosophy of science and their significance in the design of effective learning experiences." More recently, Gallagher, (1991) indicated that science teacher education programs do not seem to value nature of science instruction, and so, not surprisingly, teachers do not know how to teach about the nature of science. Loving, (1991) surveyed 17 science educators whose institutions have undergraduate and/or graduate science education programs and found that only "13% of undergraduate science education majors and 19% of graduate students have a philosophy of science course in their degree plan . . ." (p. 828). Sadly, these numbers are likely optimistic. Matthews, (1994) reports that in the United States, only the state of Florida requires prospective science teachers to complete a course in the history and philosophy of science. The lesson is clear that preservice science teachers arrive with largely unexamined beliefs about the nature of science, and, too often, graduate with such beliefs unchallenged in their teacher education programs (Haggerty, 1992; O'Brien and Korth, 1991). What this means is that teachers' ideas about the nature of science are "picked up indirectly" (Matthews, 1994) rather than deliberately during their formal science teacher education experience.

The Effectiveness of Nature of Science Instruction

The lack of formal nature of science instruction in science teacher education programs is particularly disappointing when the evidence indicates that explicit instruction in the social studies of science does succeed in changing teachers' views concerning the nature of science. For example, Carey and Strauss, (1968) investigated the influence of a science methods course on prospective secondary science teachers' views of science. The methods course started with an introduction to the nature of science through lectures, discussion, and outside readings. The pretests and posttests indicate significant enhancement in understanding the nature of science as measured by the *Wisconsin Inventory Science Process* (WISP). Similar results were obtained with experience science teachers using WISP (Carey and Strauss, 1970).

Billeh and Hasan, (1975) conducted a four-week summer program for Jordanian secondary science teachers. The program consisted of lectures and demonstrations in teaching science, laboratory investigations emphasizing a guided discovery approach, twelve fifty-minute lectures on the nature of science, and enrichment activities designed to further promote understanding of science concepts and scientists at work. The findings indicated a significant positive difference when the posttest results were compared with the pretest measures of understanding the nature of science as measured by the *Nature of Science Test* (NOST).

Lavach, (1969) studied an inservice program emphasizing the historical development of selected physical science concepts and noted that participants made statistically significant gains in their understanding regarding the nature of science as measured by the total score on TOUS. King, (1991) surveyed and interviewed thirteen student teachers at Stanford University about their beliefs, knowledge, and attitudes toward the history and philosophy of science. He found that the three participants who were formally exposed to the history or philosophy of science had more reasoned responses to the philosophical questions. Ogunniyi, (1983) and Akindehin, (1988) report on the effect of an independent NOS course in their preservice science teacher programs. They found a significant difference in the understanding of science between students who took such a course and those who did not.

While the knowledge gains reported in these studies are encouraging, sadly, little attention or research has been directed to the role science teacher education should play in ensuring such knowledge is forcefully implemented in science curricula and teaching strategies. Fortunately, exceptions to the rule do exist. In Australia, Matthews, (1990) designed a course in the history and philosophy of science (HPS) to stimulate teachers' interest in science and help them to identify the HPS issues that arise in science teaching. In Nigeria, Ogunniyi, (1983) devoted an independent course in history and philosophy of science in the science teacher preparation program for his students. Recently, Eichinger, Abell, and Dagher, (1997) presented their experience in developing and teaching a NOS course to science education graduate students at both Purdue and the University of Delaware. At the University of Iowa, two semester-length courses in the nature of science have been required of all preservice teachers and science education graduate students for more than two decades.

Description of a Desired State for Nature of Science

While acknowledging the existence of institutional constraints that thwart efforts to incorporate the nature of science in science teacher education programs, significant explicit attention to the social studies of science and relevant teaching strategies are imperative to ensure such instruction is extensively integrated in school science programs. Without this attention, the overwhelming demands placed on science teachers will prolong the current sorry state of affairs. For example, Bell et al., (1997) recently followed several preservice teachers through their student teaching experience to determine how extensively they implemented the nature of science. Most of these teachers did not show significant explicit attention to teaching the nature of science, and one subject in their study made the statement, "I don't plan to teach the nature of science . . . I don't think it is something that I would spend a great deal of time on."

While many science teachers likely take this same position, ironically, some view of the nature of science will always be communicated to students whether or not it is done purposefully or accurately. Consider a typical day in the life of a science teacher. Class begins with an opening statement, then an activity or demonstration occurs followed by some sort of discussion before a textbook assignment is given as homework. Throughout these daily experiences are explicit and implicit cues regarding how science works and the status of scientific knowledge.

Clearly, a rationale for the social studies of science in science education should spiral through all aspects of a science teacher education program. Nature of science concepts will likely seem esoteric to teachers so practical classroom applications must be extensive. Examining nature of science instruments is imperative to address the daunting task teachers face in identifying and tracking 125 students' views regarding the nature of science. Scrutinizing readings, audiovisual materials, and activities that accurately portray the nature of science while teaching science content is also imperative. The paucity of such materials necessitates that teachers be prepared to revise curricula so they effectively teach both the nature of science and science content. Presentations by practicing science teachers who implement nature of science instruction in their classes may be particularly useful in convincing skeptical teachers. Critical incidents illustrating the inseparable ties between science content instruction and the nature of science (Nott and Wellington, 1995) are valuable for increasing the sensitivity of preservice teachers to this subtle, but very real, juxtaposition. Alongside these efforts, science teacher educators must model behaviors, strategies, and language that accurately portray the nature of science. Literature and organizations devoted to the social studies of science and its implications for science teaching should be introduced so teachers have resources to seek out when a need arises. Finally, research and anecdotal reports indicating positive changes in students' views and actions regarding the nature of science are needed to bolster teachers confidence that attention to these issues will reap the desired effects (Clough, 1995).

THE NATURE OF SCIENCE IN THE CLASSROOM[1]

Four approaches to incorporating the nature of science in science teacher education programs are often suggested. Each of the four approaches has its own strengths and limitations and these are discussed below. One approach, where nature of science instruction is blended into methods classes, is most often applicable in the preservice environment. Of course, some programs offer an advanced science methods course for experienced teachers, so the first approach to incorporating the nature of science may also work for inservice teachers. The other three strategies may, with modifications, be used in either the preservice domain or applied in general science instruction or science teacher education. Regardless whether one particular approach or a combination of approaches is used, several questions remain. These include the minimal amount of nature of science content needed by science teachers, who should teach such courses, and whether they should be required for both undergraduate and graduate students.

Nature of Science in Methods Courses

In this strategy, the content and pedagogical strategies of the nature of science are communicated within a science teaching methods course.

Advantages to NOS in Methods Courses

The central advantage here is that nature of science content is discussed in an environment where the curriculum and pedagogical connections can be immediately discussed. In this fashion, both the nature of science content, a rationale for its inclusion in science teaching, and the strategies for teaching that content will be conveyed to prospective science teachers. In addition, because practicum experiences should be tied to methods course, students have the needed opportunities to put these strategies into action with students in schools. In addition, students (and perhaps other education faculty members) may be more receptive to the unfamiliar topics within the nature of science by encountering them with familiar faculty discussing familiar and applicable topics. Finally, this approach may facilitate discussions of the parallels between the development of science itself and the learning of science.

Disadvantages to NOS in Methods Courses

The disadvantages associated with blending the nature of science with methods include the possibility that other topics may be neglected or that the NOS itself may get less than adequate treatment given the many other issues needing attention in methods courses. Depending on the skill and knowledge of the instructor, the suggested NOS teaching methods may illustrate how to apply some NOS concepts, but miss other fundamental ideas and their integration with science content.

Nature of Science in Science Content Classes

This approach requires that significant attention to relevant nature of science issues be intertwined with the teaching of science content. In the pre-college environment this approach may be most strongly recommended, but in the college domain, preservice science teachers may have problems deriving teaching strategies from pure nature of science content.

Advantages to NOS in Science Content Classes

One of the advantages to this approach is that students can see the application of the nature of science in context, thus legitimizing the nature of science as a useful domain. Instructors who take such an approach may cause change in the educational infrastructure itself by encouraging other instructors to include nature of science elements in their classes. Moreover, students in the class who are interested in the education have explicit modeling to draw from when designing instruction.

Disadvantages to NOS in Science Content Classes

Central among the disadvantages for this approach is the fact that science faculty members may not know how to discuss NOS issues. Also, in the college environment, methods students learning about the nature of science in their science classres may not receive useful strategies enabling them to share nature of science content with their own students. The critical challenge that remains is the question of who will provide science teachers with specific strategies permitting them to integrate NOS in their classes.

Teachers as Scientists

In this strategy, teachers of the nature of science should have had some authentic experience actually *doing* science so that they can talk with some authority about how science is done.

Advantages to the Teachers as Scientists Approach

The advantage of this plan is that if individuals learn about the nature of science by doing scientific research, those participating could speak with authority and enthusiasm about the NOS from first-hand experience. Another advantage is that those who have had such experiences would be more able to guide students in pursuing their own research and science fair projects.

Disadvantages to the Teachers as Scientists Approach

The primary disadvantage to this plan is found in the assumption that those who do science learn enough about the nature of science to communicate it accurately. The studies reported earlier in this chapter clearly show that research experience confers no guarantee that NOS knowledge is assimilated accurately simply through working in a laboratory. In addition, the necessary strategies to translate NOS content into classroom practice are missing.

Formal Courses or Units of Study in the Nature of Science

This plan would have science teachers learn about the nature of science in a discrete unit of study or in a course taught by a science educator. In higher education, some might argue against a specific course focusing on the nature of science for science teachers on the grounds that most major universities already provide formal history and philosophy of science courses taught by experts in those fields. Unfortunately, such courses tend to be abstract examinations of science usually from a highly theoretical and prescriptive perspective commonly focusing primarily on a specific science discipline replete with symbolic logic and somewhat arcane (at least for the needs of science teaching) debates. While such courses are recommended for majors in those fields, and perhaps as continuing education for science teachers, the highly focused content is unlikely to be useful to the uninitiated science student or education major.

Advantages to Formal NOS Courses

A specific nature of science course or unit of study would guarantee that students see the fruitful connections between history, philosophy, sociology and psychology of science, not as a prescriptive endeavor within one science discipline, but as an authentic description of how science is practiced generally. If such a course were taught by a science educator, experiences in the course would likely be tailored to the needs of practicing science teachers. In addition, discrete courses could be implemented without revising other aspects of the education curriculum.

Disadvantages to Formal NOS Courses

Specific courses invariably require additional time that may impact other useful courses thus prolonging the time needed to complete a degree or certification program. Also, a discrete NOS course may be disconnected from science content possibly diluting its relevance. Finally, if the course involves science teachers then it is vital that such courses connect to problems of teaching NOS.

CONCLUSIONS

This chapter has explored the dynamic arena of the nature of science by examining both its history and ways that the nature of science has informed and should guide science teaching. We have taken the position that a pragmatic consensus exists regarding some of the most important elements regarding the process of science, but have demonstrated that constructive debate exists. Research and discussion continues regarding the relationship between what teachers believe about the nature of science and what they then communicate to students. We assert that teachers must have experiences where they explore the social studies of science and contemplate the methods by which that content may be shared with students. This is the core purpose for developing this book, a book of rationales and strategies. It is vital that the science education community provide an accurate view of how science operates to students and by inference to their teachers. What follows in the accompanying chapters are tested strategies for doing just that. Whether these plans find a home in teacher education programs, in school classrooms, or simply in the minds of interested individuals, we are confident that science education will be a richer discipline and our students will be more adequately prepared for their lives as citizens when they are afforded a fuller understanding of the nature of this thing called science.

[1,3] *University of Southern California, Los Angeles, California, USA*
[2] *University of Iowa, Iowa City, Iowa, USA*

NOTE

1. The section addressing the advantages and disadvantages of various approaches for the integration of nature of science into the teacher education curriculum is adapted from the results of an informal synthesis of a large group discussion on the topic by participants attending the *Third International History and Philosophy of Science in Science Teaching Conference* held at the University of Minnesota, Minneapolis, MN, USA from October 29 to November 1, 1995. The authors sincerely acknowledge all of those whose ideas have been used here.

REFERENCES

Abimbola, O. A. (1983). 'The relevance of the "new" philosophy of science for the science curriculum', *School Science and Mathematics*, (83), 183-190.

Aikenhead, G.S., Ryan, A.G. & Fleming, R.W. (1989). *Views on science-technology-society, Form CDNmc5* Saskatoon, SK, Canada, Dept. of Curriculum Studies, College of Education, University of Saskatoon.

Aikenhead, G. & Ryan, A. (1992). 'The development of a new instrument : "Views on Science-Technology-Society" (VOSTS)', *Science Education*, (76), 477-491.

Akindehin, F. (1988). 'Effect of an instructional package on preservice science teachers, understanding of the nature of science and acquisition of science-related attitudes,' *Science Education*, (72), 73-82.

American Association for the Advancement of Science (1989). *Project 2061: Science for all Americans*, Washington, D.C.

American Association for the Advancement of Science (1993). *Benchmarks for science literacy*, New York, Oxford University Press.

Association for Science Education (1981). *Education Through Science: An ASE Policy Statement*, Hatfield, England.

Anderson, K.E. (1950). 'The teachers of science in a representative sampling of Minnesota schools', *Science Education*, (34), 57-66.

Anderson, H.O., Harty, H. & Samuel, K.V. (1986). 'Nature of science, 1969 and 1984: Perspectives of preservice secondary science teachers', *School Science and Mathematics*, (86), 43-50.

Bell, R., Lederman, N.G., & Abd-El-Khalick, F. (1997, January). *Developing and acting upon one's conception of science: The reality of teacher preparation*, paper presented at the 1997 Association for the Education of Teachers in Science (AETS) meeting, Cincinnati, OH.

Benhke, F. L. (1961). 'Reactions of scientists and science teachers to statements bearing on certain aspects of science and science teaching', *School Science and Mathematics*, (61), 193-207.

Bentley, M.L & Garrison, J. W. (1991). 'The role of philosophy of science in science teacher education', *Journal of Science Teacher Education*, (2), 67-71.

Billeh, V.Y. & Hassan, O. E. (1975). 'Factors affecting teachers gain in understanding the nature of science', *Journal of Research in Science Teaching* (12), 209-219.

Brickhouse, N. W. (1989). 'The teaching of the philosophy of science in secondary classrooms: Case studies of teachers: personal theories', *International Journal of Science Education*, (11), 437-449.

Brush, S.G. (1989). 'History of science and science education', *Interchange*, (20), 60-71.

Burbules, N.C. (1991). 'Science education and philosophy of science: Congruence of contradiction?', *International Journal of Science Education*, (13), 227-241.

Carey, L. R. & Stauss, A. N. (1968). 'An analysis of the understanding of the nature of science by prospective secondary science teachers', *Science Education*, (52), 358-363.

Carey, L. R. & Stauss, A.N. (1970). 'An analysis of experienced science teachers' understanding of the nature of science', *School Science and Mathematics*, (70), 366-376.

Cawthron, E.R. & Rowell, J.A. (1978). 'Epistemology and science education', *Studies in Science Education*, (7), 279-304.

Clark, C. (1988). 'Asking the right questions about teachers preparation: Contributions of research on teaching thinking', *Educational Researcher*, (17), 5-12.

Clough, M.P. (1997). 'Strategies and activities for initiating and maintaining pressure on students' naive views concerning the nature of science', *Interchange*, (28),191-204.

Clough, M.P. (1994). 'Diminish students resistance to biological evolution', *The American Biology Teacher*, (56),409-415.

Clough, M.P. (1995). 'Longitudinal understanding of the nature of science as facilitated by an introductory high school biology course', in F. Finley, D. Allchin, D. Rhees & S. Fitfield (eds.), *The Proceedings of the Third International History, Philosophy and Science Teaching Conference, Vol. II.*, Minneapolis, MI, University of Minnesota, 202-211.

Clough, M.P. & Clark, R.L. (1994). 'Creative constructivism: Challenge your students with an authentic science experience', *The Science Teacher*, (61), 46-49.

Conant, A.B. (1951). *On understanding science: An historical approach*, New York, New American Library.

Connelly, F.M., Wahlstrom, M.W., Finegold, M., & Elbaz, F. (1977). *Enquiry teaching in science: A handbook for secondary school teachers*, Toronto, Ontario Institute for Studies in Education.

Craven, G. F. (1966). 'Critical thinking abilities and understandings of science by science teacher candidates at Oregon State University', *Dissertation Abstracts International* (27), 125A.

Dagher, Z.R. and BouJaoude, S. (1997). 'Scientific views and religious beliefs of college students: The case of biological evolution', *Journal of Research in Science Teaching*, (34),429-445.

DeBoer, G. (1991). *A history of ideas in science education: Implications for practice*, New York, Teachers College Press.

Dibbs, D.R. (1982). 'An investigation into the nature and consequences of teachers implicit philosophies of science', Unpublished Doctoral Dissertation, University of Aston, Birmingham, UK.
Driver, R., Leach, J., Miller, & Scott, P. (1996). *Young peoples images of science,* Bristol, PA, Open University Press.
Duffee, L. & Aikenhead, G. (1992). 'Curriculum change, student evaluation, and teacher practical knowledge', *Science Education,* (76), 493-506.
Duschl, R. A. (1994). 'Research on the history and philosophy of science', in L.G. Dorothy (ed), *Handbook of research on science teaching and learning,* New York, MacMillan, 445- 455.
Duschl, R. A. (1990). *Restructuring science education: The importance of theories and their development* New York, Teachers College Press.
Duschl, R. A. (1988). 'Abandoning the scientistic legacy of science education', *Science Education,* (72), 51-62.
Duschl, R.A. (1987). 'Improving science teacher education programs through inclusion of history and philosophy of science, in J.P. Barufaldi (ed.) *Improving Preservice/Inservice Science Teacher Education: Future Perspectives, The 1987 AETS Yearbook.* Association for the Education of Teachers in Science.
Duschl, R.A. & Wright, F. (1989). 'A case study of high school teachers' decisions -- Making models for planning and teaching science', *Journal of research in science in science teaching,* (26), 467-502.
Eichinger, D.C., Abell, S.K., and Dagher, Z.R. (1997). 'Developing a Graduate Level Science Education Course on the Nature of Science', *Science & Education,* (6), 417-429.
Einstein, A. (1933). 'On the method of theoretical physics', Herbert Spencer lecture delivered at Oxford, June 10, 1933. Published in Mein Welbild, Amsterdam: Querido Verlag, 1934. Published in "Ideas and Opinions" New York, Crown (1982).
Eisner, E. W. (1985). *Educational imagination: On the design and evaluation of school programs,* Second Edition, New York, MacMillan.
Elbaz F. (1981). 'The teacher's "practical knowledge": Report of a case study', *Curriculum Inquiry,*(11), 43-1.
Elkana, Y. (1970). 'Science, philosophy of science and science teaching', *Educational Philosophy and Theory,* (2), 15-35
Elmer-Dewitt, P. (1994). 'Don't tread on my lab', *Time,* (143), 44-45.
Ennis, R. H. (1979). 'Research in philosophy of science bearing on science education', in *Current Research in Philosophy of Science,* P. D. Asquith & H. E. Kyburg (eds.), East Lansing, MI, Philosophy of Science Association, 138-170.
Evans, J.D. & Baker, D. (1977). 'How secondary pupils see the sciences', *School Science Review,* (58), 771-774.
Eylon, B. & Linn, M. (1989). 'Learning and instruction: An examination of the research perspective in science education', *Review of Educational Research,* (58), 251-301.
Fleury, S.C. & Bentley, M.L. (1991). 'Educating elementary science teachers : Alternative conceptions of the nature of science', *Teaching Education,* (3), 57-67.
Fullan, M.G. (1991). *The new meaning of educational change, 2nd ed.,* New York, Teachers College Press
Gallagher, J.J. (1991). 'Prospective and practicing secondary school science teacher's knowledge and beliefs about the philosophy of science', *Science Education,* (75), 121-133.
Giddings, J.G. (1982). 'Presuppositions in school science textbooks', Unpublished doctoral dissertation, University of Iowa, Iowa City, Iowa.
Gill, W. (1977). 'The editor's page', *The Australian Science Teacher's Journal,* (23), 4.
Ginev, D. (1990). 'Towards a new image of science: Science teaching and non-Analytical philosophy of science', *Studies in Philosophy and Education,* (10), 63-71.
Good, T.L. & Brophy, J. E. (1987). *Looking in Classrooms, 4th Edition,* New York, Harper and Row.
Haggerty, S. M. (1992). 'Student teachers perceptions of science and science teaching', in S. Hills (ed.), *The history and philosophy of science in science education,* Kingston, Queen's University, 483-494.
Harms, H. & Yager, R. E. (1981). *What research says to the science teacher, Vol. 3,* Washington D.C., National Science Teachers Association.
Haukoos, G.D. & Penick, T. E. (1983). 'The influence of classroom climate on science process and context achievement of community college students', *Journal of Research in Science Teaching,* 20, 629-637

Herron, M.D. (1969). 'Nature of science: Panacea or Pandora's box', *Journal of Research in Science Teaching*, (6), 105-107.
Hodson, D. (1988). 'Toward a philosophically more valid science curriculum', *Science Education*, (72), 19-40.
Hodson, D. (1991). 'The role of philosophy in science teaching', in M. R. Matthews (ed.). *History, philosophy and science teaching: Selected reading*. New York, Teachers College Press.
Hodson, D. (1986). 'Rethinking the role and status of observation in science curriculum', *Journal of Curriculum Studies*, (18), 381-396.
Hodson , D. (1993). 'Philosophical stance of secondary school science teachers, curriculum experiences, and children's understanding of science :Some preliminary findings', *Interchange*, (24), 41-52.
Hollon, R., Roth, K.J. & Anderson, C.W. (1991). 'Science teachers' conceptions of teaching and learning', in J. Brophy (ed.), *Advances in research on teaching: Teachers' knowledge of subject matter as it related to their teaching practice, Vol. 2*. Greenwich, CT, Jai Press, Inc., 145-185.
Hollon, R.E., & Anderson, C.W. (1987). 'Teachers' beliefs about students' learning processes in science: Self-reinforcing belief systems', paper presented at the annual meeting of the American Educational Research Association, Washington, DC.
Hurd, P.DeH. (1969). *New directions in teaching secondary school science*. Chicago, IL, Rand McNally.
Hurd, P.DeH. (1960). 'Summary', in N.B. Henry (ed.) *Rethinking Science Education: The fifty-ninth yearbook of the National Society for the Study of Education*, Chicago, IL, University of Chicago Press, 33-38.
Jacoby, B.A. & Spargo, P.E. (1989). 'Ptolemy revisited? The existence of a mild instrumentalism in some selected high school physical science textbooks', *Interchange* (20), 33-53.
Jaffe, B.(1938). 'The history of chemistry and its place in the teaching of chemistry', *Journal of Chemical Education*, (15), 383-389.
Johnson, R. L. & Peeples, E. E. (1987). 'The role of scientific understanding in college', *The American Biology Teacher*, (49), 93-96.
Jungwirth, E. (1971). 'The pupil, the teacher, and the teacher's image: Some second thoughts of BSCS biology in Israel', *Journal of Biological Education*, (5), 165- 171.
Kilborn, B. (1982). 'Curriculum materials, teaching, and potential outcomes for students: A qualitative analysis', *Journal of Research in Science Teaching*, (19), 675- 688.
Kimball, M. (1967) 'Understanding the nature of science: A comparison of scientists and science teacher', *Journal of Research in Science Teaching*, (5), 110-120.
King, B. (1991). 'Beginning teachers' knowledge of and attitude toward history and philosophy of science', *Science Education*, (75), 135-141.
Klopfer, L. (1969). 'The teaching of science and the history of science', *Journal of Research in Science Teaching*, (6), 87-95.
Klopfer, L. (1964-66). *History of Science Cases (HOSC)*, Chicago, IL, Science Research Associates.
Knorr-Cetina, K.D. (1981). *The manufacture of knowledge: An essay on the constructivist and contextual nature of science*, New York, Pergamon Press.
Koballa, T. & Crawley, F. (1985). 'The influence of attitude on science teaching and learning', *School Science and Mathematics*, (85), 222-32.
Kuhn, T. S. (1970). *The structure of scientific revolutions*, Chicago, IL, The University of Chicago Press.
Laforgia, J. (1988). 'The affective domain related to science and its evaluation', *Science Education*, (72), 407-421.
Lakin, S. and Wellington, J. (1994). 'Who will teach the 'nature of science'?: teachers views of science and their implications for science education', *International Journal of Science Education*, (16),175-190.
Langer, J. A. & Applebee, A. N. (1987). *How writing shapes thinking: A study of teaching and learning*, (Research Report, Number 22), National Council of Teachers of English, Urbana, IL, National Council of Teachers of English.
Lapoint, A., Mead, N. A., & Phillips, G. W. (1989). *A world of differences: An international assessment of mathematics and science*, Princeton, NJ, Educational Testing Service.
Lantz, O. & Kass, H. (1987). 'Chemistry teachers' functional paradigms', *Science Education*, (71), 117-134.
Latour, B. (1987). *Science in action*, Cambridge, MA, Harvard University Press.

Latour, B. & Woolger, S. (1986). *Laboratory life: The construction of scientific facts*, Princeton, NJ, Princeton University Press.

Lavach, J. (1969). 'Organization and evaluation of an in-service program in the history of science', *Journal of Research in Science Teaching*, (6), 166-170.

Lederman, N.G. (1992). 'Students' and teachers' conceptions of the nature of science: A review of the research', *Journal of Research in Science Teaching*, (29), 331-359.

Lederman, N.G. (1983). 'Delineating classroom variables related to students' conception of the nature of science', *Dissertation abstracts international*, (45), 483A (University Microfilms No. 84-10, 728).

Lederman, N.G. (1986). 'Students and teacher's understanding the nature of science: A reassessment', *School Science and Mathematics*, (86), 91-99.

Lederman, N.G. & Zeidler, D. (1987) Science teachers conceptions of the nature of science: Do they really influence teaching behavior, *Science Education*, (71), 721- 734.

Loving, C. (1991). 'The scientific theory profile: A philosophy of science model for science teachers', *Journal of Research in Science Teaching*, (28), 823-838.

Manuel, D.E. (1981). 'Reflections on the role of history and philosophy of science in school science education', *School Science Review*, (62), 769-771.

Martin, M.R. (1972). *Concepts of science education: A philosophical analysis*, Greenview, IL, Scott, Foresman.

Matthews, M.R. (1997). Editorial, *Science & Education*, (6),323-329.

Matthews, M.R. (1994). *History, philosophy, and science teaching: A useful alliance*, New York, Routledge.

Matthews, M.R. (1994). *Science teaching; The role of history philosophy of science*, New York, Routledge.

Matthews, M.R. (1990). *History, philosophy, and science Teaching*, New York, Teachers College Press.

Matthews, M.R. (1990). 'History, philosophy and science teaching: What can be done in an undergraduate course?', *Studies in Philosophy and Education*, (10), 93-97.

Matthews, M.R. (1989). 'A role for history and philosophy in science teaching', *Interchange*, (20), 3-15.

Mellado, V. (1997). 'Preservice teachers' classroom practice and their conceptions of the nature of science', *Science & Education*, (6), 331-354.

Mendelsohn, E., Weingart, P. & Whitley, R. (eds.) (1977). *The social production of scientific knowledge*, Boston, MA, D. Reidel Publishing Co.

Meyling, H. (1997). 'How to change students' conceptions of the epistemology of science', *Science & Education*, (6),397-416.

Miller, P.E. (1963). 'A comparison of the abilities of secondary teachers and students of biology to understand science', *Proceedings of the Iowa Academy of Science*, (70), 510-513.

Moore, J. (1983). 'Evolution, education, and the nature of science and scientific inquiry, in J.P. Zetterberg, (ed.), *Evolution versus creationism: The public education controversy*, Phoenix, AZ, Oryx Press.

Mulkay, M. & Gilbert, N. (1982). 'Accounting for error: how scientists construct their social world when they account for correct and incorrect belief', *Sociology*, (16), 165-183.

Munby, A.H. (1976). 'Some implications of language in science education', *Science Education*, (60), 115-124.

National Research Council (1996). *National science education standards*, Washington, D.C., National Academy Press.

National Science Board (1996). *Science and engineering indicators: 1996*, Washington, D.C., United States Government Printing Office (NSB 96-21).

National Science Teachers Association (1995). *A high school framework for national science education standards*, Arlington, VA, National Science Teachers Association.

Nott, M. & Wellington, J. (1995). 'Critical incidents in the science classroom and the nature of science', *School Science Review*, (76), 276, 41-46.

Nunan, E. (1977). 'History and philosophy of science and science teaching: A revisit', *The Australian Science Teachers' Journal*, (23), 65-71.

O'Brian, G.E. & Korth, W.W. (1991). 'Teachers' self-examination of their understanding of the nature of science: A history and philosophy course responsive to science teachers' needs', *Journal of Science Teacher Education*, (2), 94-100.

Ogunniyi, M.B. (1982). 'An analysis of prospective science teacher's understanding of the nature of science', *Journal of Research in Science Teaching*, (19), 25-32.
Ogunniyi, M.B. (1983). 'Relative effects of a history/philosophy of science course on students teachers' performance on two models of science', *Research in Science and Technological Education*, (1), 193-199.
Olsted, R. (1980). 'Innovated doctrines and practical dilemmas: A case study of curriculum translation.', Unpublished doctoral dissertation, University of Birmingham, England.
Olson, J.K. (1981). 'Teacher influence in the classroom', *International Science*, 10, 259-75.
Olsted, R. (1969). 'The effects of science teaching methods on the understanding of science', *Science Education*, (53), 9-11.
Penick, J.E. Yager, R.E. & Bonnstetter, R. (1986). 'Teachers make exemplary programs', *Educational Leadership*, (44), 14-20.
Pomeroy, D. (1993). 'Implications of teachers' beliefs about the nature of science: Comparison of the beliefs of the scientists, secondary science teachers, and elementary teachers', *Science Education*, (77), 261-278.
Robinson, J.T. (1968). *The nature of science and science teaching*, Belmont, CA, Wadsworth Publishing Company.
Robinson, J.T. (1969). 'Philosophy of science: Implications for teacher education', *Journal of Research in Science Teaching*, (6), 99-104.
Rowell, J.A. & Cawthron, E.R. (1982). 'Images of science: An empirical study', *European Journal of Science Education*, (4), 79-94.
Rubba, P.A., Horner, J.K., Smith, J.M.. (1981). 'A study of two misconceptions about the nature of science among junior high school students', *School Science and Mathematics*, (81), 221-226
Ryan, A.G. & Aikenhead, G.S. (1992). 'Students' preconceptions about the epistemology of science', *Science Education*, (76), 559-580.
Rymonda, M.A. (1986). 'Analysis of the prospective secondary school science teacher's understanding of the nature of scientific knowledge', Unpublished doctoral dissertation, Bloomington, IN, Indiana University.
Scharmann, L.C. & Harris, W.M. (1992). 'Teaching evolution: Understanding and applying the nature of science', *Journal of Research in Science Teaching*, (29), 375-388.
Scheffler, I. (1973). *Reason and teaching*, Indianapolis, IN, Bobbs-Merrill.
Schmidt, D.J. (1967). 'Test on understanding science: A comparison among school groups', *Journal of Research in Science Teaching*, (4), 365-366.
Schwab, J.J. (1964). 'The teaching of science as enquiry', in J. J. Schwab & P. F. Brandwein (eds.), *The teaching of science*, Cambridge, MA, Harvard University Press, 31-102.
Shamos, M. (1995). *The myth of scientific literacy*, New Brunswick, NJ, Rutgers University Press.
Shapin, S. (1982). 'History of science and its sociological reconstruction', *History of Science*, 20, 157-211.
Shavelson, R.J. & Stern, P. (1981). 'Research on teachers' pedagogical thoughts, judgments, decisions, and behavior', *Review of Educational Research*, (51), 455-498.
Shulman, L.S. (1986). 'Those who understand: Knowledge growth in teaching', *Educational Researcher*, (15), 4-14.
Shymansky, J. & Penick, J. (1988). 'Teacher behavior does make a difference in hands-on science classrooms', *School Science and Mathematics*, (81), 412-422. .
Smith, H.A. (1980). 'A report on the implications for the science community on three NSF-sponsored studies of the state of precollege science education', in *What are the needs in precollege science, math, and social science education? Views of the field*, Washington, DC, National Science Foundation, 55-78.
Smith E.L. & Anderson C.W. (1984). 'Plants as producers: A case study of elementary science teaching', *Journal of Research in Science Teaching*, 21, 685- 698.
Smith, M.U, Lederman, N.G., Bell, R.L., McComas, W.F. and Clough, M.P. (1997). 'How great is the disagreement about the nature of science? A response to Alters (1997)', *Journal of Research in Science Teaching* (34), 1101-1103.
Songer, N. & Linn, M. (1991). 'How do students' views of science influence knowledge integration?', *Journal of Research in Science Teaching*, (28), 761-784.
Stake, R.E. & Easley, J.A. (1978). *Case studies in science education, Volumes I & II*, Washington, DC, United States Government Printing Office.

Summers, M. (1982). 'Philosophy of science in the science teacher education curriculum', *European Journal of Science Education,* (4), 19-27.

Thompson, A.G. (1982). 'Teacher's conceptions of mathematics and mathematics teaching: Three case studies', Unpublished doctoral dissertation, University of Georgia. Abstract from proQuest File, Dissertation abstracts item, ADD82-28729.

Tobias, S. (1990). *They're not dumb: They're different: Stalking the second tier,* Tucson, AZ, Research Council.

Tobin, K. and McRobbie, C.J. (1997). 'Beliefs about the nature of science and the enacted science curriculum', *Science & Education,* (6), 355-371.

Tuan, H. (1991, April). 'The influence of preservice secondary science teacher beliefs about science on pedagogy on their planning and teaching', paper presented at the annual meeting of the National Association of Research in Science Teaching, Lake Geneva, Wisconsin.

Visavateeranon, S. (1992). 'Effect of research experiences on teachers' perceptions of the nature of science', Abstract from proQuest File, Dissertation Abstract Item, 9239162.

Wandersee, J. (1986). 'Can the history of science help science educators anticipate students' misconceptions', *Journal of Research in Science Teaching,* 23, 581-597.

Welch, W. (1979). 'Twenty years of science education', *Review of Research in Education,* 7, 282-306.

Welch, W. W. (1984). 'A science-based approach to science learning', in D. Holdzkom & Lutz (eds.), *Research within reach: Science education,* Washington, D.C., National Science Teachers' Association, 161-170.

Weiss, I. R. (1993). 'Science teachers rely on the textbook', in R.E. Yager (ed.), *What research says to the science teacher, Volume Seven: The science, technology, society movement.* Washington, D.C., National Science Teachers Association.

Yager, R. E. (1966). 'Teacher effects upon the outcomes of science instruction', *Journal of Research in Science Teaching,* (4), 236-242.

Zeidler, D. & Lederman, N.G. (1987). 'Science teacher's conceptions of the nature of science: Do they really influence teaching behavior', *Science Education,* (71), 721-734.

Zeidler, D. & Lederman, N.G. (1989). 'The effects of teachers' language on students' conceptions of the nature of science', *Journal of Research in Science Teaching,* (26), 771-783.

WILLIAM F. McCOMAS AND JOANNE K. OLSON

2. THE NATURE OF SCIENCE IN INTERNATIONAL SCIENCE EDUCATION STANDARDS DOCUMENTS

This chapter provides one solution to the problem of what elements best represent a description of how science operates at a level that is both authentic and appropriate for science learning contexts. A qualitative analysis of recent science education standards documents from several countries has demonstrated that there is a high degree of agreement about the elements of the nature of science that should be communicated to students. Furthermore, there is evidence that four disciplines (the philosophy, history, sociology and psychology of science) add to our description of how science operates leading us to the conclusion that the nature of science (NOS) as applied by science educators is not a synonym for the philosophy of science alone.

The goal of this project was to produce a definition of the content of the nature of science (NOS) useful in informing science teaching and learning. With the proliferation of science education standards documents, this seems an ideal time to perform a rigorous qualitative analysis of those documents to provide both a matrix of the nature of science elements, look for points of central agreement, and gauge disciplines such as history and philosophy that provide the content of the nature of science.

We reviewed a number of leading science education standards documents for their recommendations relative to the nature of science. In many instances these recommendations are contained within a discrete section or chapter of the document, but where possible, the entire document was reviewed to extract NOS elements contained elsewhere in the text. We acknowledge that this procedure of analyzing text following its extraction from supporting narrative somewhat decontextualizes the statements involved. However, in keeping with a similar method applied successfully by Cossman (1989), we feel that the sacrifice of some context is reasonable because it permits cross-document comparisons to be made efficiently.

The documents reviewed include several national science education standards directed at the K-12 environment, one leading state framework document from the United States, and one U.S. publication (The Liberal Art of Science) targeting the undergraduate learner. In four of the documents (Curriculum Corporation [Australia], 1994; AAAS [Benchmarks], 1993; California Framework, 1990; Council of Ministers of Education [Canada], (1996) and AAAS [The Liberal Art of Science], 1990) an entire chapter or discrete section is devoted to the nature of science. In the National Science Education Standards (1996) the NOS elements are contained in each chapter following a discussion of content and in the New Zealand (1993) publication no section is specifically devoted to the nature of science but such material is embedded in the narrative.

We began this analysis with a thorough reading of the section of each standards document addressing issues specifically labeled as the nature of science, the history and philosophy of science, or those that appeared to represent nature of science elements applying the qualitative methods pioneered by Glaser and Strauss (1967). In cases where the entire document was available, it was reviewed completely to provide an overall feeling of how the nature of science is integrated with other standards.

Explicit individual nature of science statements from all documents were extracted from the accompanying text and affixed to cards. We both agreed on the statements to be included in the final data set. Statements were chosen if they included a specific definition, a description of what science is or how science functions, what impacts or has impacted science, issues of knowledge generation, processes involved in the intellectual (rather than mechanical) aspects of science and other facts that provide information about science and its nature. These cards held the raw data which were subjected to the next level of analysis.

Individual statements were sorted into piles based on the degree to which the statements were similar. Initially, thirty such piles were formed and labeled with categories including "observational bias," "science as a human activity," "ethics in science," "clear reporting of results," "the tentative aspect of science knowledge." This process continued with all grouped statements, but in keeping with the method of constant comparison, some categories were combined and some statement cards were shifted from one group to another.

As an example of our method, consider the statements contained in Table I. Although each uses different words, they all contain the central idea that scientific knowledge is tentative. Therefore, a statement ("scientific knowledge is tentative") was written to subsume all of these individual statements.

TABLE I
Statements from various standards documents illustrating the tentative nature of scientific knowledge

USA; Benchmarks for Science Literacy (AAAS, 1993)

". . . both incremental growth and occasional radical shifts are part of science." (p. 4)

USA; Science Framework for California Public Schools (California Department of Education, 1990)

> "But science never commits itself irrevocably to any fact or theory, no matter how firmly it appears to be established in the light of what is known." (p. 17)

USA; National Science Education Standards (NRC, 1996)

> "Because all scientific ideas depend on experimental and observational confirmation, all scientific knowledge is, in principle, subject to change as new evidence becomes available." (p. 201)

Australia; A Statement on Science (Curriculum Corporation, 1994)

> "Our scientific understanding of the world is continually changing, sometimes incrementally, sometimes in gigantic, revolutionary bounds" (p. 3)

USA; The Liberal Art of Science (AAAS, 1990)

> "Scientific knowledge is not absolute; rather, it is tentative, approximate, and subject to revision." (p. 20)

England/Wales; Science in the National Curriculum (Department of Education, 1995)

> "Students should be given opportunities to relate social and historical contexts to scientific ideas by studying how at least one scientific idea has changed over time." (p. 14)

New Zealand; Science in the New Zealand Curriculum (Ministry of Education, 1993)

> ". . . understanding in science changes." (p. 24)

Canada; Common Framework (Council of Ministers of Education, 1996)

> "While there can be major shifts in our understanding of the world through scientific revolutions, most of our understanding is the result of a steady accumulation of knowledge through a constant critical analysis of scientific work." (p. 3)

We ended the first phase of data analysis and categorization having developed approximately thirty unique statements about the nature of science. At this point, we grouped the individual statements into larger categories. Statements that were related to the history of science were associated together while statements related to the qualities of scientists were likewise grouped. A third category was formed around the behaviors of scientists and a fourth group included definitions and qualities of the discipline of science itself. Interestingly, these larger groupings conform nicely to the disciplines of the philosophy, sociology, history and psychology of science. A fifth cluster included pedagogical recommendations and strategies related to the instruction of the nature of science. For purposes of this study, only the first four groups were analyzed in detail.

This process of grouping is not without its dilemmas. For instance, the tentativeness issue itself seemed to belong within the philosophy category, but evidence that scientific ideas are tentative and whether such ideas are transformed gradually or quickly clearly comes from an analysis of the history of science. In spite of such issues, each statement seemed logically and exclusively to fit into a single category.

Our analysis of these eight standards documents has revealed that the nature of science elements they contain may be included within the following thirty individual statements. These thirty statements may, in turn, be subsumed into four larger groups related to the discipline most focused on the study of the issues involved (see Table II).

TABLE II

A matrix of insights relative to the nature of science extracted from seven international science standards documents. The number of documents containing each statement is indicated in parentheses.

	1. USA - Benchmarks	2. USA - California	3. USA - Nat'l Sci. St	4. Australia	5. USA - Undergradu	6. England / Wales	7. New Zealand	8. Canada

Philosophical Insights, Statements & Assumptions

	1	2	3	4	5	6	7	8
1. Scientific knowledge is stable (3)		x	x					x
2. Scientific knowledge is tentative (8)	x	x	x	x	x	x	x	x
3. Science will never be finished (3)		x		x	x			
4. Science relies on empirical evidence (6)		x	x	x	x		x	x

INTERNATIONAL SCIENCE EDUCATION STANDARDS

		1. USA - Benchmarks	2. USA - California	3. USA - Nat'l Sci. St	4. Australia	5. USA - Undergradu	6. England / Wales	7. New Zealand	8. Canada
5.	Science relies on logical arguments (4)	x		x	x				x
6.	Science relies on skepticism (5)			x	x	x		x	x
7.	Science aims to be objective (2)		x		x				
8.	Science aims to be testable (4)		x		x		x		x
9.	Science aims to be consistent (2)		x	x					
10.	Science aims to be precise (2)				x		x		
11.	Scientific knowledge is based on observation (5)		x	x	x		x		x
12.	Scientific knowledge is based on experimental evidence (5)		x	x	x		x		x
13.	Scientific knowledge is based on careful analysis (4)	x			x		x		x
14.	Change in science results from information of better theories (4)	x	x	x		x			
15.	There are many ways to do scientific investigations (4)	x					x	x	x
16.	Science has inherent limitations (4)	x				x	x		x
17.	Science is an attempt to explain phenomena (7)	x		x	x	x	x	x	x

	1. USA - Benchmarks	2. USA - California	3. USA - Nat'l Sci. St	4. Australia	5. USA - Undergradu	6. England / Wales	7. New Zealand	8. Canada
18. To learn about how science operates, vocabulary is vital (3)			x			x		x
Observation (5)	x	x	x			x		x
Hypothesis (4)	x	x	x					x
Law (4)	x		x		x			x
Theory (5)		x	x	x		x		x
Inference (1)		x						
Models (2)			x		x			

Sociological Insights, Statements and Assumptions

	1	2	3	4	5	6	7	8
1. All cultures (can) contribute to science (6)	x	x	x	x			x	x
2. Science is a human endeavor (5)	x		x	x			x	x
3. New knowledge must be reported clearly and openly (6)		x	x	x	x	x	x	
4. Scientists make ethical decisions (8)	x	x	x	x	x	x	x	x
5a. Scientists require: accurate record keeping (5)	x	x	x			x		x
5b. peer review (4)	x	x	x			x		
5c. replicability (7)	x	x	x		x	x	x	x
5d. truthful reporting (7)	x	x	x	x	x	x		x
6. Scientists work collaboratively (4)	x		x		x			x

INTERNATIONAL SCIENCE EDUCATION STANDARDS

Psychological Insights, Statements and Assumptions

	1. USA - Benchmarks	2. USA - California	3. USA - Nat'l Sci. St	4. Australia	5. USA - Undergradu	6. England / Wales	7. New Zealand	8. Canada
1. Observations are theory-laden (4)	x	x			x	x		
2. Scientists must be open to new ideas (5)			x	x	x	x		x
3. Scientists must be intellectually honest (3)				x		x		x
4. Scientists are creative (6)	x		x	x	x	x		x

Historical Statements and Assumptions

1. New scientific ideas have frequently been rejected (4)	x		x		x		x	
2. The past illuminates current scientific practice (2)					x	x		
3a. Change in science occurs gradually (7)	x	x	x	x	x		x	x
3b. Change in science occurs through revolutions (5)	x	x				x	x	x
4a. Science research is dictated by prevailing paradigms (1)					x			
4b. Science research is dictated by national and/or corporate interests (2)	x				x			
5. Science has global implications (7)	x		x	x	x	x	x	x
6. Technology has impacted science (4)		x	x	x				x

Historical Statements and Assumptions

	1. USA - Benchmarks	2. USA - California	3. USA - Nat'l Sci. St	4. Australia	5. USA - Undergradu	6. England / Wales	7. New Zealand	8. Canada
7a. Science is part of intellectual tradition (2)		x			x			
7b. Science is part of social tradition (8)	x	x	x	x	x	x	x	x
7c. Science is part of cultural tradition (5)		x	x	x	x			x
8. Science has played an important role in technology (6)	x	x		x		x	x	x
8. Science has been at the center of many controversies (4)		x		x		x	x	
9. Scientific ideas are affected by their social & historical milieu (6)	x			x	x	x	x	x
10. Science builds on what has gone on before (4)	x	x	x					x

[1] USA; Benchmarks for Science Literacy (AAAS, 1993)
[2] USA; Science Framework for California Public Schools (California Department of Education, 1990)
[3] USA; National Science Education Standards (NRC, 1996)
[4] Australia; A Statement on Science (Curriculum Corporation, 1994)
[5] USA; The Liberal Art of Science (AAAS, 1990)
[6] England/Wales; Science in the National Curriculum (Department of Education, 1995)
[7] New Zealand; Science in the New Zealand Curriculum (Ministry of Education, 1993)
[8] Canada; Common Framework (Council of Ministers of Education, 1996)

CONCLUSIONS

One of many conclusions that may be reached from this analysis is that there is clearly consensus regarding the nature of science issues that should inform science education. We did not assume that each document was created with the same rigor or completeness and we believe that it would be illogical to vote, in a sense, on what should be the content of the nature of science. Therefore, we have rejected the impulse to create a new statement of the NOS content by listing those elements appearing on the majority of the documents, for instance. There are some items that appear in relatively few documents that provide important insights into the way in which science and scientists operate. In our view, items such as the notion of the paradigm, the objectivity aspect of science and the idea that science has inherent limitations, are important to students' understanding although they do not appear on a majority of these documents. The fact that they appear at all indicates that a holistic view of the nature of science as suggested by a review of all of the documents seems the most logical approach to the question of what ought to be the nature of science for science education purposes.

Another interesting conclusion is that the standards documents have failed to include those prescriptive and/or speculative aspects that are of most interest to philosophers of science but say little to those who would like to provide students an authentic view of how science operates. As an example, philosopher of science Karl Popper (1963) has stated that falsification is an epistemologically useful enterprise that scientists should practice if they want to secure valid knowledge. Few would argue with his assertion, but there is very little evidence to indicate that scientists as a group actively work to demonstrate that their ideas are invalid. This issue could certainly be discussed with students, but it would be unwise to describe that science works in this fashion. Certainly some students and teachers might find the marginal debates about the NOS interesting, but the degree to which these tangential issues help learners appreciate and evaluate the scientific enterprise is minimal.

It is also useful to conclude that four major disciplines seem to provide insights regarding the nature of science. These disciplines include the philosophy, history, sociology and psychology of science. We freely admit that our prior expectation was that philosophy alone is not the sole discipline contributing to a description of how science operates, but this study has provided an empirical basis for that supposition. Related to this issue is the degree to which each of these disciplines guide our understanding. Not surprisingly, philosophy and history of science have the greatest impact on our knowledge of science, but sociology and psychology add important elements. It is clear from a review of the science education literature that the term "nature of science" is not simply a synonym for the philosophy of science, but is a hybrid domain informed primarily by descriptive scholarship from a variety of disciplines. Figure 1 represents an approximation of the degree to which each of these four disciplines adds to knowledge of how science operates.

To turn our attention to the specific generalizations seen following a review of these documents, we will briefly discuss each of the major clusters. Within the philosophy of science category, eleven specific statements were extracted which were found in at least two documents. All of the standards expressly state that scientific knowledge changes over time and seven stated that this change is usually gradual (see the section on historical statements). Five documents mention scientific revolutions as an additional agent of change. Also relating to scientific knowledge are statements about how knowledge is derived. Five documents referred to observation and experimental evidence. Another area within the philosophy category includes statements which describe what science is and how it is done. Such statements include that science is an attempt to explain phenomena, science is creative, science has inherent limitations and that science relies on empirical evidence, logical arguments and skepticism. Together, these statements provide a broad view of philosophical assumptions underlying the nature of science.

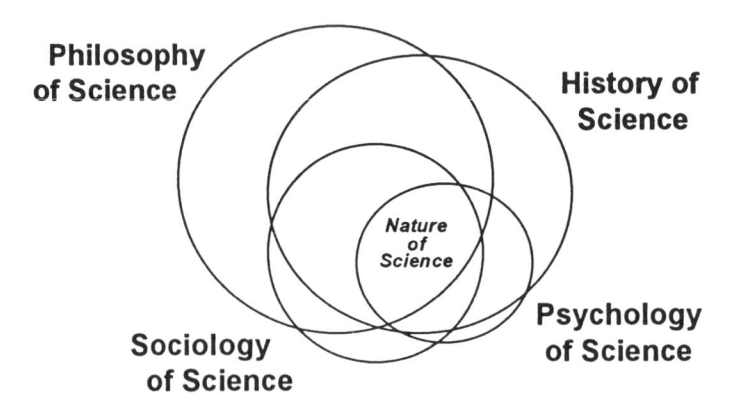

Figure 1. A proposal for the disciplines that add to our understanding of the nature of science based on a content review of various science education standards documents. The approximate extent to which each discipline contributes is represented by the relative sizes of the circles. A description of science -- the nature of science -- is found at the intersection of these various disciplines.

The sociology category comprises statements related to who scientists are and how scientists work. All documents include a statement about the ethical decision-making of scientists, and six state that new knowledge must be clearly and openly reported by scientists. It is not clear, however, if these statements come from a view of how scientists work or from an expectation that science should function in this fashion. Statements in this category also included peer review, replication of procedures and accurate record keeping.

The third category, psychology, includes statements about the characteristics of scientists. Six documents mention that scientists generate knowledge in a creative fashion. Four documents refer to the inherent bias that exists with any observation. The final group of statements in this category includes the necessity of scientists to be open to new ideas and intellectually honest.

The final category we examined includes elements from the history of science. All documents refer to science as a social tradition. Seven documents state that science has global implications, and five mention the important role that science has played in the development of technology. Also stated in six documents is the effect that social and historical contexts have on the development of scientific ideas. Interestingly, four documents include a statement which refer to the frequent rejection of new ideas by the rest of the scientific community.

We also observed that those standards documents which include aspects of NOS vocabulary (i.e., law, theory, etc.) generally fail to define these terms. This may be an oversight on the part of the authors, but given the varying interpretations held by members of the general public regarding these terms, we feel strongly that the standards must not only use the terms, but define them correctly.

It may be more than a coincidence that over half of the documents surveyed include the nature of science discussion at the outset. From this finding, we may assume that standards writers believe that student understanding of the nature of science is, or should be, a vital foundation for all future science learning. The California Framework (1990), for instance, states that "[t]his chapter [nature of science] is instrumental to understanding the rest of the framework" (p. 12).

We began this analysis with several assumptions. First, we agree that at the level of the fine details, no precise and completely agreed upon description exists concerning the structure of science or the means by which knowledge in science is produced (Herron, 1969; Laudan, et al, 1986; Lederman, 1992; Duschl, 1994). Second, we presume that some relationship between the standards documents is inevitable because developers may have inspected earlier such standards statements. However, we assume that the various teams consulted the original literature as they worked to define the nature of science.

As in any ongoing project, the conclusions presented here will be strengthened or contradicted only by examining additional standards documents and add their NOS elements to the existing data set. We look forward to including a review of science education standards in languages other than English and particularly from non-Western cultures. Not only would such an approach permit us to make firmer generalizations, but if as we suspect, our conclusions are upheld, we would be able to present evidence that a single scientific tradition unites humankind.

REFERENCES

AAAS (1993). *Benchmarks for Science Literacy,* Washington, D.C., American Association for the Advancement of Science.

Council of Ministers of Education (1996) *Common framework of science learning outcomes K-12 (Draft).* Victoria, BC, Ministry of Education, Skills and Training.

California State Department of Education (1990) *Science framework for California public schools* Sacramento.

Cossman, G. W. (1989) 'A comparison of the image of science found in two future oriented guideline documents for science education', *Proceedings of the First International History and Philosophy of Science and Science Teaching Conference*, University of Florida, Tallahassee, 83-104.

Curriculum Corporation (1994) *A statement of science for Australian schools: A joint project of the states, territories and the Commonwealth of Australia initiated by the Australian Education Council*, Carlton, Victoria.

Department of Education (1995) *Science in the national curriculum*, London.

Duschl, R. A. (1994) 'Research on the history and philosophy of science', in L.G. Dorothy (ed), *Handbook of research on science teaching and learning*, New York, MacMillan, 445-455.

Herron, M. D. (1969). 'Nature of science: Panacea or Pandora's box', *Journal of Research in Science Teaching, (6)*, 105-107.

Glaser, B. G. and Strauss, A. L. (1967). *The discovery of grounded theory: Strategies for qualitative research*, New York, Aldine de Gruyter.

Lauden, et al. (1986). 'Philosophical models and historical research', *Synthese 69*, 141-223.

Lederman, N.G. (1992). 'Students' and teachers' conceptions of the nature of science: A review of the research', *Journal of Research in Science Teaching,* (29), 331-359.

National Research Council (NRC) (1996) *National science education standards*, Washington, D.C., National Academy Press.

Ministry of Education (1993). *Science in the New Zealand Curriculum*, Wellington, New Zealand, Learning Media.

Popper, K. R. (1963) *Conjectures and refutations: The growth of scientific knowledge*, New York, Harper and Row.

WILLIAM F. McCOMAS

3. THE PRINCIPAL ELEMENTS OF THE NATURE OF SCIENCE: DISPELLING THE MYTHS

In the first chapter we explored rationales for the inclusion of the nature of science in science education and in chapter two we demonstrated the degree of consensus that exists relative to the elements of the nature of science expressed in international science education documents. This chapter features a discussion of fifteen major issues related to the NOS that seem most problematic in the experience of many science educators. These fifteen issues, described here as "myths of science," do not represent all of the important issues that teachers should consider when designing instruction relative to the nature of science, but may serve as starting points for evaluating current instructional foci while enhancing future curriculum design.

The "myths of science" discussed here are commonly included in science textbooks, in classroom discourse and in the minds of adult Americans. Misconceptions about science are most likely due to the lack of philosophy of science content in teacher education programs and the failure of such programs to provide real science research experiences for preservice teachers while another source of the problem may be the generally shallow treatment of the nature of science in the textbooks to which teachers might turn for guidance. Some of these myths, such as the idea that there is a scientific method, are most likely caused by the explicit inclusion of faulty ideas in textbooks while others, such as lack of knowledge of the social construction of scientific knowledge, are the result of omissions in texts.

As Steven Jay Gould points out in *The Case of the Creeping Fox Terrier Clone* (1988), science textbook writers are among the most egregious purveyors of myth and inaccuracy. The "fox terrier" refers to the classic comparison used to express the size of the dawn horse, tiny precursor to the modern horse. This comparison is unfortunate for two reasons. Not only was this horse ancestor much bigger than a fox terrier, but the fox terrier breed of dog is virtually unknown to American students. The major criticism leveled by Gould is that once this comparison took hold, no one bothered checking its validity or utility. Through time, one author after another simply repeated the inept comparison and continued a tradition making many science texts virtual clones of each other on this and countless other points.

In an attempt to provide a more realistic view of science and point out issues on which science teachers should focus, this chapter presents and discusses fifteen widely-held, yet incorrect ideas about the nature of science. There is no implication that all students, or most teachers for that matter, hold all of these views to be true, nor is the list meant to be the definitive catalog. Cole (1986) and Rothman (1992) have suggested additional misconceptions worthy of consideration. However, years of science teaching and the review of countless texts has substantiated the validity of the following inventory presented here.

MYTH 1: HYPOTHESES BECOME THEORIES THAT IN TURN BECOME LAWS

This myth deals with the general belief that with increased evidence there is a developmental sequence through which scientific ideas pass on their way to final acceptance (see Figure 1) as mature laws. The implication is that hypotheses and theories are less secure than laws. A former U.S. president expressed his misunderstanding of science by saying that he was not troubled by the idea of evolution because it was, in his words, "just a theory." The president's misstatement is the essence of this myth; an idea is not worthy of consideration until "lawness" has been bestowed upon it.

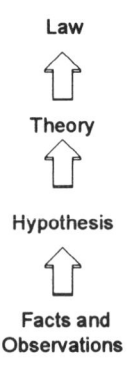

Law
⇧
Theory
⇧
Hypothesis
⇧
Facts and Observations

Figure 1. The false hierarchical relationship between facts, hypotheses, theories and laws.

Theories and laws are very different kinds of knowledge, but the misconception portrays them as different forms of the same knowledge construct. Of course there is a relationship between laws and theories, but it is not the case that one simply becomes the other -- no matter how much empirical evidence is amassed. Laws are generalizations, principles or patterns in nature and theories are the explanations of those generalizations (Rhodes and Schaible, 1989; Horner and Rubba, 1979; Campbell, 1953). Dunbar (1995) addresses the distinction in a very useful fashion by calling laws "cookbook science," and the explanations "theoretical science." He labels the multiple examples of the kind of science practiced by traditional peoples as "cookbook" because those who apply the rules after observing the patterns in nature, do not understand why nature operates in the fashion that it does. The rules work and that is enough.

Even in more sophisticated settings, cookbook science is occasionally practiced. For example, Newton described the relationship of mass and distance to gravitational attraction between objects with such precision that we can use the law of gravity to plan spaceflights. During the Apollo 8 mission, astronaut Bill Anders responded to the question of who was flying the spacecraft by saying, "I think Isaac Newton is doing most of the driving right now" (Chaikin, 1994, p. 127). His response was understood to mean that the capsule was simply following the basic laws of physics described by Isaac Newton centuries earlier.

The more thorny, and many would say more interesting, issue with respect to gravity is the explanation for why the law operates as it does. At this point, there is no well-accepted theory of gravity. Some physicists suggest that gravity waves are the correct explanation, but with clear confirmation and consensus lacking, most feel that the theory of gravity still eludes science. Interestingly, Newton addressed the distinction between law and theory with respect to gravity. Although he had discovered the law of gravity, he refrained from speculating about its cause. In *Principia*, Newton states ". . . I have not been able to discover the cause of those properties of gravity from phenomena, and I frame no hypothesis . . ." ". . . it is enough that gravity does really exist, and act according to the laws which we have explained . . ." (Newton, 1720/1946, p. 547).

MYTH 2: SCIENTIFIC LAWS AND OTHER SUCH IDEAS ARE ABSOLUTE

This myth involves two elements. First, even if individuals understand that scientific laws are equal in importance to theories, they rarely appreciate that all knowledge in science is tentative, occasionally seeing "proof" in science equal to proof in mathematics. The issue of tentativeness is part of the self correcting aspect of science but one that those who fault science frequently ignore. Creationists, for instance, are quick to criticize science by pointing to the discovery of several teeth found in Nebraska early in this century (Gould, 1991). Initially, these teeth were considered to have come from a primitive human, but were later found to be those of an extinct pig. Scientists made both the initial identification and the later revision, but those who would like to fault science only discuss the error, while rarely mentioning the inevitable correction.

Another aspect of this myth stems from the realization that there are several basic kinds of laws — deterministic and probabilistic. Although both types of laws are as tentative as any scientific knowledge, the laws of the physical sciences are typically deterministic in that cause and effect are more securely linked while the laws in biology usually have a probability factor associated. In the life sciences it is typical to see limitations placed on the application of laws. For example, Mendel's laws of inheritance work only with single gene pairs and not even with all such pairs. This issue has called some to question if there are really laws in biology. My response would be that there are laws in the life sciences, but the rules for their application are somewhat distinct from those applied in the physical sciences.

MYTH 3: A HYPOTHESIS IS AN EDUCATED GUESS

The definition of the term hypothesis has taken on an almost mantra-like life of its own in science classes. If a hypothesis is always an educated guess as students typically assert, the question remains, "an educated guess about what?" The best answer for this question must be, that without a clear view of the context in which the term is used, it is impossible to tell.

The term hypothesis has at least three definitions, and for that reason, should be abandoned and replaced, or at least used with caution. For instance, when Newton said that he framed no hypothesis as to the cause of gravity he was saying that he had no speculation about an explanation of why the law of gravity operates as it does. In this case, Newton used the term hypothesis to represent an immature theory.

As a solution to the hypothesis problem, Sonleitner (1989) suggested that tentative or trial laws be called generalizing hypotheses with provisional theories referred to as explanatory hypotheses. Another approach would be to abandon the word hypothesis in favor of terms such as speculative law or speculative theory. With evidence, generalizing hypotheses may become laws and speculative theories become theories, but under no circumstances do theories become laws. Finally, when students are asked to propose a hypothesis during a laboratory experience, the term now means a prediction. As for those hypotheses that are really forecasts, perhaps they should simply be called what they are, predictions.

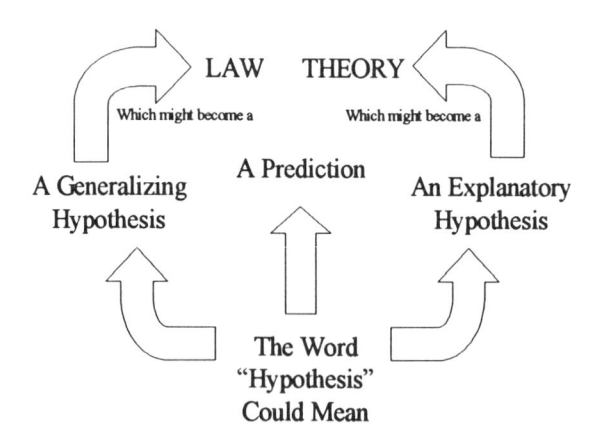

Figure 2. "Family tree" of hypotheses, illustrating the multiple definitions of the term.

MYTH 4: A GENERAL AND UNIVERSAL SCIENTIFIC METHOD EXISTS

The notion that a common series of steps is followed by all research scientists must be among the most pervasive myths of science given the appearance of such a list in the introductory chapters of many precollege science texts. The steps listed for the scientific method vary somewhat from text to text but usually include: a) defining the problem, b) gathering background information, c) forming a hypothesis, d) making observations, e) testing the hypothesis and f) drawing conclusions. Some texts conclude their list of the steps by listing communication of results as the final ingredient as illustrated in Figure 3.

7. Report Results

6. Form Conclusions

5. Test the Hypothesis

4. Make Relevant Observations

3. Form a Hypothesis

2. Gather Information

1. Define the Problem

Figure 3.   The typical steps associated with the so-called scientific method.

The universal scientific method is one of science educations' most pervasive "creeping fox terriers." The multi-step list seems to have started innocently enough when Keeslar (1945a b) prepared a list of a number of characteristics associated with scientific research such as establishing controls, keeping accurate records, making careful observations and measurements. This list was refined into a questionnaire and submitted to research scientists for validation. Items that were highly ranked were put in a logical order and made part of the final list of elements associated with the investigation of scientific problems. This list was quickly adopted by textbook writers as *the* description of how science is done. In time the list was reduced from ten items to those mentioned above, but in the hands of generations of textbook writers, a simple list of characteristics associated with scientific research became a description of how all scientists work.

Another reason for the widespread belief in a general scientific method may be the way in which results are presented for publication in research journals. The standardized style makes it appear that scientists follow a standard research plan. Medawar (1991) reacted to the common style exhibited by research papers by calling the scientific paper a fraud since the final journal report rarely outlines the actual way in which the problem was investigated.

Those who study scientists at work have shown that no research method is applied universally (Carey, 1994; Gibbs and Lawson, 1992; Chalmers, 1990 and Gjertsen, 1989). The notion of a single scientific method is so pervasive that many students must be disappointed when they discover that scientists do not have a framed copy of the steps of the scientific method posted above each laboratory workbench.

Close inspection will reveal that scientists approach and solve problems with imagination, creativity, prior knowledge and perseverance. These, of course, are the same methods used by all effective problem-solvers. The lesson to be learned is that science is no different from other human endeavors when puzzles are investigated. Fortunately, this is one myth that may eventually be displaced since many newer texts are abandoning or augmenting the list in favor of discussions of *methods* of science.

MYTH 5: EVIDENCE ACCUMULATED CAREFULLY WILL RESULT IN SURE KNOWLEDGE

All investigators, including scientists, collect and interpret empirical evidence through the process called induction. This is a technique by which individual pieces of evidence are collected and examined until a law is discovered or a theory is invented. Useful as this technique is, even a preponderance of evidence does not guarantee the production of valid knowledge because of what is called the problem of induction.

Induction was first formalized by Frances Bacon in the 17th century. In his 1620 book, *Novum Organum*, Bacon advised that facts should be assimilated without bias to reach a conclusion. The method of induction he suggested is in part the principal way by which humans traditionally have produced generalizations that permit predictions. Baconian induction, and the related process of deduction (or hypothetico-deductivism) is illustrated in Figure 3. Without the creative leap (shown later in Figure 4), the process of Baconian induction is most accurately characterized as naive induction.

The problem with induction is that it is both impossible to make all observations pertaining to a given situation and illogical to secure all relevant facts for all time, past, present and future. However, only by making all relevant observations throughout all time, could one say that a final valid conclusion had been made. On a personal level, this problem is of little consequence, but in science the problem is significant. Scientists formulate laws and theories that are supposed to hold true in all places and for all time but the problem of induction makes such a guarantee impossible. This problem is particularly acute in biology and to some extent in geology. The laws of biology for instance, are confined at the moment to the only planet on which they have been tested. It is unlikely that the rules of the life sciences, as we know them, would, in fact, operate on other planets where life has evolved.

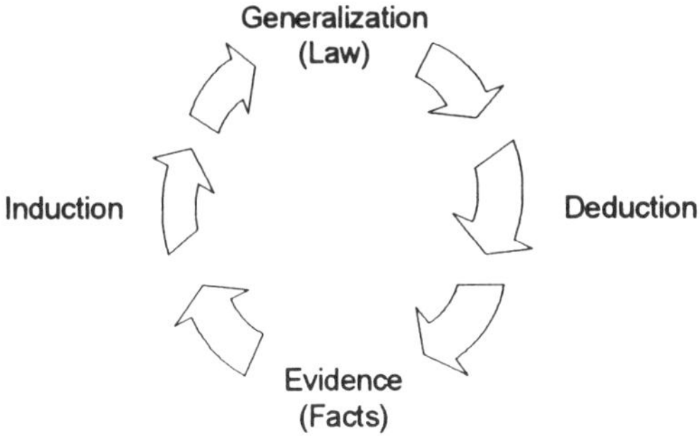

```
         Generalization
            (Law)
    ↗                  ↘
Induction              Deduction
    ↖                  ↙
          Evidence
          (Facts)
```

Figure 4. A typical view of Baconian knowledge production. Bacon's view (on the left) of the production of new generalizations and deduction, or hypothetico-deductivism (on the right) for the testing of such generalizations. The diagram does not imply that the laws produce new facts, but rather that a valid law would permit the accurate prediction of facts not yet known.

The proposal of a new law begins through induction as facts are heaped upon other relevant facts. Deduction is useful in checking the validity of a law. For example, if we postulate that all swans are white, we can evaluate the law by predicting that the next swan found will also be white. If it is, the law is supported (but not proved as we discuss). Locating a black swan will cause the law to be questioned.

The nature of induction itself is another interesting aspect associated with this myth. If we set aside the problem of induction momentarily, there is still the issue of how scientists make the final leap from the mass of evidence to the conclusion. In an idealized view of induction, the accumulated evidence will simply result in the production of a new law or theory in a procedural or mechanical fashion. In reality, there is no such method. The issue is far more complex and interesting than that. The final creative leap from evidence to scientific knowledge is the focus of another myth of science. (See Figure 4).

MYTH 6: SCIENCE AND ITS METHODS PROVIDE ABSOLUTE PROOF

The general success of the scientific endeavor suggests that its products must be valid. However, a hallmark of science is that it is subject to revision when new information is presented. Tentativeness resulting in a lack of dogmatism is one of the points that differentiates science from other forms of knowledge. Accumulated evidence can provide support, validation and substantiation for a law or theory, but will never prove those laws and theories to be true. This idea has been addressed well by Horner and Rubba (1978) and Lopushinsky (1993).

The problem of induction argues against proof in science, but there is another element of this myth worth exploring. In actuality, the only truly conclusive knowledge produced by science results when a notion is falsified. What this means is that no matter what scientific idea is considered, once disconfirming evidence begins to accumulate, at least we know that the notion is untrue. Consider the example of the white swans discussed earlier. One could search the world and see only white swans, and arrive at the generalization that "all swans are white." However, the discovery of one black swan has the potential to overturn, or at least result in modifications of, this proposed law of nature. Finding yet another white swan does not prove anything, its discovery simply provides some comfort that the original idea has merit. Whether scientists routinely try to falsify their notions as has been recommended by philosopher of science Karl Popper, and how much contrary evidence it takes for a scientist's mind to change are fascinating issues (Lakatos, 1972).

MYTH 7: SCIENCE IS PROCEDURAL MORE THAN CREATIVE

We accept that no single guaranteed method of science can account for the success of science, but realize that induction, the collection and interpretation of individual facts providing the raw materials for laws and theories, is at the foundation of most scientific endeavors. This awareness suggests a paradox. If induction itself is not a guaranteed method for arriving at conclusions, how do scientists develop useful laws and theories? Induction makes use of individual facts that are collected, analyzed and examined. Some observers may perceive a pattern in these data and propose a law in response, but there is no logical or procedural method by which the pattern is suggested. With a theory, the issue is much the same. Only the creativity of the individual scientist permits the discovery of laws and the invention of theories. If there truly was a single scientific method, two individuals with the same expertise could review the same facts and likely reach identical conclusions. There is no guarantee of this because the range, nature, and application of creativity is a personal attribute. See Figure 5 for an illustration of the role of creativity in the knowledge generation process.

Unfortunately, many common science teaching orientations and methods serve to work against the creative element in science. The majority of laboratory exercises, for instance, are verification activities. The teacher discusses what is going to happen in the laboratory, the manual provides step-by-step directions and the student is expected to arrive at a particular answer. Not only is this approach the antithesis of the way in which science actually operates, but such a portrayal must seem dry, clinical and uninteresting to many students. In her 1990 book, *They're Not Dumb, They're Different*, Tobias argues that many capable and clever students reject science as a career because they are not given opportunities to see it as an exciting and creative pursuit. The moral in Tobias' thesis is that science may be impoverished when students who feel a need for a creative outlet eliminate it as a potential career because of the way it is taught.

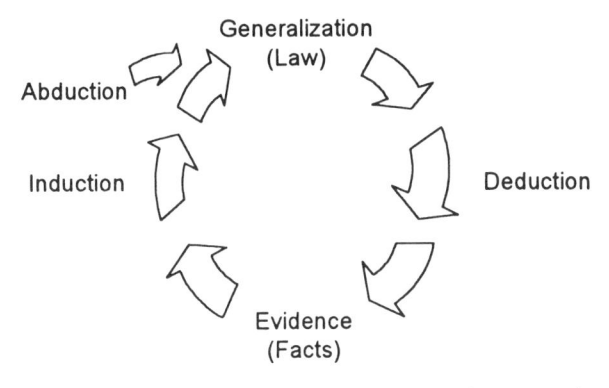

Figure 5. A more accurate illustration of the knowledge generation process in science. Here the creative leap (sometimes called abduction) is shown as a necessary element leading from the evidence to the generalization.

MYTH 8: SCIENCE AND ITS METHODS CAN ANSWER ALL QUESTIONS.

Philosophers of science have found it useful to refer to the work of Karl Popper (1968) and his principle of falsifiability to provide an operational definition of what counts as science. Popper suggested that only those ideas that are potentially falsifiable are scientific ideas.

For instance, the law of gravity states that more massive objects exert a stronger gravitational attraction than do objects with less mass when distance is held constant. This is a scientific law because it could be falsified if newly-discovered objects operate differently with respect to gravitational attraction. In contrast, the core idea among creationists is that species were placed on earth fully-formed by some supernatural force. Obviously, there is no scientific method by which such a belief could be shown to be false. Since this special creation view is impossible to falsify, it is not scientific and the term "creation science" is an oxymoron. Creation science is a religious belief and as such, does not require that it be falsifiable. Hundreds of years ago thoughtful theologians and scientists carved out their spheres of influence and expertise and have coexisted since with little acrimony. Today, only those who fail to understand the distinction between science and religion confuse the rules, roles, and limitations of these two important world views.

It should now be clear that some questions simply must not be asked of scientists. During one of the recent creation science trials for instance, science Nobel laureates were asked to sign a statement about the nature of science to provide some guidance to the court. Seventy-two of these famous scientists responded resoundingly to support such a statement; after all they were experts in the realm of science (Klayman, Slocombe, Lehman & Kaufman, 1986). Later, those interested in citing expert opinion in the abortion debate asked scientists to issue a statement regarding their feelings on this issue. Wisely, few participated. Science as a discipline cannot answer the moral and ethical questions engendered by the matter of abortion. Of course, scientists as individuals have personal opinions about many issues, but as a group, they must remain silent if those issues are outside the realm of scientific inquiry. Science simply cannot answer moral, ethical, aesthetic, social and metaphysical questions, although it can provide some insights which might be illuminating. For instance, science and resulting technology may be able to clone mammals, but only society can decide whether such cloning is moral and ethical.

MYTH 9: SCIENTISTS ARE PARTICULARLY OBJECTIVE

Scientists are no different in their level of objectivity than are other professionals. They are careful in the analysis of evidence and in the procedures applied to arrive at conclusions. With this admission, it may seem that this myth is valid, but contributions from both the philosophy of science and psychology reveal that complete objectivity is impossible for at least three major reasons.

Many philosophers of science support Popper's (1963) view that science can advance only through a string of what he called conjectures and refutations. In other words, Popper recommends that scientists should propose laws and theories as conjectures and then actively work to disprove or refute those ideas. Popper suggests that the absence of contrary evidence, demonstrated through an active program of refutation, will provide the best support available. It may seem like a strange way of thinking about verification, but the absence of disproof is considered support. There is one major problem with the idea of conjecture and refutation. Popper seems to have proposed it as a recommendation for scientists, not as a description of what scientists do. From a philosophical perspective the idea is sound, but there are no indications that scientists actively practice programs to search for disconfirming evidence.

Another aspect of the inability of scientists to be objective is found in theory-laden observation, a psychological notion (Hodson, 1986). Scientists, like all observers, hold myriad preconceptions and biases about the way the world operates.

These notions, held in the subconscious, affect the ability of everyone to make observations. It is impossible to collect and interpret facts without any bias. There have been countless cases in the history of science in which scientists have failed to include particular observations in their final reports. This occurs, not because of fraud or deceit, but because of the prior knowledge possessed by the individual. Certain facts either were not seen at all or were deemed unimportant based on the scientists' prior expectations. In earlier discussions of induction, we postulated that two individuals reviewing the same data would not be expected to reach the same conclusions. Not only does individual creativity play a role, but the issue of personal theory-laden observation further complicates the situation.

This lesson has clear implications for science teaching. Teachers typically provide learning experiences for students without considering their prior knowledge. In the laboratory, for instance, students are asked to perform activities, make observations and then form conclusions. There is an expectation that the conclusions formed will be both self-evident and uniform. In other words, teachers anticipate that the data will lead all pupils to the same conclusion. This could only happen if each student had exactly the same prior conceptions and made and evaluated observations using identical schemes. The does not happen in science nor does it occur in the science classroom.

Related to the issue of theory-based observations is the allegiance to the paradigm. Thomas Kuhn (1970), in his ground-breaking analysis of the history of science, suggested that scientists work within a research tradition called a paradigm. This research tradition, shared by those working in a given discipline, provides clues to the questions worth investigating, dictates what evidence is admissible and prescribes the tests and techniques that are reasonable. Although the paradigm provides direction to the research it may also stifle or limit investigation. Anything that confines the research endeavor necessarily limits objectivity. While there is no conscious desire on the part of scientists to limit discussion, it is likely that some new ideas in science are rejected because of the paradigm issue. When research reports are submitted for publication, they are reviewed by other members of the discipline. Ideas from outside the paradigm are liable to be eliminated from consideration as crackpot or poor science and thus will not appear in print.

Examples of scientific ideas that were originally rejected because they fell outside the accepted paradigm include the sun-centered solar system, warm-bloodedness in dinosaurs, the germ-theory of disease, and continental drift. When the idea of moving continents was first proposed early in this century by Alfred Wegener it was vigorously rejected. Scientists were simply not ready to embrace a notion so contrary to the traditional teachings of their discipline. Continental drift was finally accepted in the 1960's with the proposal of a mechanism or theory to explain how continental plates move (Hallam, 1975 and Mendard, 1986). This fundamental change in the earth sciences, called a revolution by Kuhn, might have occurred decades earlier had it not been for the strength of the prevailing paradigm.

It would be misleading to conclude a discussion of scientific paradigms on a negative note. Although the examples provided do show the contrary aspects associated with paradigm-fixity, Kuhn would likely argue that the blinders created by allegiance to the paradigm help keep scientists on track. His review of the history of science demonstrates that paradigms are responsible for far more successes in science than delays.

MYTH 10: EXPERIMENTS ARE THE PRINCIPAL ROUTE TO SCIENTIFIC KNOWLEDGE

Throughout their school science careers, students are encouraged to associate science with experimentation. Virtually all hands-on experiences that students have in science class are called experiments even if they would more accurately be labeled as technical procedures, explorations or activities. True experiments involve carefully orchestrated procedures accompanied by control and test groups. Usually experiments have as a primary goal the establishment of a cause and effect relationship. Of course, true experimentation is a useful tool in science, but is not the sole route to knowledge.

Many noteworthy scientists have used non-experimental techniques to advance knowledge. In fact, in a number of science disciplines, true experimentation is not possible because of the inability to control variables. Many fundamental discoveries in astronomy are based on extensive observations rather than experiments. Copernicus and Kepler changed our view of the solar system using observational evidence derived from lengthy and detailed observations frequently contributed by other scientists, but neither performed experiments.

Charles Darwin's investigatory regime was frequently more similar to qualitative techniques used in the social sciences than the experimental techniques associated with the natural sciences. For his most revolutionary discoveries, Darwin recorded his extensive observations in notebooks annotated by speculations and thoughts about those observations. Although Darwin supported the inductive method proposed by Bacon, he was aware that observation without speculation or prior understanding was both ineffective and impossible. In fact he stated this view clearly by saying, "I could not help making hypotheses about everything I saw." (Darwin, 1958). The techniques advanced by Darwin have been widely used by scientists such as Goodall and Fossey in their primate studies. Scientific knowledge is gained in a variety of ways including observation, analysis, speculation, library investigation *and* experimentation.

MYTH 11: SCIENTIFIC CONCLUSIONS ARE REVIEWED FOR ACCURACY

When preparing school laboratory reports, students are frequently told to present their methods clearly so that others could repeat the investigation. The conclusion that students will likely to draw from this requirement is that professional scientists are also constantly reviewing each other's experiments to check up on each other. Unfortunately, while such a check and balance system would be useful, the number of findings from one laboratory checked by others is small. In reality, most scientists are simply too busy and research funds too limited for this type of review.

It is interesting to note that when scientific experiments are repeated it is usually because a scientific conclusion attacks the prevailing paradigm. In the recent case of cold fusion, scientists worldwide dropped what they were doing to try to repeat the findings provided by Fleishman and Pons. In fairness, these two scientists not only assailed the conventional wisdom but presented their results in a press conference rather than in a peer-reviewed journal. Therefore, the community of scientists had two reasons to be suspicious. One can infer a measure of the disdain exhibited by the scientific community toward cold fusion and its "discoverers" in the titles of several new books on the subject. *Bad Science: The Short Life and Weird Times of Cold Fusion* (Taubes, 1993) and *Cold Fusion: The Scientific Fiasco of the Century* (Huizenga, 1992) both tell the tale of what happens when a new idea is too far outside scientific norms – at least as far as those norms are presently perceived. The fact that cold fusion did not exist likely vindicated those who quickly attacked it, but the more interesting lesson is that is was attacked because the idea was so distant from the expectation on the part of the scientific community.

The result of the lack of oversight has recently put science itself under suspicion. The pressures of achieving tenure, accruing honors, and gaining funds do result in instances of outright scientific fraud, but fortunately such cases are quite rare. However, even without fraud, the enormous amount of original scientific research published, and the pressure to produce new information rather than reproduce others' work dramatically increases the possibility that errors will go unnoticed.

An interesting corollary to this myth is that scientists rarely report valid, but negative results. While this is understandable given the space limitations in scientific journals, the failure to report what did *not* work is a problem. Only when those working in a particular scientific discipline have access to all information regarding a phenomenon can the discipline progress most effectively.

MYTH 12: ACCEPTANCE OF NEW SCIENTIFIC KNOWLEDGE IS STRAIGHTFORWARD

This misconception addresses the belief that when a more accurate interpretation for the evidence is produced it will immediately be accepted by the scientific community.

Nothing could be farther from the truth as we have seen in at least one previous myth. A new idea that is not too far from the expectations of scientists working in a particular field would probably gain entry into scientific journals without much trouble — particularly if it comes from someone working in that field. However, if the idea is a significant breakthrough or revolution in Kuhn's use of the term, particularly if it is counterintuitive or comes from outside the discipline, its acceptance is by no means quick and easy.

The lesson to be learned from this myth, is that science is at its heart a human activity. Humans are the producers of new knowledge and also the arbiters of what counts as new knowledge. While nothing like a vote takes place when a new idea is proposed, the peer review system acts as a gatekeeper to new ideas. Those notions that cannot find a place in the journals will never have a chance to be accepted or denied. Even those new visions of reality that do make it into the journals still have to pass what might best be called the "conference test" if they are to be accepted. Discrepant notions are the talk of professional conferences where they are debated both in the meeting halls but also during dinner and over drinks. As an example, consider the current debate about the origin of modern humans. One view suggests that modern humans arose in various places around the world from ancestral stock while a competing story places the origin of modern humans squarely in Africa from which they migrated to displace the more primitive human forms living elsewhere. The story is told well in a wonderful book, *The Neandertal Enigma*. In this book, Shreeve (1995) discusses the evidence, the personalities and the politics that have directed the conversation about which view should prevail. The final result in the case of human origins is still unsettled, but in many cases, the acceptance of a new scientific idea might be as much a matter of the dynamics of the personalities involved as the strength of the arguments.

MYTH 13: SCIENCE MODELS REPRESENT REALITY

This may be one myth that is shared by both scientists and lay persons alike and is related to the distinction between the philosophical views of realism and instrumentalism. Realism is a position that what science produces not only works and permits the production of accurate predictions but really does represent and/or describe the actual situation in nature as known by some omniscient entity. Of course, one of the central limitations of science is that the "true" nature of reality can never be known because there is no omniscient entity to ask. Science was invented, at least in part, to answer questions about the natural world and get as close to "the truth" as possible, but no bell rings or light blinks to tell scientists that they have found the truth. Another philosophical precept is that as long as the scientific ideas function properly and are consonant with all of the evidence it does not matter whether they correspond with reality or not. The ideas are useful and descriptive and that should be the end of it.

With this distinction between realism and instrumentalism in mind we can now turn to the idea of a scientific model. Although no survey has ever been taken on this issue, it seems logical that scientists do believe that they are not just producing useful ideas but that their ideas and descriptions correspond to a reality external to the scientists themselves. Certainly the average person believes this to be true. It is doubtful that anyone seriously questions the model suggested by the kinetic molecular theory of matter which pictures atoms and molecules as tiny discrete balls that have elastic collisions, thus explaining whole ranges of phenomena. Never mind that no one has ever seen these tiny balls or witnessed their impacts, but the model works; it permits both predictions and explanations and therefore must be true. A realist would say that it is true while an instrumentalist would say it does not matter as long as there is something to be gained from keeping the idea in mind.

The story may be apocryphal, but it is commonly repeated among science educators that when students were once asked what color atoms were, their answer was closely linked to the textbook in use by those students. If the book illustrated atoms as blue, then blue was the color students would assign to atoms when asked. It would probably serve us well to think of models as "useful fictions," but it is doubtful that more than a few keep this warning in mind. After all, what caused Galileo trouble was not that he adopted and supported the sun-centered universe, but that he taught it as the truth in an age when the church felt it had authority over what was considered the truth.

MYTH 14: SCIENCE AND TECHNOLOGY ARE IDENTICAL

A common misconception is the idea that science and technology are the same. In fact, many believe that television, rockets, computers and even refrigerators are science, but one of the hallmarks of science is that it is not necessarily practical while refrigerators certainly are. The pursuit of knowledge for the sake of knowledge alone is called pure science while its exploitation in the production of a commercial product is applied science or technology.

Today, most investigators are working on problems that are at least in part directed from outside their laboratories. Scientists typically blend the quest of pure science in order to solve a technology challenge. In many ways the distinction between pure and applied science is not crucial, but it is interesting to explore what motivates scientists to work on their problems. Few scientists have the luxury to pursue any goal they choose since most scientific work is funded by organizations with an agenda. This funding relationship is not necessarily damaging, but the freedom experienced by the pure scientists of the Victorian age is long gone.

MYTH 15: SCIENCE IS A SOLITARY PURSUIT

Most would likely accept the premise that science builds on prior work, but that essentially great scientific discoveries are made by great scientists. Even the Nobel prizes recognize the achievements of individual scientists rather than research teams. Therefore, science must be a solitary and individual pursuit. Sociologists of science who study scientists at work have shown that only rarely does a scientific idea arise in the mind of a lone individual which is then validated by that individual alone and accepted by the scientific community. The process is much more like a negotiation than the revelation of truth. Scientists work in research teams within a community of like-minded investigators. Many problems in science are simply too complex for a sole individual to pursue alone due to constraints of time, intellectual capital and financing.

CONCLUSIONS

The message from the Science and Engineering Indicators Study (National Science Board, 1996) discussed in the first chapter, and from an evaluation of the myths of science presented here is simple. We must rethink the goals for science instruction. Both students and those who teach science must focus on the nature of science itself rather than just its facts and principles. School science must give students an opportunity to experience science and its processes, free of the legends, misconceptions and idealizations inherent in the myths about the nature of the scientific enterprise. There must be increased opportunity for both beginning and experienced teachers to learn about and apply the real rules of the game of science accompanied by careful review of textbooks to remove the "creeping fox terriers" that have helped provide an inaccurate view of science and its nature. Only by clearing away the mist of half-truths and revealing science in its full light, with knowledge of both its strengths and limitations, will all learners appreciate the true pageant of science and be able to judge fairly its processes and products.

University of Southern California, Los Angeles, USA

NOTE

This chapter is an expanded and modified version of those originally published in *School Science and Mathematics* under the title, *Myths of Science: Reexamining What We Think We Know About the Nature of Science* (1996), 96, 10-16, and in *Skeptic*, *15 Myths of Science: Lessons of Misconceptions and Misunderstandings from a Science Educator* (1997), 5, 88-95.

REFERENCES

American Association for the Advancement of Science (1993). *Benchmarks for science literacy,* New York, Oxford University Press.

Bacon, F. 1952 (1620). 'The new organon', in R. M. Hutchins, (ed.), *Great books of the western world, Vol. 30. The works of Francis Bacon,* Chicago, Encyclopedia Britannica, Inc., 107-195.

Campbell, N. (1953). *What is science?,* New York, Dover Publications.

Carey, S. S. (1994). *A beginners guide to scientific method,* Belmont, CA, Wadsworth Publishing Company.

Chaikin, A. (1994). *A man on the Moon: The voyages of the Apollo astronauts.* New York, Viking Press.

Chalmers, A. (1990). *Science and its fabrication,* Minneapolis, MN, University of Minnesota Press.

Cole, K.C. (1986). 'Things your teacher never told you about science: Nine shocking revelations!', *The Newsday Magazine,* March 23, 21-27

Darwin, C.R. (1958). *The autobiography of Charles Darwin, Nora Barlow (ed.),* New York, W.W. Norton & Company.

Dunbar, R. (1995). *The trouble with science,* Cambridge, MA, Harvard University Press.

Gibbs, A. and Lawson, A.E. (1992). 'The nature of scientific thinking as reflected by the work of biologists and by biology textbooks', *The American Biology Teacher,* (54), 137-152.

Gjertsen, D. (1989). *Science and philosophy past and present,* New York, Penguin Books.

Gould, S.J. (1991). 'An essay on a pig roast', in Bully for brontosaurus, New York, W.W Norton, 432-447.

Gould, S.J. (1988). 'The case of the creeping fox terrier clone', *Natural History,* (96), 16-24.

Hallam, A. (1975). 'Alfred Wegener and the hypothesis of continental drift', *Scientific American, (232)*, 88-97.

Hodson, D. (1986). 'The nature of scientific observation', *School Science Review, (*68), 17-28.

Horner, J.K. & Rubba, P.A. (1979). 'The laws are mature theories fable', *The Science Teacher,*(46), 31.

Horner, J.K. & Rubba, P.A. (1978). 'The myth of absolute truth', *The Science Teacher,* (45), 29-30.

Huizenga, J.R. (1992). *Cold fusion: The scientific fiasco of the century,* Rochester, NY, University of Rochester Press.

Keeslar, O. (1945a). 'A survey of research studies dealing with the elements of scientific method', *Science Education,* (29), 212-216.

Keeslar, O. (1945b). 'The elements of scientific method', *Science Education,* (29), 273-278.

Klayman, R. A., Slocombe, W. B., Lehman, J. S. and Kaufman, B.S. 1986. 'Amicus curiae brief of 72 Nobel laureates, 17 state academies of science, and 7 other scientific organizations, in support of appellees', in support of Appellees. Edwards v. Aguillard, 85-1513.

Kuhn, T. S. (1970). *The structure of scientific revolutions.* 2nd edition, Chicago, IL, University of Chicago Press.

Lopushinsky, T. (1993). 'Does science deal in truth?', *The Journal of College Science Teaching, 23*, 208.

Medawar. P. B. (1963). 'Is the scientific paper a fraud?', in P B. Medawar (1991), *The threat and the glory.* New York, Harper Collins. (228-233)

Menard, H. W. (1986). *The ocean of truth: a personal history of global tectonics,* Princeton, NJ, Princeton University Press.

National Research Council (1996). *The national science education standards,* Washington, D.C., National Academy Press.

National Science Board (1996). *Science and engineering indicators: 1996,* Washington, D.C., United States. Government Printing Office (NSB 96-21).

Newton, I. (1946). *Sir Isaac Newton's mathematical principles of natural philosophy and his system of the world,* A Motte, trans; revised and appendix supplied by F. Cajori, Berkeley, CA, University of California Press, (Original work published 1720).

Popper, K.R. (1968). *The logic of scientific discovery,* 2nd ed. revised, New York, Harper Torchbooks.

Popper, K.R. (1963). *Conjectures and refutations: the growth of scientific knowledge,* New York, Harper and Row.

Rhodes, G. and Schaible, R. (1989). 'Fact, law, and theory, ways of thinking in science and literature', *Journal of College Science Teaching,* 18, 228-232 & 288.

Rothman, M.A. (1992). *The science gap,* Buffalo, NY, Prometheus Books.

Schwab, J.J. (1964). The teaching of science as enquiry, in J.J. Schwab & P.F. Brandwein (eds), *The teaching of science*, Cambridge, MA, Harvard University Press, 31-102.

Sonleitner, F.J. (1989). 'Theories, laws and all that', *National Center for Science Education, Newsletter, 9*, 3-4.

Shreeve, J. (1995). *The Neandertal enigma: Solving the mystery of modern human origins*, New York, William Morrow and Company.

Taubes, G. (1993). *Bad science: The short life and weird times of cold fusion*, New York, Random House.

Tobias, S. (1990). *They're not dumb, they're different: Stalking the second tier*, Tucson, AZ, The Research Corporation.

SECTION II

COMMUNICATING THE NATURE OF SCIENCE: PLANS, APPROACHES AND STRATEGIES

WILLIAM W. COBERN AND CATHLEEN C. LOVING

4. THE CARD EXCHANGE: INTRODUCING THE PHILOSOPHY OF SCIENCE

The nature of science is an important though difficult subject to teach meaningfully and effectively to preservice teachers. To engage the students' minds in this subject that many find obscure and esoteric, a good introduction is a necessity. This chapter presents a learning game called *The Card Exchange* which has been found effective in arousing student interest in the philosophy of science. The chapter presents a brief description of how the game is set up and played and how it relates to the authors' instruction on the philosophy of science. The chapter includes a list of card statements. The statements as well as the text of the chapter have been revised and updated from an earlier publication (Cobern, 1991a).

There are a number of thoughtful articles in the literature stressing the need for philosophically literate teachers of science at all school levels (e.g., Andersen, Harty & Samuel, 1986; Hodson, 1985; Martin, 1979) and for many years the textbooks used in science methods courses have contained at least some material on the philosophy and nature of science. Nevertheless, science educators have been concerned that an acceptable level of philosophical sophistication was not being reached within the ranks of science teachers, and consequently are concerned about views toward the nature of science promoted in the classroom (e.g., Schmansky & Kyle, 1986). Duschl (1988, p. 51) summarizes the classroom situation by saying that "the prevailing view of the nature of science in our classrooms reflects an authoritarian view; a view in which scientific knowledge is presented as absolute truth and as a final form." This view has been called scientism. This is a problem first because as we learn more about the world views that students bring to the classroom we begin to understand how the scientific view extinguishes students nascent interest in science (Cobern, 1991b; 1996). Secondly, those students who do accept the scientific view are likely to become disenchanted with science at a later date as science fails to achieve the unrealistic expectations accompanying a scientism orientation. The challenge is how to teach the philosophy of science be taught to teachers with greater effectiveness?

THE CARD EXCHANGE

Sometime early each year in many schools, lessons are taught addressing the nature of science. Often instruction in the process of science is nothing more than a method listed on the board and provided as *the* way all scientists work. Or it may be suggested that students will be following various aspects of this method in numerous activities throughout the year. Students are told, therefore, they will be doing real science. We take the view that students' understanding of: a) what science is, b) just how human the endeavor really is and -- perhaps equally important -- c) what science is not, can be enriched and made more engaging by showing that those who do science, and those who write about it, hold varying views as to just what is authentic science (Martin, Kass, and Brouwer, 1990). If we can find ways to determine what individual students currently think, we at least can acknowledge their varying views--whether they come from ignorance, first impressions, or an extensive knowledge base about science. If necessary, teachers can then try to help them construct meanings more in line with a balanced view of science. Our purpose in this chapter is to present an activity, a learning game, which acts as a powerful set induction for subsequent instruction in the philosophy of science. We have found that this activity engages our students' minds and precipitates enthusiastic discussion on the question, "what is science all about?"

We have used the game successfully in a variety of settings. Elementary and secondary preservice methods classes are one example. Here we found our challenge to be how much time we can spend on the nature of science versus all the pedagogical and content issues one must deal with for a variety of science disciplines and a variety of grades preservice students will teach. We found that if students have only one science methods class, it is difficult to find the necessary time to do a good job with nature of science issues. It is always the struggle between our desire to give them the necessary background and their desire to know "what can I do in my classroom tomorrow." The card game does, however, serve as a highly effective entry into a world many students do not know exists.

Another group with whom we have used the card game are veteran classroom teachers, either during summer workshops or at state science teacher meetings in workshop settings. They love the activity, the engagement and, for many, the discovery that there is a whole area about science for which they have not had much background or experience. "Light bulbs" often go on in these settings and some teachers crave more. We both have had, from time to time, this activity result in teachers later enrolling in our graduate courses which concentrate on the nature of science and science teaching. There is little indication that most teachers who have become familiar with this strategy use the cards immediately with their students, although the high school teachers were more likely to see this as a possibility for their students in tenth to twelfth grade. Instead, it appeared that they were seeing this as a self-enriching experience that might enable them to teach from a different perspective.

When graduate students in science education play the card game, they are potentially the best prepared to get the most out of this activity. These students tend to have good backgrounds in science, have taught for a number of years, and have combined that experience with recent course work and, for some, active research in current issues of science education reform. Not only do they tend to have the most intense and detailed conversations, but their resultant paragraphs about science tend to be what we would consider the most perceptive and balanced. Later in the course, they go back to some of these statements to design exhibitions for their peers about how they would teach this principle about science to students. For example, two graduate students designed five different posters depicting five well-known models of classification systems throughout the history of science from Aristotle to the present. Their peers loved it because it was such a vivid way to teach something all children learn in a developmental way. It so clearly showed these systems to be human constructions that were later replaced with what the scientific community decided were more authentic models. What better way to show that "science builds on what has gone on before and refines its conclusions" or "theory and observation interact" or "theories help scientists interpret their observations."

Finally, interesting results occurred when we used the card game with some university scientists. Scientists are diverse in their views about science -- some holding rather strong empiricist views, others seeming theoretically driven, and others appearing balanced. The cultural component was minimally referred to by our scientists. The research piece to the card game -- looking more closely at the relationship between composition (race, culture, gender) of our various card-playing groups and our results, what they do with the cards, how they respond to the activity initially and in retrospect, what they propose to do differently when they leave us, and what they end up doing back in their schools -- is richly layered and ongoing.

The activity we are writing about is called a "card exchange," a learning game developed by Bergquist and Phillips (1975) for classes of 20 students or more. We use the game very much as it was originally developed except that we have changed the game content to the philosophy of science. The game works well because at the beginning students are encouraged to move around and talk with each other, things almost all students like to do! The subject of conversation is the content of the cards. This works as a set induction because in the course of their conversation students quite naturally begin considering what they believe about science and how those beliefs may or may not coincide with what others believe. Later in the game students form groups based on the content of the cards they hold and then corporately produce a written summary. Both of these later acts require compromise which forces the students to give a rough rank order to their beliefs about science. The result is that when we begin our part of the instructional process, our students are not only keenly aware that many of them hold quite different views on the nature of science, and many of them now have doubts about the validity of their own views. They are engaged.

PLAYING THE GAME

To prepare for this game the teacher must develop a set of science statements related to what that teacher later wishes to accomplish with his or her philosophy of science instruction. A single statement is placed on each card. The statements should be succinct and easily understood. They should represent a broad range of viewpoints including specific views to be expressed in the course. The set of card statements may be redundant. In fact, redundancy as well as diversity is necessary so that students can avoid being trapped with statements that they cannot affirm. We personally use a set of more than 200 cards containing 40 unique statements representing six categories (See Appendix A for the actual statements):

1. Theoretical Emphasis: Science is primarily a rationalistic, theory-driven endeavor (for example, see California Department of Education, 1990, p. 14-18).

2. Empirical Emphasis: Science is primarily and a data gathering, experimental endeavor in pursuit of physical evidence (for example, see Braithwaite, 1955).

3. Anti-Science View: Science is overrated. One should not give much credence to the aims, methods or results of science (for example, see Appleyard, 1992; Sale, 1995; Skolimowski, 1974).

4. Scientism: Science is the way of knowing; it is the perfect discipline. For a good introduction to this topic see Poole (1995) and Settle (1990).

5. Cultural View: Science is embedded in a social, historical, and psychological context which affects all that goes on in science (for example, see Cobern, 1991b; Fuller, 1991; Harding, 1993; Hodson, 1993).

6. Balanced View: This view point, which reflects our own aversion to extremism, takes science to be a complicated affair that cannot easily be reduced to one or even a few simple descriptions (for example, see Loving 1991; 1992).

The statements (see Appendix) used in the activity reflect the diversity found in current thought. They allow for comparison and contrast with our objective which simply put is, science viewed as both empirical and theoretical; science as a powerful though limited way of knowing; science as a human, not mechanical endeavor; science as a dynamic process. Depending on the instructor's objectives, other statements can be used. Our statements were drawn from a number of different sources. In addition to those listed above, we refer the reader to AAAS (1993), Aicken (1984), Eastman (1969), Kimball (1967), Matthews (1994), and National Research Council (1996).

The game begins with the instructor giving to each student a randomly drawn set of six cards (six to eight cards usually works best). The students evaluate their cards according to what they can most and least affirm. They then have a period of time in which to mill about examining each other's statements and making trades. Sufficient time should be allowed for each student to examine every other student's cards. The goal is to improve one's hand by trading cards one for one, in other words the students' goal is to trade cards they like less for ones they like more. There is no discarding. We typically allow our classes of 30 to 40 students a minimum of ten minutes for this phase of the game. At the end of the period we have everyone sit down while we give the next set of instructions. Instructions for each phase should not be given in advance.

In the second phase students are again to mill about, but this time seeking someone with whom they can pair. The pairing rules are that each pair must hold eight cards on which they have relative agreement. Each member of a pair must contribute at least three cards. This is important if the pairs are to be truly formed by compromise. The pair's remaining four cards are discarded.

Phase three of the game is a repeat of phase two, except now the pairs form quadruplets. Each foursome is to hold eight cards with each pair contributing at least three cards. Once the foursome has been established, the students are asked to rank order their cards. Then if they wish they may discard the two bottomed-ranked cards. Based on this final set of cards the students cooperate to write a statement of paragraph length on the nature of science. At the conclusion of the game we ask the various groups to share their paragraphs and to say why they accepted some statements while rejecting others. Generally this is enough to precipitate vigorous discussion. We facilitate the discussion by writing on the board a few phrases that characterize the views being presented.

We follow up the discussion with a presentation of two case studies from the history of science. Typically we use Ignaz Semmelweiss' work with childbed fever and Newton's exploration of the phenomena of colors (Mannoia, 1980). In these case studies we look for examples of the statements on the nature of science that the students have advocated in their card exchange summaries. The case studies can be presented orally in a recitation format by the professor or in the form of a printed handout. The advantage of using a handout is that the groups working individually at comparing and contrasting their card exchange summaries with the case studies do a more thorough job. The disadvantage is the amount of time required. The discussion of the card exchange summaries vis-a-vis the case studies concludes the set induction. From this point we begin the main body of instruction on the nature of science.

CONCLUSION

We personally have found the card exchange activity to be an effective method of drawing our students into the philosophy of science, a subject they heretofore resisted. It capitalizes on the innate gregariousness of students and the diversity of opinion among students. A set induction is, however, only the beginning of a lesson. The effectiveness of what happens afterwards depends on how well one can hold the attention captured during the set induction. Obviously there is a need for many creative instructional strategies if the philosophical preparation of preservice science teachers is to be effective.

ACKNOWLEDGMENTS

We gratefully acknowledge the assistance of Dr. Jennifer Helms of the University of Colorado, Boulder. Dr. Helms has used variations of the activity in her own teaching and provided us with valuable insight as we revised the original activity. Her assistance was especially helpful in the development of the "Cultural View" category.

[1] *Western Michigan University, Kalamazoo, Michigan, USA*
[2] *Texas A&M University, College Station, Texas, USA*

REFERENCES

Aicken, F. (1984). *The nature of science*, London, UK, Heinemann Educational Books.
American Association for the Advancement of Science (AAAS). (1993). *Benchmarks for science literacy: Project 2061*, New York, Oxford University Press.
Andersen, H.O., Harty, H., & Samuel, K.V. (1986). 'Nature of science, 1969 and 1984: Perspectives of preservice secondary science teachers', *School Science and Mathematics*, (86), 43-50.
Appleyard, B. (1992). *Understanding the present - Science and the soul of modern man*, New York, Anchor Books Doubleday.
Bergquist, W. H., & Phillips, S. R. (1975). *A handbook for faculty development*, Danville, NY, Danville Press.
Braithwaite, R.B. (1955). *Scientific explanation: A study of the function of theory, probability and law in science*, New York, Cambridge University Press.
California Department of Education (1990). *Science framework for California public schools: Kindergarten through grade twelve*, Sacramento, CA, California State Board of Education.
Cobern, W.W. (1991a). 'Introducing teachers to the philosophy of science', *Journal of Science Teacher Education*, (2), 45-47.
Cobern, W.W. (1991b). *'World view theory and science education research'*, NARST Monograph No. 3. Manhattan, KS, National Association for Research in Science Teaching.
Cobern, W.W. (1996). 'Worldview theory and conceptual change in science education', *Science Education*, (80), 579-610.
Duschl, R.A. (1988). 'Abandoning the scientific legacy of science education', *Science Education*, (72), 51-62.
Eastman, G. (1969). 'Scientism in science education', *The Science Teacher*, (36), 19-22.
Fuller, S. (1991). *Social epistemology*, Bloomington, IN, Indiana University Press.

Harding, S. (ed.) (1993). *The "racial" economy of science: Toward a democratic future'*, Bloomington, IN, Indiana University Press.

Hodson, D. (1985). 'Philosophy of science, science and science education', *Studies in Science Education*, (1), 25-57.

Hodson, D. (1993). 'In search of a rationale for multicultural science education', *Science Education*, (77), 685-711.

Kimball, M. E. (1967-1968). 'Understanding the nature of science: a comparison of scientists and science teachers', *Journal of Research in Science Teaching*, (5), 110-120.

Loving, C. C. (1991). 'The scientific theory profile: A philosophy of science models for science teachers', *Journal of Research in Science Teaching*, (28), 823-838.

Loving, C. C. (1992). 'From constructive realism to deconstructive anti-realism: Helping science teachers find a balanced philosophy of science', *proceedings of the Second International Conference on the History and Philosophy of Science and Science Teaching*, Vol. II, Kingston, Queen's University.

Mannoia, V. J. (1980). *What is science?: An introduction to the structure and methodology of science*, Washington D.C., University Press of America.

Martin, B., Kass, H., & Brouwer, W. (1990). 'Authentic science: a diversity of meanings', *Science Education* (74), 541-554.

Martin, M. (1979). Connections between philosophy of science and science education, *Studies in Philosophy and Education*, (9), 329-332.

Matthews, M.R. (1994). *Science teaching: The role of history and philosophy of science*, New York, Routledge.

National Research Council (1996). *National science education standards*, Washington, D.C., National Academy Press.

Poole, M.W. (1995). *Beliefs and values in science education*, Philadelphia, PA, Open University Press.

Sale, K. (1995). *Rebels against the future*, New York, Addison-Wesley Publishing Co.

Settle, T. (1990). 'How to avoid implying that physicalism is true: A problem for teachers of science', *International Journal of Science Education*, (12), 258-264.

Shymansky, J.A., & Kyle, W.C. Jr. (1986). 'A summary of research in science education - 1986', *Science Education*, (72), 254-275.

Skolimowski, H. (1974). 'The scientific world view and the illusions of progress', *Social Research*, (41), 52-82.

APPENDIX A
CARD EXCHANGE STATEMENTS

Theoretical Emphasis

1. Science is open-ended, but scientists operate with expectations based on the predictions of theory.

2. A theory is what scientists strive for: a large body of continually refined observations, inferences, and testable hypotheses.

3. Theories help scientists interpret their observations: facts do not speak for themselves.

4. In general, scientists plan investigations by working along the lines suggested by theories, which in turn are based on previous knowledge. Theories serve to give direction to observations, i.e., they tell one where to look.

5. A theory is a logical construct of facts and hypotheses that attempts to explain a range of natural phenomena and that can be tested in the natural world.

6. Good science cannot be done without good theories.

Empirical Emphasis

1. Observation is central to all of science, i.e., seeing is believing.

2. A scientist should not allow preconceived theoretical ideas to influence observation and experimentation.

3. Unless an idea is testable it is of little or no use; thus, scientists attempt to convert possible explanations into testable predictions.

4. Careful, repeatable observation and experiment give the facts about the world around us.

5. Good science always begins with observations.

6. Science is never dogmatic; it is pragmatic--always subject to adjustment in the light of solid, new observations.

7. A phrase such as "Many scientists believe.." misrepresents scientific inquiry because scientists deal in evidence.

Anti-Science View

1. Science is always changing and therefore is not very reliable.

2. Scientists should be held responsible for harm their discoveries have caused, e.g., pollution, nuclear weapons.

3. Earning recognition from other scientists is really the main motivation of more scientists.

4. Most of what scientists do will never be of much practical value.

5. Money spent on projects such as NASA space flights would be better spent on healthcare for the needy.

6. Science destroys values and morality by disparaging the unique nature of men and women.

7. Science and religion are fundamentally at odds.

Scientism

1. The scientific method should be followed in all fields of study.

2. Scientists and engineers should make the decisions about things like types of energy to use because they know the facts best.

3. Science is the most important way of gaining knowledge open to humanity.

4. Science knowledge is of much greater value than any other type of knowledge.

5. Only science can tell us what is really true about the world.

6. Science knowledge is always objective and self-correcting.

7. Credit for our advanced way of life must go to science and scientific progress.

Cultural View

1. Funding influences the direction of science by virtue of the decisions that are made on which research to support.

2. The scientific enterprise is situated in specific historical, political, cultural, and social settings; thus, scientific questions, methods, and results vary according to time, place, and purpose.

3. The predominance of men in the sciences has led to bias in the choice and definition of the problems scientists have addressed. This male bias is also one factor in the under representation of women in science.

4. Scientific facts are manufactured through social negotiations. Nature has nothing to say on its own behalf.

5. Scientists in one research group tend to see things alike, so even groups of scientists may have trouble being entirely objective.

6. The Early Egyptians, Greeks, Chinese, Hindu and Arabic cultures are responsible for many scientific and mathematical ideas and technological inventions.

7. Until recently, some racial minorities, because of restrictions on their education and employment opportunities, were essentially left out of the formal work of the science establishment. The remarkable few who overcame these obstacles were even then likely to have their work disregarded by the science establishment because of their race.

Balanced View

1. Science is one of several powerful ways of knowing and understanding the natural world, however, some matters cannot be examined usefully in a scientific way.

2. Science leads to generalizations based on observations or theories. Science always aims to be testable, objective and consistent.

3. As with all human endeavors science is subject to many influences both good and bad.

4. Science builds on what has gone on before and refines its conclusions, but scientific work does not result in infallible propositions, such as the word "proof" implies to a nonscientist.

5. Scientific progress has made possible some of the best things in life and some of the worst.

6. Theory and observation interact. Each contributes to the other: If theory without observation is empty, then observation without theory is blind.

NORMAN LEDERMAN AND FOUAD ABD-EL-KHALICK

5. AVOIDING DE-NATURED SCIENCE: ACTIVITIES THAT PROMOTE UNDERSTANDINGS OF THE NATURE OF SCIENCE

In spite of a renewed interest in establishing the nature of science as a major element of the science curriculum we fear that the current reform emphasis relative to the nature of science (NOS) will have as little impact as earlier efforts. It seems that two critical omissions noted in previous reform efforts are again apparent. There has not been a concerted effort to communicate clearly what is meant by the NOS and how a functional understanding of this crucial aspect of science can be communicated to K-12 students (Lederman & Niess, 1997).

The focus of this chapter is to address this second issue by providing strategies to communicate important aspects of the nature of science to students in a hands-on fashion. As far as communicating proper understandings of the NOS to students is concerned, teachers have been led to believe that their students will come to understand the NOS simply through the performance of scientific inquiry and/or investigations. This advice is no more valid than assuming that students will learn the details of cellular respiration by watching an animal breathe. Too many students and teachers believe that scientific knowledge is provable in an absolute sense, objective, and devoid of creativity and human imagination. It is also just as common for students and teachers to believe that laws are theories that have been proven and that there exists a single step-wise scientific method which characterizes scientific investigations (Lederman, 1992).

It is highly unlikely that students and their teachers will come to understand that science is tentative, empirically-based, partly the product of human imagination and creativity, and is influenced by social and cultural factors solely through learning about the content of science or its processes. We believe that a concerted effort on the part of science educators and teachers to explicitly guide learners in their attempts to develop proper understandings of the nature of the scientific enterprise is essential. The notion of explicitness is imperative. It is critical that we target teaching the NOS if the desired impact on learners' conceptions is to be achieved.

ABOUT THE ACTIVITIES

This chapter introduces a set of activities designed to model an explicit approach to teaching crucial aspects of the NOS. These activities have been successfully used with students and science teachers. Science educators and science teachers can use the following activities to convey to students and preservice or inservice science teachers adequate notions of the NOS. The notions advocated in these activities are designed at a level of generality that renders them virtually non-controversial. Moreover, the scientific knowledge prerequisite is minimal to remove the constraints on using the activities.

Science educators can use these activities in either science content or science methods courses. In case your audience consists of precollege students, the appropriate grade level(s) for using a certain activity are pointed out, but most activities can be adapted for audiences of varying levels of sophistication. Where appropriate, extensions that make an activity more amenable for use with older students are included. Additionally, by not requiring specialized scientific knowledge, the activities free the learners from having to struggle with complex scientific concepts as they try to internalize certain aspects of the NOS. The following activities are presented with no sequence in mind; each can stand on its own. The activities can be presented in a nature of science unit or can be infused throughout a science content course of study.

The activities are grouped in three sections relative to important aspects of the NOS. These sections include observation versus inference, creativity, and tentativeness, subjectivity and social and cultural context in science, and black-box activities. Each section starts with a general discussion of related NOS notions. Each activity specifies student objectives, materials, setup and/or procedure in addition to a possible scenario. The overheads that you need for the activities can be conveniently reproduced from the ones included. The procedure/scenario section provides an explicit idea about the kind of questions and answers that can be expected and the notions that need emphasis during a certain activity. These scenarios are not meant to be prescriptive in any respect. They are drawn from classroom experiences and genuine student reactions to the activities and are meant to give you a better idea about one possible discourse among many equally fruitful classroom interactions. The approach you choose to undertake to a certain activity is totally up to your professional judgment.

Throughout the discussions and activities, examples and ideas from science are infused. These are intended to enrich classroom discourse and support the notions presented with instances from science. Moreover, references to several scientific articles and/or books are included for your own use. Most of these readings are not appropriate for precollege students but should be adequate for preservice and inservice secondary science teachers. They are intended to provide a background that makes the ideas presented more contextual and embedded in scientific practice.

SECTION I: OBSERVATION, INFERENCE, CREATIVITY, AND TENTATIVENESS

Among the inappropriate conceptions of the NOS frequently portrayed in textbooks is the notion that for every question posed about the natural world, scientists will eventually find 'the correct and absolute' answer. This idea is reinforced when students are expected to come up with 'the' correct answer to end-of-chapter textbook exercises; choose the *one* correct answer on multiple choice tests; or reach *the* right conclusion in 'cook-book' laboratory sessions.

These notions and experiences are not consistent with the way scientific knowledge is produced. There seldom is, at least initially, and most often indefinitely, one answer to the questions that scientists investigate. This is because scientific knowledge is partly a product of human inference, imagination, and creativity, even though it is, at least partially, supported by empirical evidence. As such, scientific knowledge is never absolute or certain. This knowledge (including theories and laws) is tentative and subject to change.

The following activities are designed to help students make the crucial distinction between observation and inference, and appreciate the tentativeness of scientific knowledge and the role of creativity in science.

Tricky Tracks!

This activity can be typically used to introduce students to the NOS. You can use the activity to establish an atmosphere that supports your students' active participation in classroom discussion. This is crucial if you are to derive the benefit from these activities.

'Tricky Tracks!' conveys to students the message that every idea counts irrespective of it being the 'correct' answer. Students completing this activity will gain experience in distinguishing between observation and inference and realizing that, based on the same set of evidence (observations, or data), several answers to the same question may be equally valid.

Level: Upper elementary, middle, or high school.

Materials: Overheads (Figures 1, 2 & 3)

Procedure / Scenario:

1. Show Figure 3 to students and ask them to write down an account of what they think might have happened as indicated by what they see. A typical

story line is that "two birds approached each other over the snow, had a fight, and the big bird ate the smaller one and went on its way." Make sure that each student writes his/her own account. This written record will render students' dissatisfaction with their accounts greater and facilitate their attending to the ideas being presented.

2. Place Figure 1 on overhead. Ask students: "What do you observe?" Typically students would answer: "Bird (or any other animal) tracks" or "Tracks left by birds (or other animals) as they walked toward the same spot," etc. Accept all answers at this point and avoid making any judgment. You can list those answers on the board.

3. To continue with the bird scenario, at this point you may ask: "Can you see the birds?" or "How can you tell that these tracks are left by birds?" The fact that we can not see birds makes the statement "bird tracks" an inference rather than an observation. A possible observation would be: "Two sets of black marks of different shapes and sizes left on a transparency!" It is the case that based on this observation and probably on our familiarity with the kind of tracks that some animals leave behind, we inferred that birds made these tracks, but they may be something else. The tracks may be those of two different species of dinosaurs, or a mother (or father!) and a baby dinosaur of the same species. The tracks may as well be those of two different kinds of birds, or a large and a small bird of the same species. Even our claim that larger tracks are left by the larger animal is an inference. The important point to emphasize is that student statements similar to the ones above are inferences as contrasted to observations.

4. You may ask your students: "Why were the two animals heading toward the same spot?" Again the answers may vary: Students may say that the animals were aiming at a common prey, or moving toward a source of water. One animal may have been attacking the other, or the two had to move to the same spot by virtue of the nature of the terrain, etc. It is important to point out that all of these statements are inferences and that all those inferences are equally plausible. Emphasize that based on the same set of observations or evidence, you and your students were able to come up with several, but equally plausible answers (inferences) to the same question: "What has happened?"

5. Place Figure 2 on overhead. Ask your students: "What do you observe?" Some may answer: "The two sets of marks now appear to be close and randomly mingled," which is a possible observation. Others may say: "The

two birds are having a fight," which is an inference. Point out to students the difference between the two. Again note that many inferences are possible: The two animals are fighting, or engaged in a mating ritual, or battling over a prey that one of them has captured, etc.

6. Now place Figure 3 on overhead and ask students what they observe. By now the answer should typically be: "The set of the larger marks is left on the transparency. The smaller marks are not visible any more." Ask them: "What do you infer?" Again the possibilities are many: One animal may have eaten the other, one may have grabbed the other and moved on, one animal may have flown while the other kept walking, etc. Again stress the point that all these inferences are equally justified by the evidence available.

7. Now ask each pair of students to compare their written accounts and what they think of them after the class discussion. (You can ask younger students to write in their journals whether and how the discussion made them change their mind about their own accounts). Next, ask students whether we can ever know, based on the evidence available, what has "really" happened (see extensions, below).

8. Conclude by making explicit the two main points: a) the difference between observation and inference, and b) based on the same set of evidence many equally warranted answers to the same question can be inferred. Continue that scientists make similar inferences as they attempt to derive answers to questions about natural phenomena. And even though their answers are consistent with the evidence available to them, no single answer (or story) may solely account for that evidence. Several answers are often plausible. And similar to the case of our tracks, scientists may simply never find the answer as to what has really happened. Alternatively, this activity can be presented in an inductive manner. You can show students Figures 1, 2 and 3 respectively, each time asking them to make observations and draw inferences as to what might have happened. Thus, as you proceed, students would be provided with additional data or evidence that they need to incorporate into their accounts.

Upper middle and high school extensions: Half way through the activity, encourage the student to rule out those inferences that seem to be inconsistent with the observations. For example that the tracks were caused by a car! Or a fish! Take all student responses seriously even if they were meant to be humorous. It is important to convey the idea, and then to explicitly explain, that inferences must be consistent with the evidence. Even though a certain collection of evidence may equally justify several inferences, not all inferences can be based on that evidence.

Eventually, scientific knowledge should be based on and consistent with empirical evidence.

High school extensions:

Another possibility to pursue in the scenario with older students is the time frame. What guarantees, or what evidence do we have that both sets of marks were made at the same time. It could equally well be that each animal made its tracks at a different time, and that both were never actually present in the same place at the same time. Moreover, the whole thing may not have ever happened.

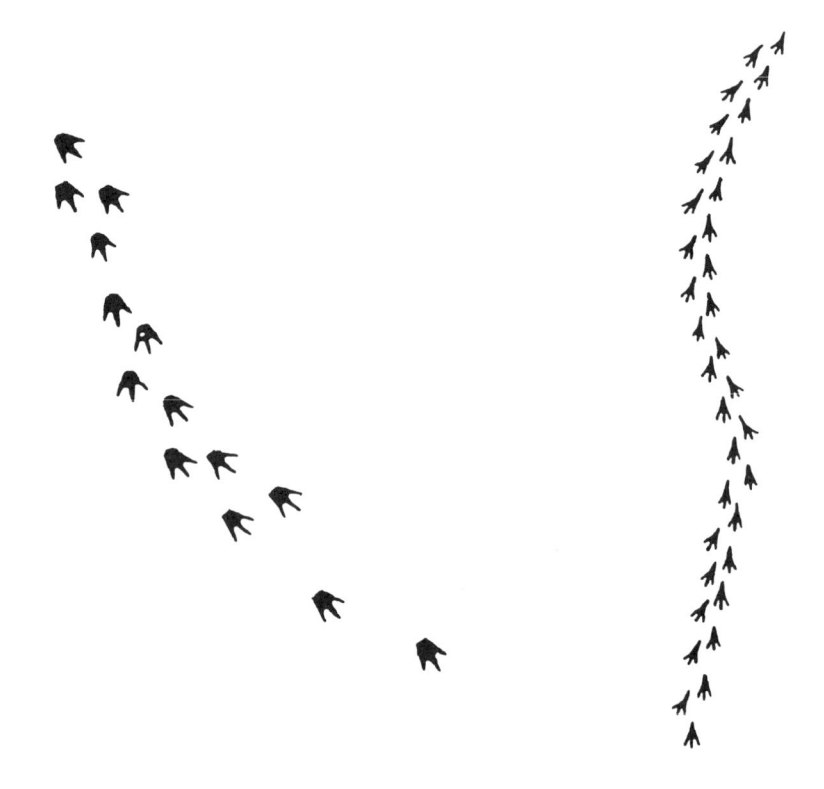

Figure 1. Tricky Tracks, Set #1

AVOIDING DE-NATURED SCIENCE 89

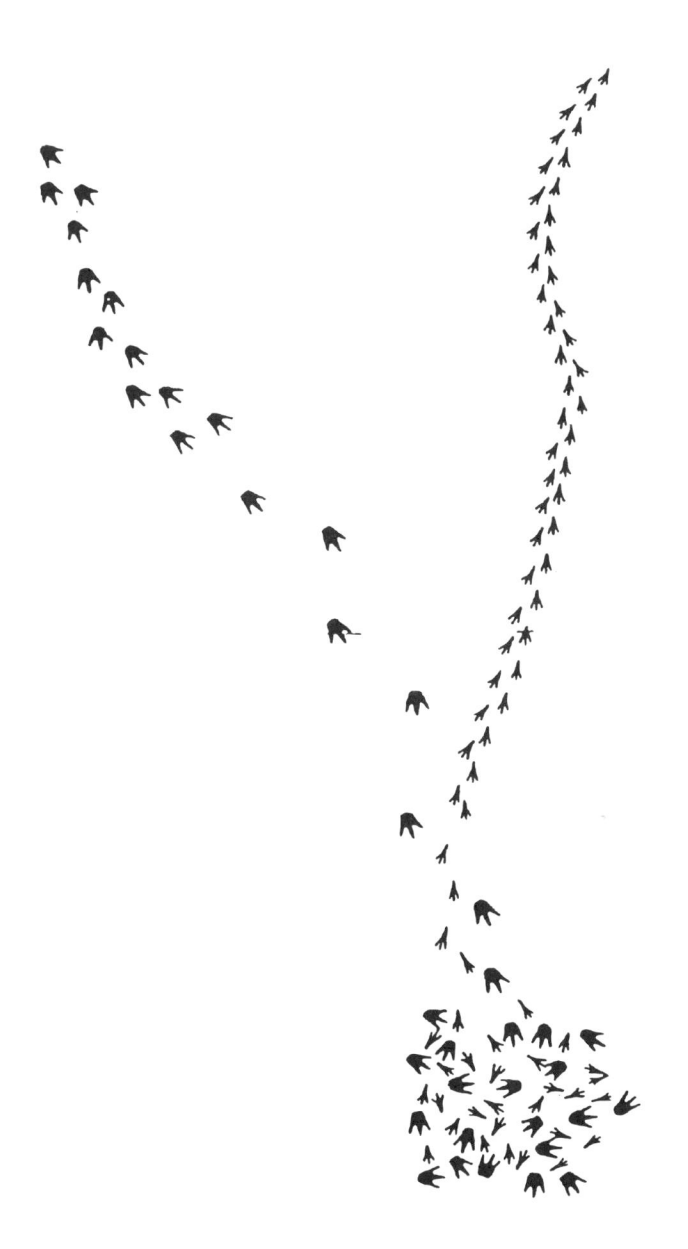

Figure 2. Tricky Tracks, Set #2

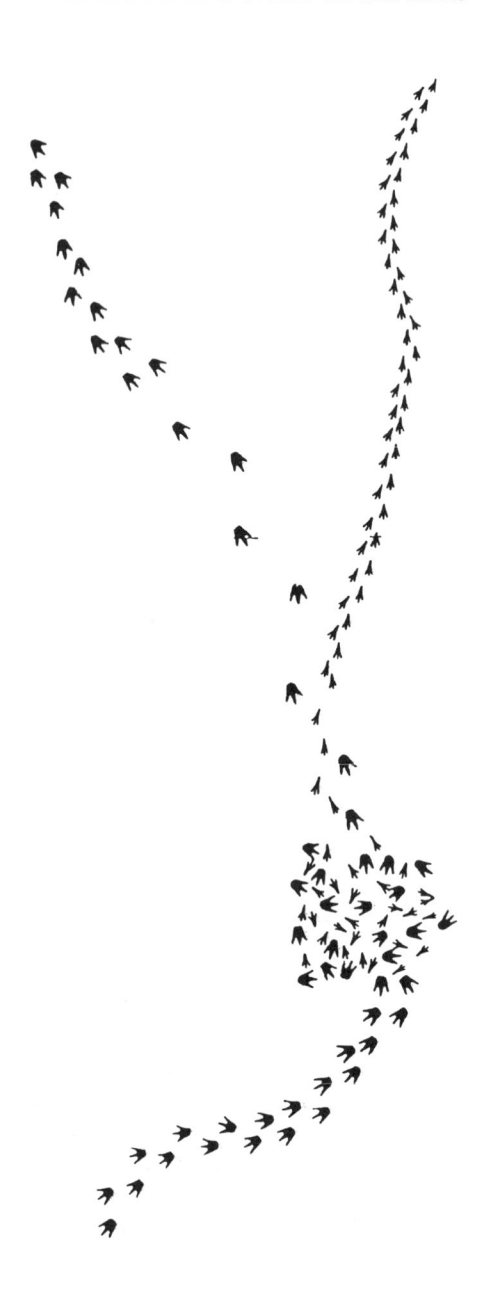

Figure 3. Tricky Tracks, Set #3

It may simply be a set of marks on a transparency! There is no evidence to rule out the possibility that these marks were simply left by the touch of a pen on paper. This can be pursued all the way to the notion that certain concepts in science, even though based on some evidence, may only exist in the scientists' minds (for example, see Lewin, 1983, for an account of the recent controversy about the actual viability of the concept of 'competition' in community ecology). The issue of whether we can ever know through inference only what has actually happened can be further pursued. A case in point is the disclosure by the American space agency of evidence that supports the claim that life once existed on Mars. Ask students to research the issue and argue whether inference is at play, and to what extent. Of course, remember that there is no single answer to this question!

The Hole Picture!

This activity is intended to reinforce students' understanding of observation and inference in science and to introduce them to the notion of creativity and its role in constructing scientific knowledge. Students should be able to appreciate that scientific knowledge is partly a product of human inference, imagination, and creativity, even though it is, at least partially, supported by empirical evidence. Moreover, students should be helped to realize that science does not produce absolute or certain knowledge. All scientific knowledge is subject to change (i.e. tentative) as more evidence is accumulated or as already available evidence is re-interpreted in light of newly formulated hypotheses, theories, and/or laws.

Level: Upper elementary and middle school

Materials: One manila file folder, a sheet of construction paper that will fit inside the folder, one overhead transparency (per student or group of students). Paper sheets of various colors, adhesive tape, glue.

Construction: On the large or tab side of the manila file folder, punch a number of holes randomly (Figure 4). It may be necessary to cover the folder with an opaque material to prevent seeing the shapes in the folder. Next, tape the folder leaving the tab end open creating an envelope (Figure 5). Create an insert by gluing differently colored, randomly shaped pieces of paper to the sheet of construction paper (Figure 6). The inserts need not be the same for all prepared folders. Insert the construction paper into the envelope with the colored pieces of paper facing the holes (Figure 7).

92   N. LEDERMAN AND F. ABD-EL-KHALICK

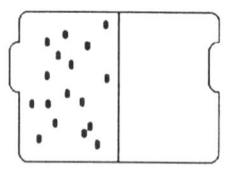

Figure 4.   Open file folder with randomly-spaced holes on one side.

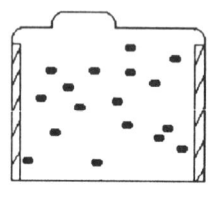

Figure 5.   Closed file folder showing the sides taped.

Figure 6.   Illustration of the construction paper shapes glued to the insert.

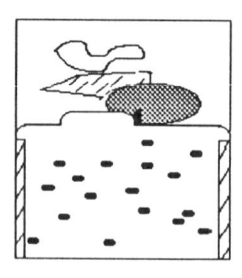

Figure 7.   Illustration of the insert being placed into the file folder.

AVOIDING DE-NATURED SCIENCE 93

At this point, tape an overhead transparency over the side of the folder with the holes. With non-permanent pen, students can draw on the transparency and erase their drawing.

Procedure / Scenario

1. Hand each student, or group of students, a manila folder (with the insert inside) and an erasable or non-permanent pen.

2. Inform students that the inserts have certain colored shapes glued to them. Students, without removing the inserts, are instructed to figure out those shapes and colors. The only available information to the students is what they see of the colored paper through the holes.

3. Have students trace their proposed shapes on the overhead transparency. Using transparencies and erasable pens allows you to re-use the folders in other or future classes.

4. The activity aims to put students in a situation similar to the ones that scientists face when going about their work. When your students finish their proposed shapes, take the time to make explicit the similarities between what they are doing and what scientists usually do: Faced with a natural phenomenon (the insert), scientists pose certain questions to which there usually are no readily available answers (What is the shape of the colored pieces of paper on the insert?)

5. Just like your students, scientists would rather handle the phenomenon first hand, which in the present case would be to simply pull out the insert and see how it looks. This, however, is rarely possible. For example, for a few hundred years physicists have theorized the presence of atoms, formulated many atomic theories, investigated the structure of the atom, and accumulated a great deal of knowledge about the atom and its components. This knowledge in turn has allowed many advances in physics and other related fields. This was the case without the scientists being able to actually see an atom. (It is true that using super-accelerators / super-colliders physicists were able to break atoms into smaller pieces. However, another problem of 'visibility' arose. The now-famous "Higgs" particle seems to 'block the vision' of scientists who again seem not able to 'see' what they would like to see firsthand). In a similar fashion, astrophysicists have produced scientific knowledge about the inside of the sun and the kinds of reactions taking place within, all without splitting the sun open!

6. The question arises, "How do scientists produce a seemingly reliable body of knowledge about such phenomena?" Scientists collect data about the phenomena they study. The holes on the folder represent data points allowing us to view a part of the object of investigation. The data points that can be collected about different phenomena vary in several ways. Some of those ways include the:

   - Amount (e.g., number of holes) which may depend on the feasibility and practicality of collecting the data. For example, in geological surveys, it is possible, but not at all practical, to collect rock samples from every square meter of terrain. Samples are usually collected from a much larger unit area.

   - Quality (such as small versus larger holes) which relates to the accuracy, precision, etc., of the data. The quality of data depend on a multitude of factors. Technology is one. For example, the quality of the Hubble telescope photos of distant galaxies now available to astronomers are by far more informative than earlier photos taken by observatories on Earth.

   - Availability (for example, it may not be possible for us to indefinitely punch holes in the folder to see the whole insert below). For example, astrogeologists would certainly like to put their hands on rock samples from every meteorite they can observe. In reality, however, they can only hope to locate meteorite samples available on Earth as a result of collisions that took place in distant times.

7. After collecting data, scientists, much like what your students did, infer answers to their questions that are consistent with the data. Creativity and imagination are essential to this process. In much the same way that your students have literally filled in the gaps between the holes to generate a final picture of what they thought the colored pieces of paper look like, scientists engage in a creative process to make sense of the data they have collected and come up with a final picture or an answer.

8. Now ask a few of your students to remove the insert from their folders (other students should keep their inserts inside). Ask those students to compare, in front of the class, their proposed drawings with what the inserts actually look like. If you were careful to glue randomly shaped pieces of paper to the insert, your students' surprise is sure to follow. Point out that scientists, very often, do not have the luxury of 'pulling out their inserts and examining them'. Rather, they have to infer an answer from the available data.

9. Ask some of your students whose inserts are still inside their folders how certain they are about their proposed drawings! Ask your students whether they think scientific knowledge can be absolute or certain (A good discussion usually results).

10. With older students it might be a good idea not to permit they to remove their inserts. This provides an experience that is more consistent with actual scientific investigations. With younger students you might have to let all of them remove their inserts, especially if they show signs of frustration. (Note, it is true that in most cases scientists do not stop at the initial phase of collecting and inferring. Next, they derive predictions based on their hypothesized answers and test those predictions by collecting more direct or indirect data. This aspect is dealt with in the following activities.)

Real Fossils, Real Science

This activity helps students realize that scientific knowledge is partly a product of human inference, imagination, and creativity. The advantage of this activity is that students work with the same artifacts and data (i.e, fossil fragments) as paleobiologists.

Level: Upper elementary and middle school

Materials: Fossil fragments (not complete fossils), construction paper, scissors (per student or pair of students).

Procedure / Scenario:

1. Give each student (or pair of students) a fossil fragment and ask them to make a detailed diagram of it. The diagrams may be larger than the actual fragments. The students, however, must include the appropriate scale with their diagrams. If possible, obtain sets of similar or identical fossil fragments so that different students may get similar or identical fragments.

2. Ask students to trace the outline of their fossil fragment on a separate sheet of colored construction paper. This tracing is cut out and discarded to form a window so that when the construction paper border is placed over the paper containing the fossil fragment diagram, only the diagram appears.

3. Using a different color pencil instruct students to complete their fossil drawing (to scale) on the construction paper containing the fossil fragment diagram. Students should end up with a drawing of an organism from which, they believe, the fossil fragment has come. You can have your students complete their fossil diagrams in class or assign this as homework.

4. Each student ends up with a complete fossil drawing having two parts: the original fossil fragment drawing in one color and the inferred drawing of a complete organism in another color.

5. Ask students to staple together the construction paper with the previously cut window and the paper with the complete drawing. The papers should be stapled on one side such that they can be flipped open. Moreover, the fossil fragment diagram should only show through the construction paper window. This format enhances the presentation of original (fossil fragment) and completed diagrams to other students (see Figures 8 and 9).

6. Ask students to make an oral presentation in which they describe the habitat, diet, behavior, and other characteristics of the organisms they have extrapolated from the fossil fragments. At this point you might ask whether some students knew in advance what organism their fossil fragment came from (e.g., coral). Ask those students, if any, whether their knowledge affected the inferences they made about the habitat, diet, etc., of the complete organism that they inferred from the fossil fragment. You might want to explain that scientists' prior knowledge often influences their interpretations of the data and affects their conclusions.

7. It would be interesting to compare those organisms that different students have inferred from similar or identical fossil fragments. If those organisms were different, ask your students: "Can we tell for certain from which organism the original fossil fragment come?" Explain to students that we might not be able to give a definite answer. Continue by asking: "Is it possible that scientists face a similar situation?" "Can scientists differ in the inferences they derive from evidence?" "If yes. How can such differences be settled?" Explain to students that all too often scientists may reach differing conclusions based on the same evidence, just as the students have done in this activity. Scientists also often hold their views strongly and do not give them up easily.

8. Make explicit to students that what they have done is very similar to what paleobiologists and other scientists that investigate fossils often do. Point

out that much creativity is involved in extrapolating or inferring from fossils the kind, habitat, and life style of the organisms whose fossils or fossil fragments are investigated.

9. You can conclude this activity by talking about the famous case of the dinosaur Iguanodon. When it was first reconstructed, the thumb was originally placed as a spike above the nose! It is useful to remind students that any reconstruction should be considered tentative -- just like all of the products of science.

High School Extensions

You can initiate a discussion about the extent to which creativity plays a role in science with the case of hominid evolution: Telling the story of the evolution of man (*Homo sapiens*) over the course of the past five million years. Scientists have formulated several elaborate and differing story lines about this evolution. It is noteworthy that all that is available to those scientists is a few teeth, tools, and parts of skulls and skeletons that can be spread over one moderately-sized table! Inference, imagination, and creativity serve to fill in the gaps, which in this case seem to be enormous!

The same discussion can be carried further to introduce students to the notion that scientific knowledge is affected, to varying degrees, by the social and cultural context in which it is produced. The differing story lines in the above example about the evolution of humans were heavily influenced by social and cultural factors. Until recently, the dominant story was centered about 'the man-hunter' and his crucial role in the evolution of humans to the form we now know (see Lovejoy, 1981). The hunter scenario was consistent with the white-male culture that dominated scientific circles up to the 1960s and early 1970s. As the feminist movement grew stronger and women claimed recognition in the various scientific disciplines, the story about hominid evolution changed. One story more consistent with a feminist approach is centered about 'the female-gatherer' and her central role in the evolution of humans (see Hrdy, 1986). It is noteworthy that both story lines are consistent with the available evidence.

SECTION II: SUBJECTIVITY, AND SOCIAL AND CULTURAL CONTEXT

Scientists are often portrayed as being objective. As they engage in their work, scientists are thought to set aside their personal prejudices, perspectives, and beliefs.

This objectivity, among other things, is believed to allow scientists to:

- Conduct 'objective' observations. Scientists make theory-free observations: They simply describe and measure things as they are. These observations are independent of what the scientists know, believe or how they view the world.

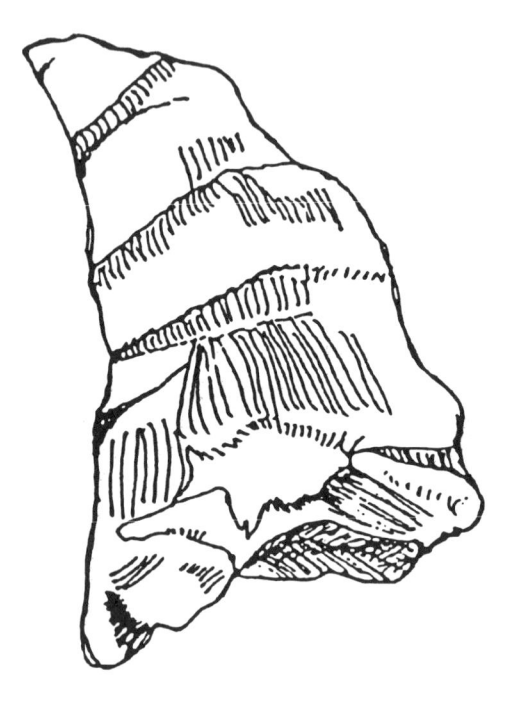

Figure 8. View of the original fossil fragment.

- Reach 'objective' conclusions. Based solely on their objective observations, scientists use the rules of logic and inference to formulate hypotheses or theories to explain the phenomenon under investigation.

- Evaluate new evidence objectively. After they formulate a hypothesis or theory, scientists collect more evidence to test the adequacy of their hypotheses or theories or to test their predictive power. Hypotheses and theories are 'objectively' evaluated against this evidence. Confirmatory evidence tends to strengthen the hypothesis or theory and eventually leads to its acceptance by scientists. However, if the hypothesis or theory is not supported by the evidence, it is rejected.

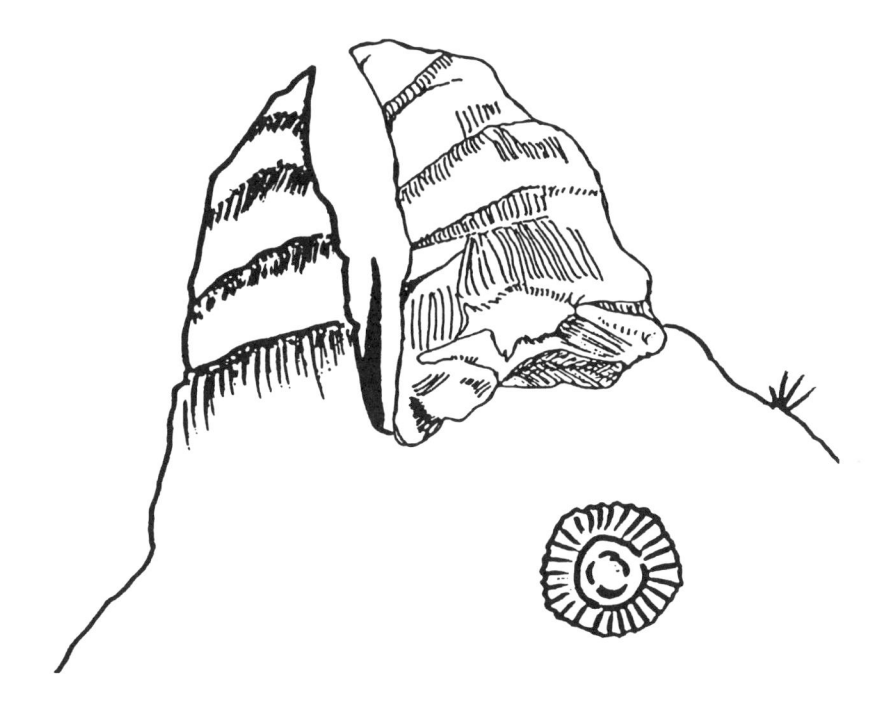

Figure 9. View of the completed fossil diagram.

It may be tempting to accept the above claims. The history of science, however, is replete with instances that counteract each of them. It is often the case that scientists interpret the same evidence differently, formulate different hypotheses to explain that evidence, and fiercely defend those explanations or hypotheses. It fact, controversies are commonplace in science. This is the case even though the various parties collect and present abundant evidence to champion their views (see the discussion of mass extinction in Chapter 16 for a recent example). The above notions of objectivity have been discounted by many philosophers and historians of science. For instance, Kuhn (1970), suggested that all scientific observations and interpretations are in some respect subjective. Kuhn advanced the notion of 'paradigm' to account for what usually happens in science. A paradigm defines, for a certain research community, the phenomena that are worth researching, acceptable questions to ask of those phenomena, appropriate research methodologies, adequate instrumentation, and the relevant and admissible evidence. For a scientist, a paradigm acts as a lens through which his/her observations are filtered. In a sense, the interpretations and explanations that a scientist formulates are consistent with that paradigm.

Although you may not want, and rightly so, to formally introduce your students to all the above notions, certain ideas are worth emphasizing to your upper middle and high school students. Scientists' beliefs, previous knowledge, training, experiences, and expectations actually influence their work. All these background factors form a mind-set that affects what scientists observe (and don't observe) and how they make sense of or interpret their observations. It is this individuality that accounts for the role of subjectivity in the production of scientific knowledge.

That's Part of Life!

Level: Upper middle and high school

Materials: Overhead (Figure 10)

Procedure / Scenario:

1. Place Figure 10 on the overhead projector and ask your students to carefully read the text and tell you what it means.

2. It is usually quite difficult to make sense of the text, even though the individual words and sentences are easily comprehensible. Tease students for possibilities. After a while, suggest that the passage may simply have no meaning at all. Ask how many students agree with that. (The idea is to make the activity as discrepant as possible.)

3. Next tell your students that this passage is about 'doing the laundry'. Ask them to read the text again and tell you whether things now make more sense to them, and whether, after all, there was a sensible meaning to that collection of words and sentences.

4. Ask students what they think is the idea behind this exercise.

5. Context is very important for making sense of what we observe, and how we interpret those observations. The individual words and sentences, even though each by itself was comprehensible, made little sense to your students. In the same manner, for a scientist, a mere collection of data or facts, lacking any context, may not make any sense.

6. In order to put things in context, we need to bring in our prior knowledge, experiences, and expectations into a situation. Consider someone who understands English very well, but who has never seen or used a washing machine, but rather hand washed his/her clothes. Ask your students whether they expect such a person to make sense of the text if he/she were told that the text is about 'doing the laundry'? It is because we know about washing machines, either by using them, or through watching others use them, that we were able to interpret the text meaningfully.

7. In conclusion you should emphasize that trying to make sense of a mere collection of data may not get a scientist anywhere. Scientists bring their prior knowledge, experiences, and expectations in order to put the data into context. Only then can scientists make sense of or interpret available data.

The procedure is actually quite simple. First arrange things into different groups. Of course, one pile may be sufficient depending on how much there is to do. If you have to go somewhere else due to lack of facilities, that is the next step, otherwise you are pretty well set. It is important not to overdo things. That is, it is better to do too few things at once than too many. In the short run this may not seem important but complications can easily arise. A mistake can be expensive as well. At first, the whole procedure will seem complicated. Soon, however, it will become just another facet of life. It is difficult to foresee any end to the necessity of this task in the immediate future, but then one never can tell. After the procedure is completed one arranges the materials into different groups again. Then they can be put into their appropriate places. Eventually they will be used once more and the whole cycle will then have to be repeated. However, that is part of life.

Figure 10. "The Instructions"

Young? Old?

Level: Upper middle and high school

Materials: Overheads (Figures 11, 12a, b and c.)

Procedure / Scenario:

1. Place Figure 12a on the overhead. Ask students what they see.

2. Students usually first recognize the face of an old lady. A few usually see the profile of an attractive young woman.

3. If students can not see the young lady, insist that it shows in the drawing, and that they can see it if they look hard enough. Do not at this stage point at the drawing to help students 'see' one image or the other.

4. Now point out, for example, how the nose of the old lady forms the cheek and chin of the young women to help students recognize the image. Many students will still not be able to see one or the other image.

5. Ask students, "How come we are looking at the very same drawing and seeing two different things?" If this was a piece of evidence, for example, a witness's recollection of a murder's face which she saw at a crime scene, then the police would end up looking for two women: an old lady and a young one!

6. Ask, "How can it be that some of us see only one face and not the other?" "Is it possible that some scientists may look at the same piece of evidence or set of data and see different things?" You can at this point discuss with students how a scientist's training, previous knowledge, and experiences dispose him/her to 'see' a certain set of evidence from a certain perspective. In the same manner that your students were not able to see the face of the young lady in the drawing, scientists sometimes fail to 'see' (or perceive of) a certain set of evidence as relevant to their questions. Scientists sometimes tend to infer different things from the same set of data in the same manner that your students inferred totally different things from the same piece of evidence: the drawing.

7. To help students see both images, show them Figure 12b of the old lady and Figure 12c of the young women. Now students can look at Figure 12a and, with some effort, see both faces. Students can now shift from one face to the

other. They, however, can never see both faces at the same time. Figure 11 can be used in the same manner, by asking, is it a rabbit or a duck?

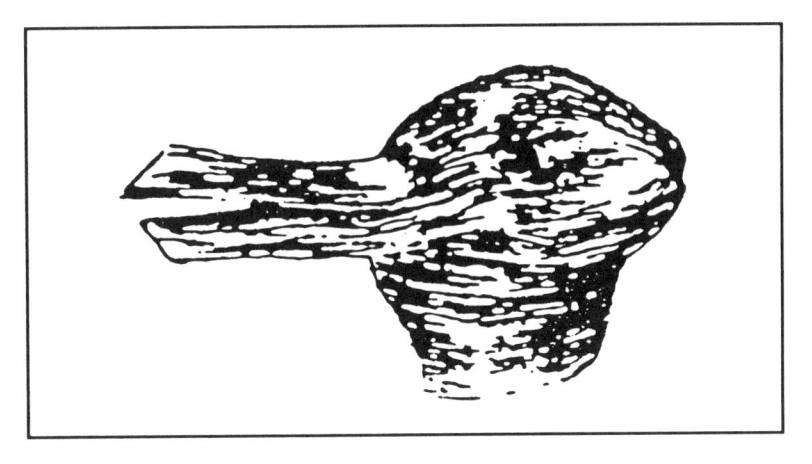

Figure 11. Is it a duck or a rabbit? From Atkinson (1975), p. 102.

The Aging President

This activity gives students a feel of what it means to approach a phenomenon with a certain paradigm or mind-set or perspective (if you do not wish to introduce the term paradigm). Even though certain facts change, a paradigm lingers on and sets expectations. A paradigm will very much make us see what we expect to see.

Level: High school

Materials: Individual overhead transparencies created from Figure 13a-h.

Procedure / Scenario:

1. Enlarge the panels from Figure 13 as individual images for use on the overhead projector or as hand outs

2. Put the first panel on the overhead projector and tell students that this is a caricature of President Regan at the beginning of his term. Next, tell students that you are going to show other caricatures of the president made at later stages in his two terms. Ask the students to observe the caricatures and describe what changes took place as the president ages.

Figure 12a. Young Woman, Old Woman[1]. From McNeil & Rubin, (1977), p. 89.

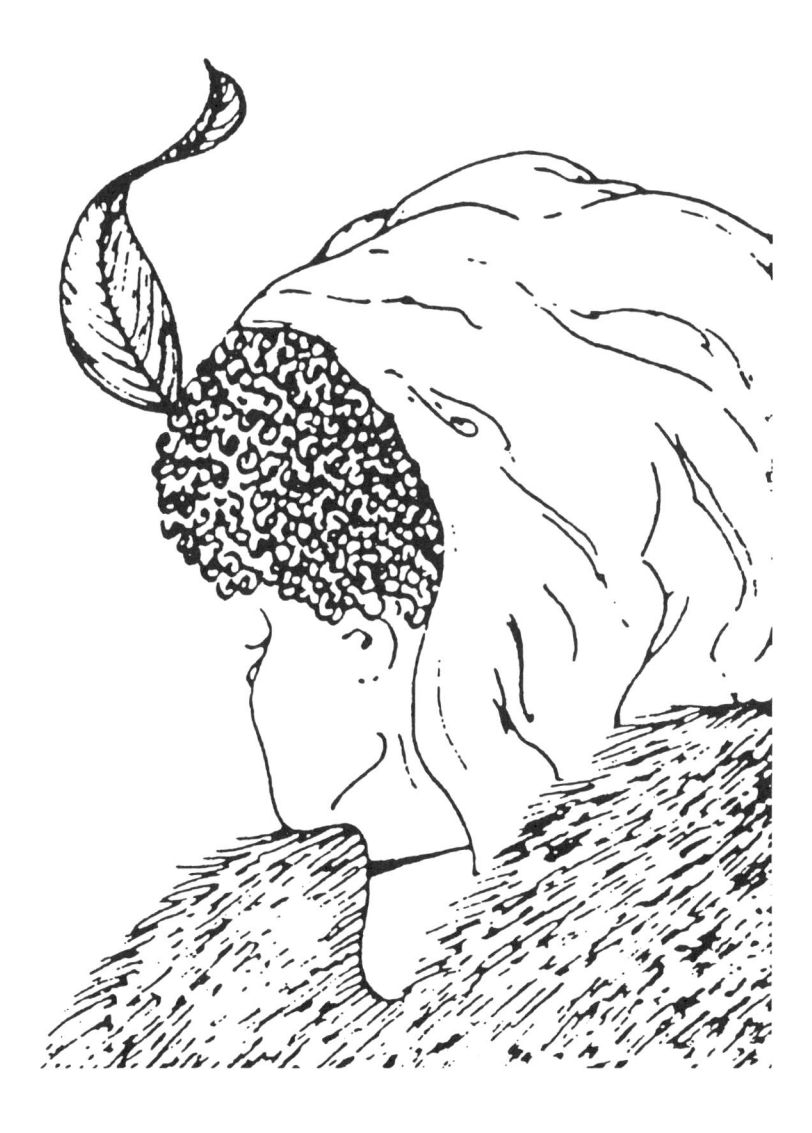

Figure 12b.   Young Woman, Old Woman[1]. From McNeil & Rubin, (1977), p. 89.

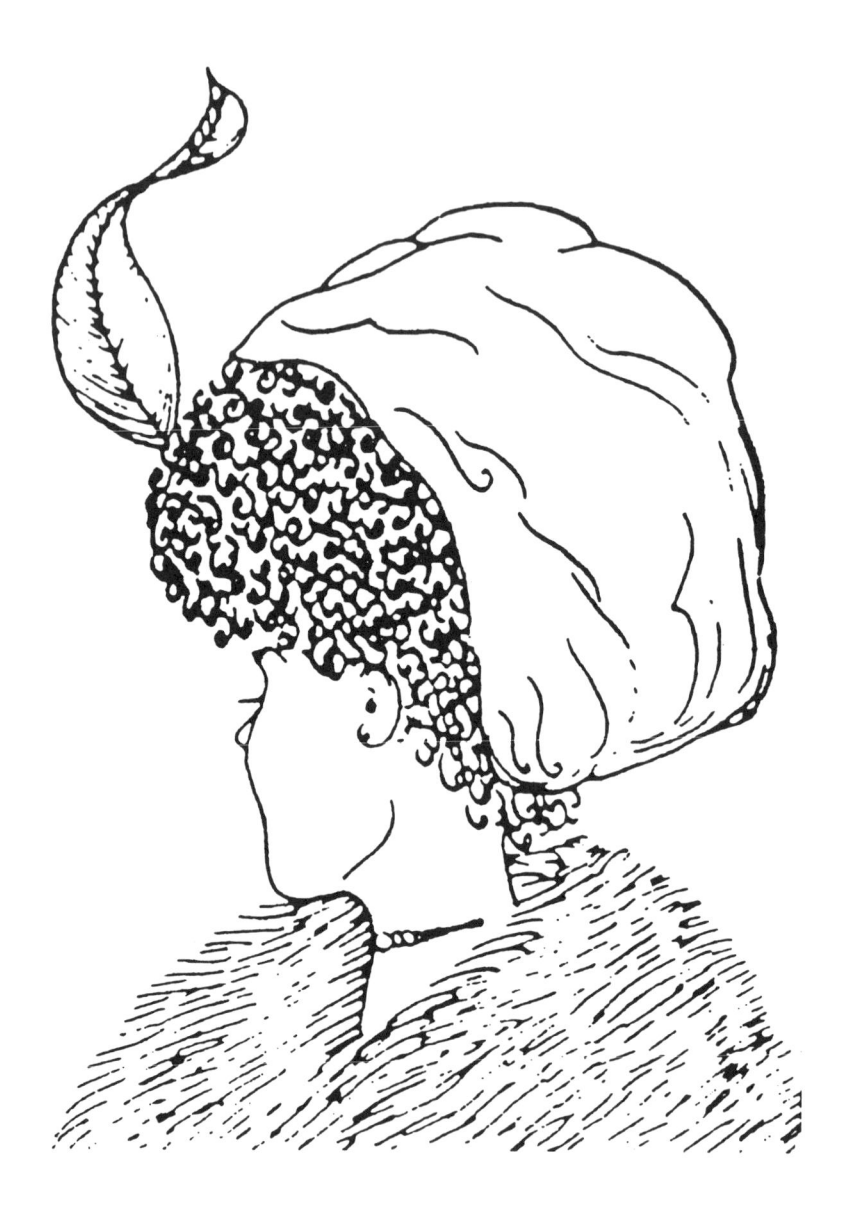

Figure 12c. Young Woman, Old Woman[1]. From McNeil & Rubin, (1977), p. 89.

AVOIDING DE-NATURED SCIENCE   107

3. Place the next panel on the overhead. Students usually note things such as the "chin has flattened," or "a piece is lost from his ear," etc.

4. Repeat the above step with the next three panels. Till now students usually keep reporting the changes they observe in the president's face.

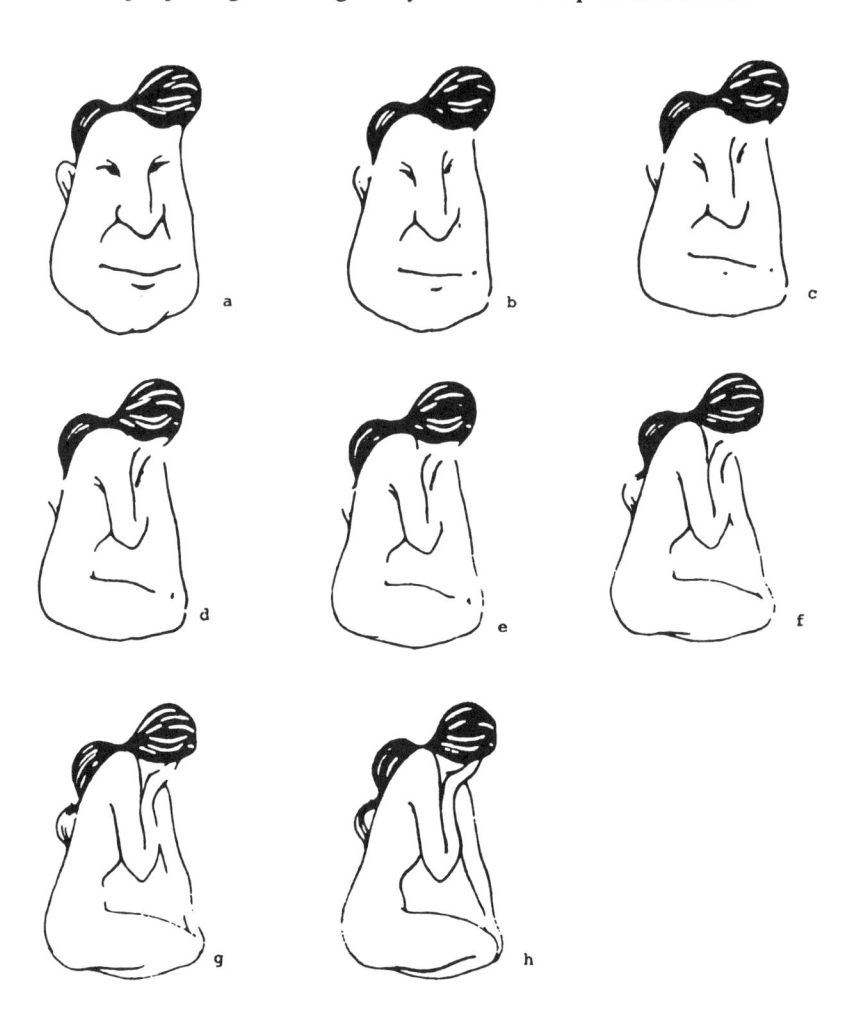

Figure 13a-h.   The Aging President? From Atkinson, (1975), p. 103.

5. Usually, it will not be until the middle of the series that students start to note something other than the face, but may still not 'see' the body of a female.

6. Place Figure 13h on the overhead. And ask students to describe what they see now. You can go back to the earlier panels and it would not be easy for students to recognize the female body.

7. Now you can show students the drawings next to each other (Figure 13) and ask them where do they start to see the female body.

8. Ask students why they were first able to see only a man's face? Did telling them that the drawing is a caricature of the president anything to do with it? Ask them whether they would have realized the female body earlier had you told them from the beginning that the drawing was that of a female body?

9. The kind of knowledge and expectations with which we approach any phenomenon may affect the way we interpret that phenomenon. A certain perspective may even determine what we see (a man's face) and what we do not (the female body) in the same data (drawing). Likewise, the kind of knowledge, training, experiences, and expectations that scientists bring into an investigation affect what they discern in the available data. And, most of the time, scientists do not give up their perspective (e.g., that the drawing was a face) even if evidence to the contrary is made available to them (let students now look at Figure 13c and ask them whether they wonder how could they have seen this drawing as a man's face!) It usually takes dramatic evidence, over a relatively long period of time (contrast for students Figure 13h with 13a) before scientists exchange their old views for new ones.

SECTION III: BLACK-BOX ACTIVITIES

Black-box activities provide students with challenges similar to those of encountered by scientists. Students examine phenomena and attempt to explain how they work. They make observations, collect data, draw inferences, and suggest hypotheses to explain their data. Next, based on those hypotheses, students make predictions and devise ways to test them (these 'ways' need not be limited to controlled experiments). Based on their tests, they judge whether their hypotheses are appropriate or not. Students finally construct models to explain the phenomena investigated and test whether their models 'work'. Black-box activities can be used to convey to students appropriate conceptions of many aspects of the nature of science.

General Learning Objectives for Black Box Activities:

Students will learn:

- The distinction between observation and inference.

- That scientific knowledge is partly a product of human inference, imagination, and creativity.

- That scientific knowledge is, eventually, empirically based (i.e., based on and/or derived from experiment and observation).

- That scientific knowledge (both theories and laws) is tentative and subject to change.

- That scientific models (e.g., atom, gene) are not copies of reality. Rather, these are inferred constructs that help to explain observable phenomena. Scientific theories are analogous to scientific models in the sense that theories are inferred explanations for observable phenomena. It should prove useful to explain to students the crucial distinction between scientific theories and laws. While theories attempt to explain observable phenomena, laws are descriptions of discernable patterns or regularities in these phenomena.

In addition, these activities provide students with opportunities to practice some science process skills. Among these processes are observing and collecting data, inferring, hypothesizing and devising 'ways' to test those hypotheses (or inferences) and constructing models.

A Generalized Model for Black Box Activities

The activities in this section are all black-box activities and share a common presentation model. Aspects of the model are illustrated by reference to one or more of the activities included in this section so you might find it useful to shim through the activities before examining the generalized model. In addition, you will find it useful to adapt many other activities to fit this model and use them to convey to students appropriate conceptions of the NOS. Black-box activities can be presented in a single class session or it can span several days. The amount of time you decide to dedicate to a certain activity depends on several factors such as time availability, grade level, and student interest.

The 'Phenomenon'

Present students with a demonstration that simulates a natural phenomenon or the universe around us. The demonstration should have three essential aspects:

a) It should be *discrepant*. This will serve to capture students' attention and arouse their curiosity. For instance, you can pour about 60 milliters of water into the water-making machine and recover about 330 milliters!

b) It should provide students with an *open-ended situation* so it is vital that the answer as to how the activity works should not be readily known to students.

This open-ended nature is essential if you wish your students to conduct inquiries the way scientists do. This feature, for instance, makes a black-box activity genuinely different from a cook-book laboratory experience. Students are often perplexed as to how the cans demonstration works. Their answers usually range from there being a small pump inside the cans, all the way to very complex chemical reactions. In the case of the tube, there are many workable possibilities: One can use a plastic ring to connect the upper and lower (or left and right) ropes, or simply loop the upper and lower (or left and right) ropes, or twist the left and right ropes, etc.

c) Students should not be able to 'see' what is going on. Most phenomena that scientists investigate are 'black' in the sense that they can not be 'directly' observed (e.g., atoms, black holes, radiation, reaction dynamics, temperature, gravity, etc.) Students are asked to explain the phenomenon without actually opening the box or demonstration to 'see' what is inside. You may elect at the conclusion of an activity to let students see how it is set up. However, not having students see the actual solution by looking inside the box is more consistent with the way science works. Scientists cannot open an atom and 'see' inside it. Despite that, scientists are able to produce relatively reliable bodies of knowledge about the phenomena they investigate.

Observing and Inferring

While demonstrating, ask students to make observations. Students can either call out or make records of their observations (data). Take time to go through student responses and differentiate observations from inferences. Make sure students understand that distinction by asking them to make further observations. You may discuss with students the extent to which their inferences are consistent with the observations they make.

- In the cans demonstration: "Water is cycling between the two cans" and "Gaseous pressure from a chemical reaction pushes the water" are inferences. Possible observations would be more like: "No water is passing through the glass-tube piece between the rubber tubes" and "There was a change in color from blue to red". In the water-making machine demo: "We poured 60 mL in funnel 1 and collected 330 mL in the recovery beaker" is an observation. "The machine made water" or "There is water inside the machine" are inferences. We can not observe that there is water inside the box, we can only claim that by inference.

Hypothesizing

a) Based on their observations and inferences, ask students (or groups of students) to suggest hypotheses to explain how the phenomenon (or demo) works. Have students present their hypotheses to the class.

b) Take time to discuss these hypotheses. Ask students to judge whether their hypotheses are consistent with the data that they have collected. Discuss how these data can support or weaken their hypotheses. Emphasize that scientists' hypotheses must be *consistent* with the available evidence.

- In the cans demonstration, students may hypothesize that a chemical reaction produces a gas that pushes the water from one can to the other. Ask students whether such a hypothesis is consistent with the evidence. For instance, a change in temperature and/or the production of a gas, are possible indications of chemical change. Students can feel the cans and observe whether gas fumes pass through the small piece of glass tubing. Discuss how these data can support or weaken their hypotheses. For example, many, but not all, chemical reactions are accompanied by a change in color. Ask students whether we can infer with confidence that the change in color in the demo is the result of a chemical reaction, etc.

c) At this point accept all hypotheses consistent with the evidence or prior knowledge, even though they do not completely explain what is going on.

- "There was water in the water-making machine at the start." This inference is based on the simple observation that the water-making works only once. This is an opportunity to discuss the role that prior knowledge plays in science. It is because we know that in the physical world "matter is not created nor destroyed," we can conclude that there is water inside the machine at the start. This knowledge has greatly reduces the number of acceptable explanations for our observations.

Testing Hypotheses

a) After making their initial hypotheses, ask students to design tests to determine whether their hypotheses are supported by evidence. Based on a hypothesis, a student (or group of students) can make a prediction and then test this prediction by collecting more evidence.

- In the water-making machine, students may hypothesize that there was water in the machine to start with. If this hypothesis is correct, then colored water poured into the machine should become diluted when collected in the recovery beaker. Students can pour water colored with food coloring into funnel 1 (you would have to prepare more than one water-making machine before class) and then observe the collected water. If the water collected has a lighter color, then this water 'must' have been diluted (see below). The prediction is thus observed and the hypothesis is *supported* (not proved).

b) Ask students whether they can, through testing, prove their hypotheses? Through testing (and experimentation) we can never prove an hypothesis for certain. Tests can only add support to a certain hypothesis. Irrespective of the amount of evidence collected, a hypothesis can never be proven with absolute certainty. However, when sufficient supportive evidence is collected, a certain hypothesis gains more acceptance as a plausible explanation. As such, scientific knowledge is never certain, and is subject to change (tentative).

- In the example given above, it was suggested that because the intensity of the water coloration decreased, we concluded that the water must have been diluted. This, however, is just another inference. There is another plausible explanation for the color dilution: The machine 'makes' water and not food coloring. As such, any colored amount of water will necessarily be diluted as the machine increases the amount of water but not that of the food coloring!

Designing and Testing Models

a) Based on their hypotheses students can now design actual models of the phenomenon they observed. A model is acceptable only as long as it 'behaves' like the phenomenon it is supposed to represent.

In the tube demonstration, each student can bring to class the core of a toilet paper roll. You can provide students with rope. Ask students to build their own model of the tube. Next ask students whether their model 'behaves' like yours. You can start pulling on one end of the rope and students can do the same step to see whether their model accords with yours.

1. When students have a working model of the phenomenon, ask whether they know what is inside the demo you showed them (which represents the natural phenomenon). If we can never 'open' the box and look inside, can we ever tell whether our model is an exact copy of what is inside? Can we ever be certain how the phenomenon 'actually' works? Scientific models are never exact copies of natural phenomena. These models are rather inferred or hypothesized from the behavior of the phenomenon. They are workable representations of those phenomena. Scientific knowledge, in this sense, is never certain. It is a product of human inference, even though it is based on empirical evidence.

- In the tube demonstration for example, students' models can never be like yours, since you do not provide them with rings, but the tube can be built without them.

- You might, later on, open the black box. This is especially helpful for students to see how far off their models or inferences might have been. This will help them get a feeling of the role of inference in science, and that models are not copies of reality.

In the cans, the hypothesis box, and the water-making machine demos, you might find it to be time consuming for students to build their own models, or that students do not show a sustained interest in going through the model construction phase. In such cases, have students draw sketches of their models on overheads which can be reproduced from the schematics provided (see Figures 14, 15, 16 and 17) by covering the 'inside' of the demonstration. Have students defend their models by explaining how each accounts for the collected data. You can tell students whether their models might work or not.

Discussion

It is essential at the conclusion of each activity you take time to *explicitly* point out to students the aspects of the NOS that the activity emphasized (these aspects were pointed out in each of the above sections). Discuss with students the implications of such aspects of the NOS on the way they view scientists, scientific knowledge, and the conduct of science.

The Tube

The way the ropes are set inside the tube will cause a seemingly complex and amazing movement pattern of the ropes. When the teacher pulls one end of the rope, another end will be pulled in with a seemingly random pattern. The teacher can pull on rope ends clockwise at one time, then across the tube at another.

Level: Elementary and middle school

Prerequisite knowledge: None

Model Construction: Individual students (See Figure 14).

Materials: 1 tube (mailing tube or PVC pipe, approx. 30 cm), 1 plastic ring (optional, you can simply loop the lower rope over the upper rope), rubber stoppers or tape (to seal tube ends), 1 roll of clothesline rope (for whole class), 1 toilet paper roll core (each student can get his/hers).

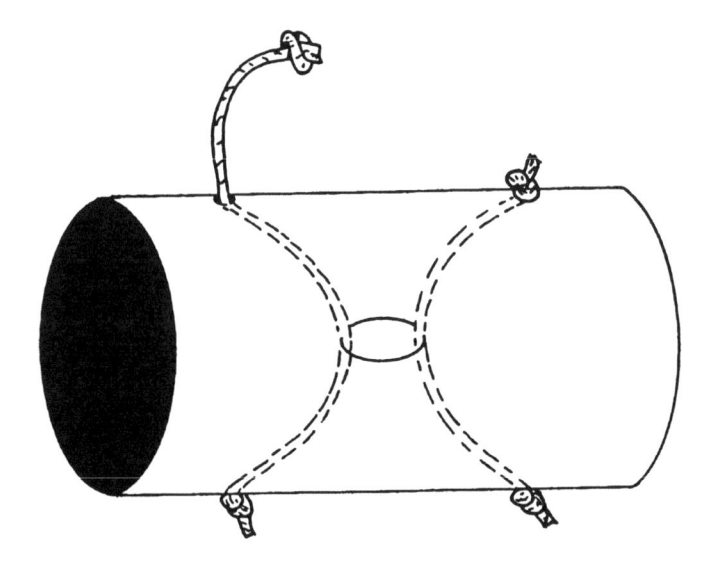

Figure 14. The construction of the mystery tube. Students see only the knotted ropes that appear on the outside of the tube.

The Hypothesis Box

Students can only see the front of the box with the top funnel and outlet tube. The teacher can pour clear water from top funnel and re-collect it from the outlet. By changing the position of the tubing inside the box to funnels with food coloring soaked paper-cones, the teacher can pour clear water but collect solutions of different colors!

Level: Upper elementary and middle school

Prerequisite knowledge: None

Model Construction: Groups of 4 or 5 students (See Figure 15).

Materials: 1 cardboard box (approx. 70 x 50 x 30 cm) with back open, 1 wood sheet (approx. 50 x 30 cm), 4 funnels, rubber tubing, 2 three-way rubber tube connectors, 2 filter-paper cones soaked with 2 different food colorings, 300 mL beakers.

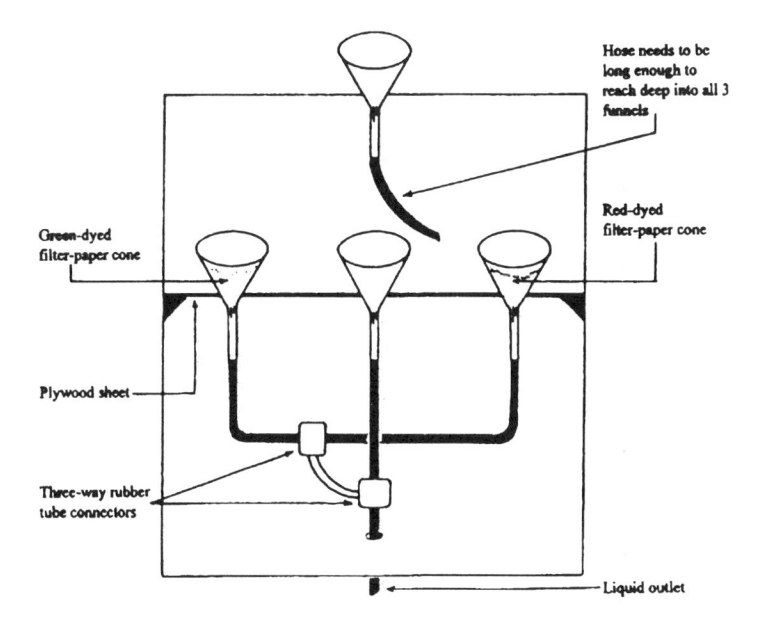

Figure 15. The is the view of the apparatus facing the instructor. On the other side, students can only see the top funnel and the liquid outlet.

The Cans

The initial levels of liquid in cans A & B appear in Figure 16. Using a beaker, pour enough water into thistle funnel until water starts flowing into the funnel from the glass tubing above. The water will now keep running for a long time. The cans will appear to be a self-perpetuated or closed system in which the liquid seems to cycle by itself! Instead of clear water in can B, fill half with colored water and pour above ethyl alcohol colored with iodine. A more complex pattern will now be apparent.

Level: Upper middle and High School

Prerequisite knowledge: Liquid Pressure

Model Construction: Groups of 4 or 5 students. (See Figure 16).

Materials: 2 ditto-master fluid or camping fuel cans (if not available use 2 500-mL Erlenmeyer flasks wrapped with aluminum foil), 2 rubber stoppers, rubber tubing, 1 thistle funnel, glass tubing, (Optional materials include ethyl alcohol, iodine and food coloring).

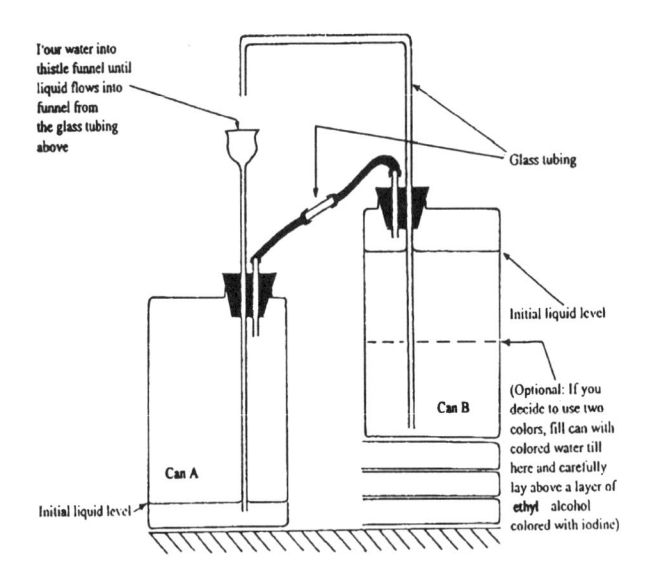

Figure 16. The design of the can apparatus.

The Water-Making Machine

Before class pour about 270 mL of water through funnel 1. Tell students you have a water making machine. You can pour approximately 60 mL of water through funnel 1 and recover approximately 330 mL ! Make sure students record the initial and final (recovered) volumes of water.

Level: Upper middle and High School

Prerequisite knowledge: Liquid pressure and the siphon effect

Model Construction: Groups of 4 or 5 students. (See Figure 17).

Materials: 1 cardboard box (approx. 35 x 25 x 15 cm), 2 empty soft-drink cans, 1 goose-necked straw, 2 500-mL beakers, 2 funnels, rubber tubing, 1 tube of silicone.

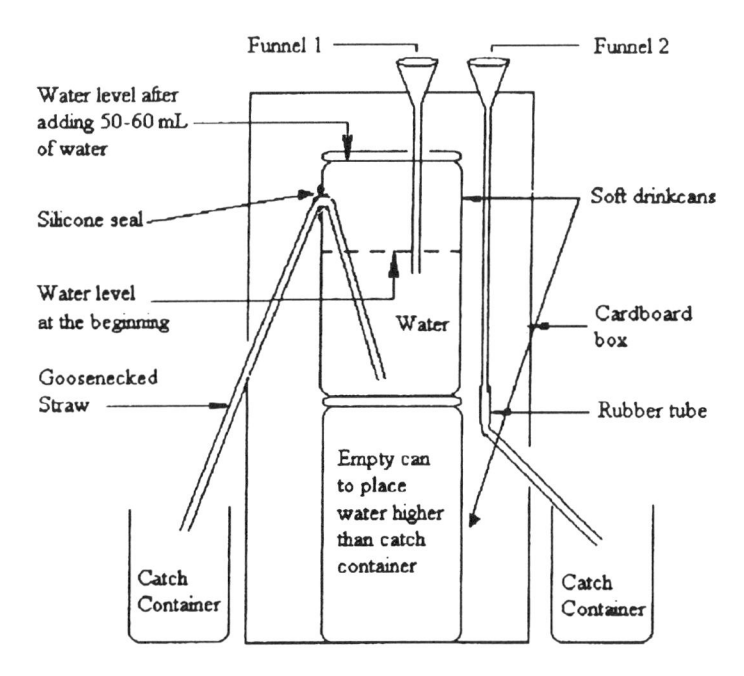

Figure 17. The design of the water making machine. The students only see the two funnels, the cardboard box and the two catch containers. From BouJaoude (1995), pp. 46-49.

The Cube Activity

This activity is similar to black-box activities in several respects. It has the additional advantage of giving students a sense of, and experience with another creative activity that scientists often undertake: The search for patterns. Scientists often search the data they collect for certain patterns or regularities. Based on these regularities, scientists can, for instance, extrapolate their data in order to predict possible future behaviors of the phenomenon under investigation. A common example comes from meteorology. Meteorologists collect data on several relevant phenomena (e.g., atmospheric pressure, temperature, humidity, wind direction and speed, cloud formations, etc.), and then attempt to single out patterns that would allow them to anticipate or predict the behavior of the 'weather' in the near future based on what they know. Such anticipation or prediction is only approximate and probabilistic (we have all regularly experienced 'unexpected' weather). Scientists can not invariably predict the future (since science does not provide absolute knowledge); they can only suggest what might happen in the near future by extrapolation from past and present evidence and experiences.

It should be noted that the patterns that scientists often single out, may or may not exist in the natural world. On the one hand, for example, planetary movements in our solar system do seem to follow a certain pattern that allows scientists to make fairly accurate predictions about the position of a certain planet at a certain point in time. On the other hand, for instance, one group of community ecologists advocates 'competition' as a pattern common to the behavior of all groups of animals. Another group of scientists claims that such 'competition' is not an actual natural phenomena, but was rather imposed on the behavior of animals by the scientists themselves. (For example, see Lewin (1983) for an account of the recent controversy regarding the concept of 'competition' in community ecology). Whether some patterns or regularities actually exist in nature, is a question similar to asking whether the models that students build to account for the black-box phenomena that they investigated are actual copies of what exists inside the boxes. The main point to emphasize to your students is that patterns are partly based on evidence, but are also partly the product of the scientists' imagination and creativity. The activity aims to convey to students the notions that scientific knowledge is partly a product of human inference and creativity, is empirically based (based on and/or derived from observation and experiment), and tentative (subject to change).

Level: Upper elementary, middle, and high school.

Materials: One cube per group of four students (black-line masters are provided as Figures 18-22). A single black-line master can be used to prepare the cubes needed for a single activity.

Procedure/ Scenario:

1. Have students form groups of four. Place the cubes on the center of the tables where the students are working. All cubes should have the same face on the bottom. Use the bottom square of the black-line masters to serve as the face on the bottom. Students should not turn or lift the cubes.

2. Tell students they have to answer the question: "What is on the bottom of the cube?" Their answers should be supported by evidence. They should also include an explanation of how an answer was reached.

3. Ask students in each group to make observations and record the data each from his/her position (for instance, what is the number or word that a student can see on the cube surface facing him/her). Then ask them to share their observations. Each student in the group can verbalize his/her observations to the 'recorder' who compiles all the data. This is intended to simulate scientists working together and sharing data.

4. Based on their observations, students in each group should be able to figure out the pattern on the cube, and consequently infer what is on the bottom. Each group should then prepare a written report of the suggested answer and corresponding pattern and explanation.

5. Put the cubes away without showing the bottom. (You can make the experience more genuine by gluing the bottom of the cubes to a piece of cardboard). Scientists often have no way of 'seeing' the phenomena they are investigating. (This makes 'The Cubes' a black-box activity.)

6. Ask one or more groups to present their suggested answer to the class, including the pattern they discerned, and the way they reasoned to formulate that pattern. If different groups come up with different answers, then all differing views should be presented.

If all groups come up with the same answer, you can initiate a discussion of the role of evidence (or observation) in deciding the pattern, and how the answers were consistent with the available data. If different groups come up with different answers, the two scenarios are possible. In some cases, some groups may have inferred patterns that are obviously inconsistent with the data. Here, you can emphasize the importance of evidence in supporting or weakening a certain conclusion by asking students to identify how a certain pattern is consistent or otherwise inconsistent with the data. In other instances, two or more patterns may be equally consistent with the data and consequently legitimate. In this latter case, you can discuss with students whether it is possible to tell who is 'right' and who is 'wrong'! If differing explanations are consistent with all available evidence, is it possible to decide which might correspond with the answer on the bottom of the cubes, which we cannot see? The patterns in the cubes vary in difficulty. You might want to match those difficulty levels to your students' grade level(s). Table I gives a brief description of possible patterns for the different cubes. (In certain cases, more than one pattern is possible, and students often come up with new patterns that are consistent with the evidence).

TABLE I
The Cubes Activity: Possible Patterns.

Cube #	Figure #	Possible pattern (with the bottom square of black-line master serving as bottom face)
1	18	First letters vary as consecutive consonant letters of the alphabet that form meaningful words with _AT: the bottom is PAT ('e' is a vowel and does not count).
2	19	2 faces are not divided; 2 faces are divided in two; 1 face is divided in four: bottom is divided in four. Number on upper left represents the number of divisions on face: bottom has 4 on upper left, etc. The lower right corner numbers follow a pattern in which sequential numbers (1, 2, 3, 4, 5) are added as follows: 1 + 1 = 2, the result is taken and added to the next number in the sequence: 2 + 2 = 4, and so on: 4 + 3 = 7, 7 + 4 = 11: bottom should have 11 + 5 = 16 on lower right corner.
3	20	No real pattern in mind. Useful with older students to discuss whether scientists may in certain cases impose a pattern on nature. This is especially pertinent if students do come up with some pattern(s).
4	21	Exposed sides have either a male or female name; opposing sides have a male name on one side and a female name on the other; names on opposite sides begin with the same letters. On this data any one of four possibilities can be on the bottom: Fran, Frances, Francene, and Francine. The number in the upper right corner of each side corresponds to the number of letters in the name on that side; the number in the lower left corner of each side corresponds to the number of the first letters the names on opposite sides have in common. This new evidence leads to put either Francene or Francine on the bottom. We still are, however, not sure what is exactly on the bottom because we can not see it. We can only infer it! Note, another option is to use this cube without the numbers.
5	22	If the alphabet is numbered, then the numbers on the cube correspond to the numbers of the 6 vowels (a is 1, e is 5, etc.): the bottom is 23.

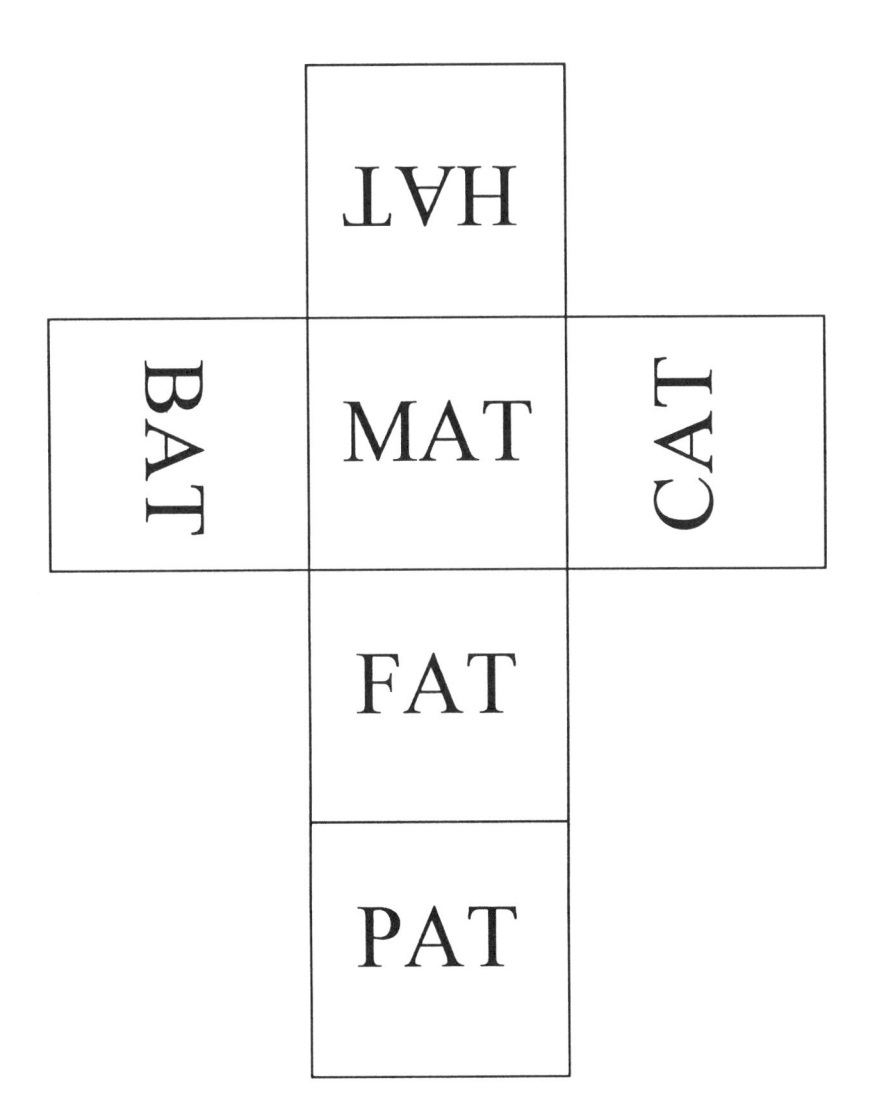

Figure 18.   Mystery cube #1.

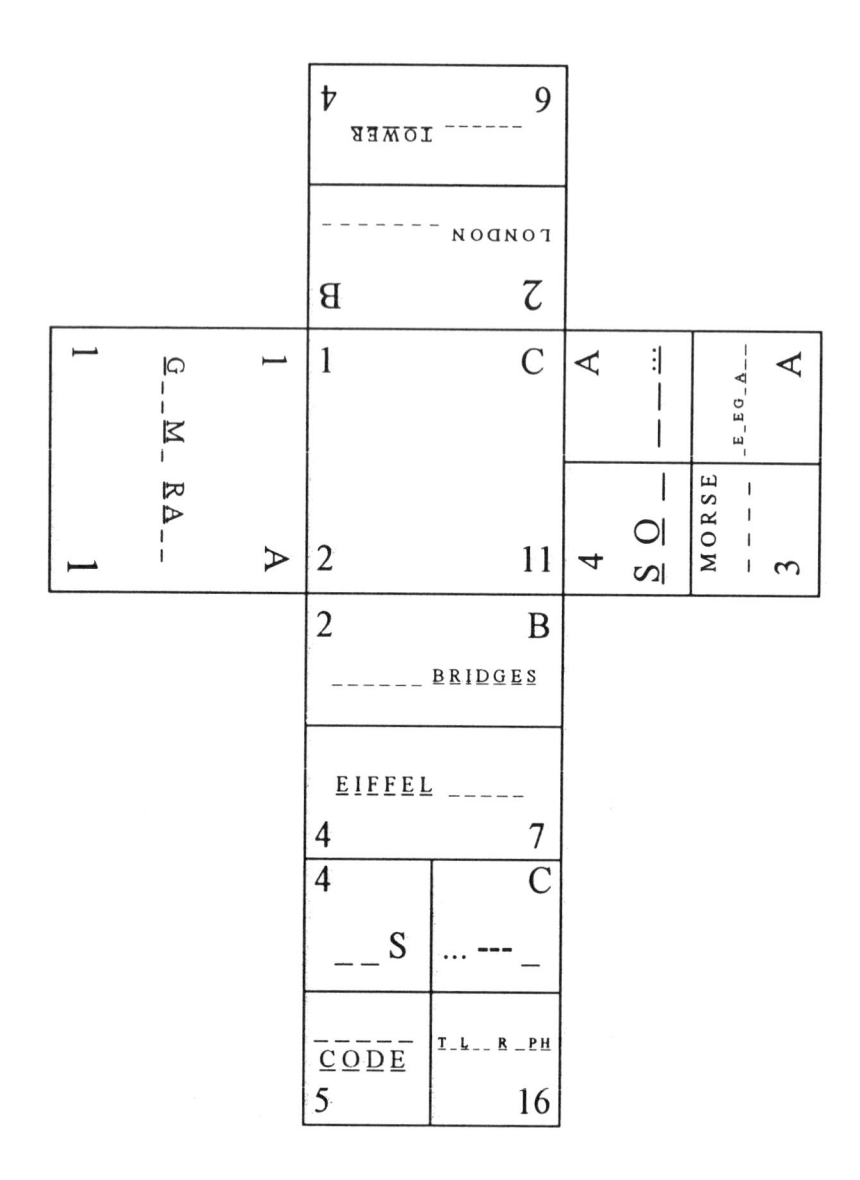

Figure 19. Mystery cube #2.

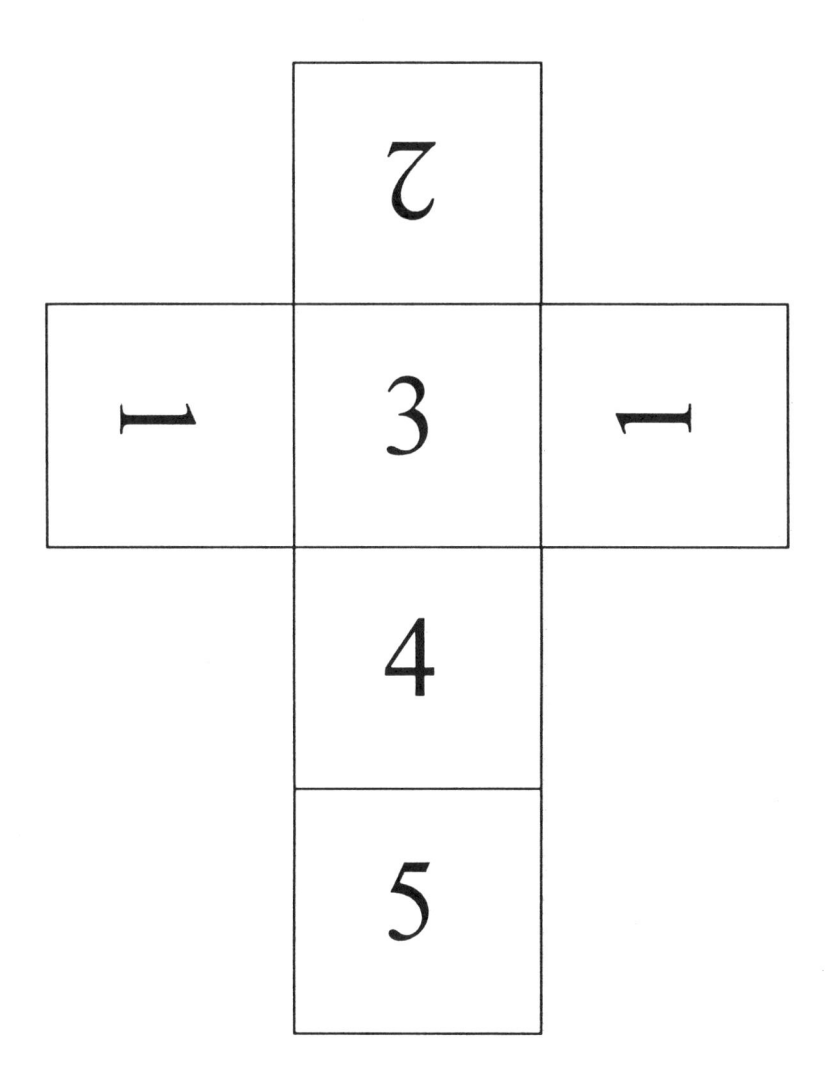

Figure 20.   Mystery cube #3.

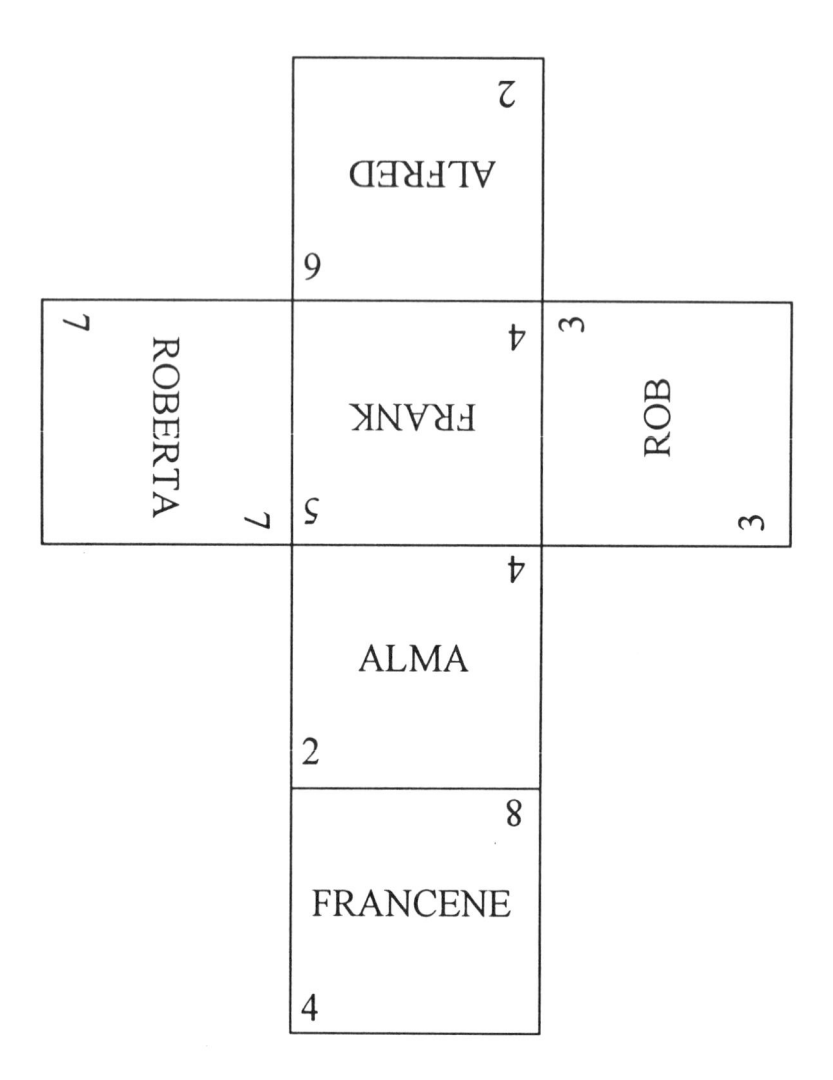

Figure 21. Mystery cube #4.

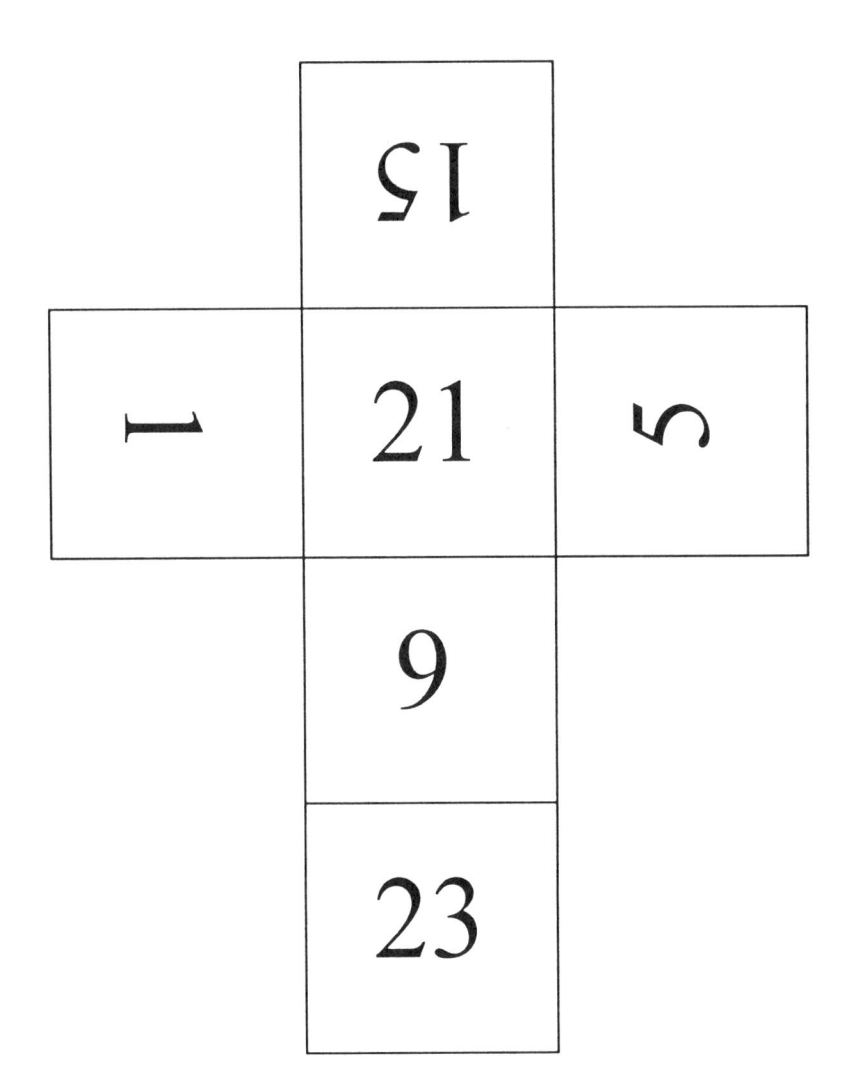

Figure 22.   Mystery cube #5.

Oregon State University, Corvallis, Oregon, USA

NOTE[1]

The image of the young woman/old woman has had a lengthy history. It was drawn by the cartoonist W. E. Hill and appeared in the November 6, 1915 issue of *Puck* titled "My wife and my mother-in-law." The psychological utility of the image seems first to have been noted by E.G. Boring (1930) in a short article appearing in the *American Journal of Psychology*, pp. 444-45.

REFERENCES

Atkinson, R.C. (1975). *Psychology in progress: Readings from Scientific American*, New York, Freeman.
BouJaoude, S. (1995). 'Demonstrating the nature of science', *The Science Teacher*, (62), 46-49.
Hrdy, S. B. (1986). 'Empathy, polyandry, and the myth of the coy female', in R. Bleier (ed.), *Feminist approaches to science*, New York, Pergamon Publishers, 119-146.
Kuhn, T. (1970). *The structure of scientific revolutions*, 2nd ed., Chicago, The University of Chicago Press.
Lederman, N.G. (1992). 'Students' and teachers' conceptions about the nature of science: A review of the research', *Journal of Research in Science Teaching*, (29), 331-359.
Lederman, N. G. (1994, April). *Scientific hypotheses, theories, laws, and other dangerous ideas*, paper presented at the National Convention of the National Science Teachers Association, Anaheim, CA.
Lederman, N.G., & Niess, M. (1997). 'The nature of science: Naturally?', *School Science and Mathematics*, (97), 1-2.
Lewin, R. (1983). 'Santa Rosalia was a goat', *Science*, (222), 636-639.
Lovejoy, C.O. (1981). 'The origin of man', *Science*, (211), 341-350.
Luchessa, K., & Lederman, N.G. (1992). 'Real fossils, real science', *The Science Teacher*, (59), 68-92.
McNeil, E.B. & Rubin, Z. (1977). *The psychology of being human, 2nd edition*, San Francisco, CA, Canfield Press.
Shoresman, P. (1965). 'A technique to clarify the nature of theories', *The Science Teacher*, (32), 53-54.
Social Science Education Consortium & Biological Sciences Curriculum Study (1994). *Teaching about the history and nature of science and technology: Teacher's resource guide* (field test ed.), Boulder, CO, Social Science Education Consortium.

PENNY L. HAMMERICH

6. CONFRONTING STUDENTS' CONCEPTIONS OF THE NATURE OF SCIENCE WITH COOPERATIVE CONTROVERSY

Have you learned lessons only of those who admired you, and were tender with you, and stood aside for you?
Have you not learned great lessons from those who braced themselves against you, and disputed the passage with you?

Walt Whitman, 1860

Among the fundamental principles shaping students' understanding of science are the conceptions they hold about the nature of science itself. Students' conceptions of the nature of science influence their thoughts, feelings and actions associated with doing, understanding, and teaching science (Smith, 1990). Constructivism implies that teachers cannot assume students have the same conception of science, nor are their conceptions necessarily accurate. It is important for teachers to find out what conceptions of the nature of science students hold so that lessons and activities can be developed which challenge conceptions of the nature of science.

One ideal instructional strategy for examining students' nature of science conceptions is cooperative controversy. Cooperative controversy exists when one's ideas, conclusions, theories, or opinions are incompatible with those of another and the two then seek to reach an agreement (Johnson & Johnson, 1979). This strategy forces students to actively engage in a debate of two opposing sides of an issue. Research indicates that cooperative controversy dramatically enhances critical thinking (Johnson et al. 1985; Scharman 1990; Solomon et al. 1992).

The conception or understanding of the nature of science as socially constructed and validated has important implications for science education. Although learning is individually processed, further learning is constructed when individuals engage socially in talk and debate about issues. Learning science thus involves being confronted with the ideas and practices of the science community and making these ideas and practices individually meaningful (Driver et al., 1994). The aim of cooperative controversy is to have students examine their conceptions of the nature of science.

Throughout their life experiences, students develop many interconnected and valid conceptions about the scientific world. Some argue there are many understandings among students, teachers, philosophers, and science educators of the nature of science (Millar, Driver, Leach & Scott, 1993). Science is both symbolic in nature and socially constructed. The role of the science educator is to mediate learners' understanding of science in order to help them make sense of the way in which scientific concepts are generated and validated. It is not the role of the science educator to present science as the only way of knowing. The cooperative controversy lesson is important because the way teachers teach science is linked to the teachers' own understanding of science. If this is the case, then it can be inferred that students' conceptions of the nature of science are linked to how they learn science material.

The concept of the nature of science is a foundation for the knowledge base for teaching and learning science. It is likely that the nature of science is a global conception that frames a learner's total scientific knowledge. It is similar to what Bohm and Peat (1987) call the infrastructure of scientific knowledge, the set of tacit beliefs and skills which allow one to understand and build scientific knowledge. If a learner's infrastructure contains misconceptions and contradictions, subsequent knowledge and concepts built upon these faulty ideas are likely to be both erroneous and fragmented. Smith (1990) has found that the conceptions learners hold about the nature of science influence the kinds of information they find relevant and therefore tend to seek and value. Thus, these conceptions have consequences in the course of the development of science teaching. An effort is made in the following cooperative controversy lesson to recognize and better understand college students' conceptions of the nature of science.

THE METHOD

The lesson presented here has been used successfully in a college science methods class. Thirty-seven upper-level elementary education majors at an urban university participated in a cooperative controversy lesson designed to reveal and challenge their conceptions of the nature of science. Students are asked to write down their conception of the nature of science before and after participating in the cooperative controversy lesson. Students are told that by writing down their conceptions of the nature of science they will be able to discover and to reflect upon their own conceptions. The question is open-ended and the responses are analyzed by measuring patterns and trends related to students definitions of the nature of science. Both the author and another science educator analyzed the responses. The author analyzed the responses twice for an agreement of 93% (intrarater reliability). The agreement between the author's and the independent science educator's analysis was 89% (interrater reliability). Results and examples of responses appear in a subsequent section.

USING A COOPERATIVE CONTROVERSY LESSON TO TEACH ASPECTS OF THE NATURE OF SCIENCE

Conflicts over conceptions within the nature of science are inevitable but these conflicts can provide a positive experience and increase learning. The cooperative controversy lesson is designed to engage students actively in personal knowledge construction. Figure 1 identifies the cooperative controversy model.

```
            Cooperative Controversy:
            Present opposing sides of
               thenature of science

New Conception:                    Discrepant Viewpoint:
Organization of                    Challange to prior or
new conception                     alternative conception

               Active Reflection:
           Conceptual conflict and a
          search for a new conceptions
```

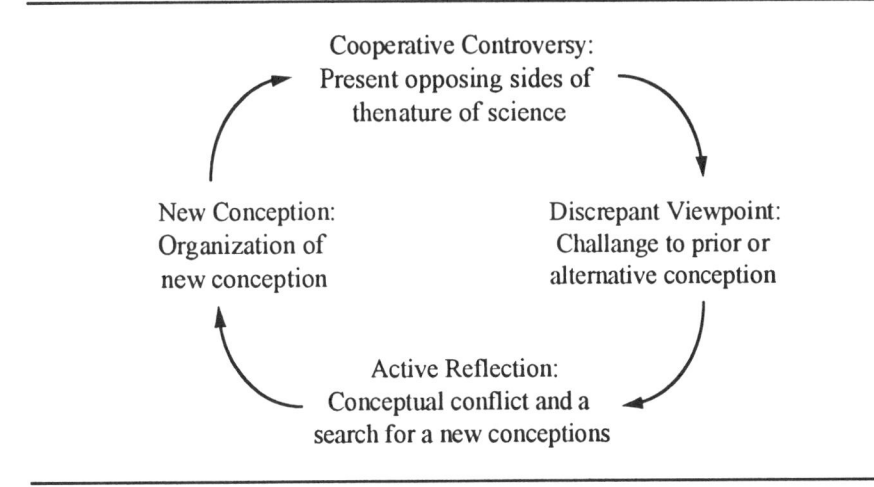

Figure 1: The Process of Confronting Students' Conceptions with Cooperative Controversy

When any controversial issue such as the nature of science is approached there tends to be a "choosing of sides" behavior where learners take a position on the issue. This is described as "cooperative controversy." If an individual conception is left unchanged, it will continue to exist and be strengthened as new knowledge is built upon it. During the lesson, individual conceptions are challenged by other ideas. The challenge happens face to face in what is called the "discrepant viewpoint." The uncertainty resulting from the "discrepant viewpoint" leads to a search for more information and the desire to determine the origin of individual conceptions. This is referred to as "active reflection" and occurs when students are focused on the goal of the exercise: to come to a consensus on the most accepted conception of the nature of science. Students may change their conceptions several times during the activity based on the ideas put forth by others.

In the following cooperative controversy lesson, the two opposing sides of the nature of science are *science as fact or scientific knowledge as etched in stone* to *science knowledge as myth and belief.* Figures 2 and 3 identify the two opposing views of the nature of science. Although neither is considered "correct," the two opposing sides are seen as opposite ends of a continuum with "science as fact" at the far left and "science as myth and beliefs" at the far right. Most students have conceptions of the nature of science that fall somewhere between the two extremes.

Science as FACT

Adopt the perspective that science is factual as your own and prepare a brief presentation of the idea for your group. Add examples and statements that will help you argue your point of view successfully. (What do you suppose the other point of view is?)

Further points to consider:

The American College dictionary defines science as "a branch of knowledge of study dealing with a body of facts or truths systematically arranged and sharing the operations of general laws . . ." From this definition it is assumed that science is a series of facts about nature that are static and fixed collections of truths about the world. These facts are discovered by scientists who make observations of the world and carefully record and organize the facts or truths in a systematic way.

Facts exist in nature, can be observed, recorded and classified to produce a body of knowledge about the world. Facts are truths in the past, present, and future. This definition implies that nature is separate from the observer and exists whether or not anyone is looking. The role of scientists in this case is to record what exists and systematically organize the resulting body of knowledge. In this interpretation of science the role of a science teacher is to point out the facts of nature and organize these facts for the student.

Figure 2.   The position of science knowledge as etched in stone.

Science as BELIEF

Adopt the perspective that science is uncertain as your own and prepare a brief presentation of the idea for your group. Add examples and statements that will help you argue your point of view successfully. (What do you suppose the other point of view is?)

Further points to consider:

The universe is seen as full of uncertainties and mysteries that individuals experience every day. In this view, science is the seeking of approximate answers to questions about nature. There are many interpretations of nature and these interpretations constantly change with new experiences, conceptions, and understandings of information discovered.

Our understandings of the universe are abandoned through scientific progress; explanations and theories of the universe are only working hypotheses that are abandoned one day when another discovery supersedes and causes new conceptions of the universe. For example, at the beginning of the 20th century the scientific conception of the atom was simply a model, which proved its truth only after it had been surpassed by another scientific explanation.

In this view, the teaching of science is seen as a process of discovering new scientific conceptions instead of focusing on the fact of content.

Figure 3. The position of science knowledge as myth and belief.

Each pair of students is given a written passage that describes one of the two opposing sides. Students are asked to read, discuss, and write a persuasive argument defending the position they are given. Then two pairs, with opposing sides, engage in the cooperative controversy activity by presenting their side to the other pair and defending their position. Students are encouraged to ask questions of clarification on any point made that they do not understand. After the discussion, the two pairs are asked to reverse roles by taking the opposite side to read and defend. The goal in the role reversal is for each pair to elaborate on what members of the other pair said previously. The final goal in the cooperative controversy is for the two pairs, representing opposing sides, to reach a group decision on their conception of the nature of science. Table I identifies the steps involved in setting up the cooperative controversy.

TABLE I
The steps involved in setting up the controversy

1. Assign cooperative groups of four students which are then further divided into pairs of two.

2. Students meet with their partner, read their position and plan how to argue effectively for their position.

3. Each pair presents their position while the other pair takes notes and asks for clarification on anything they don't understand.

4. Open discussion takes place where each group argues forcefully and persuasively for their position, presenting as many facts as they can to support their point of view. Students as an entire group are to make sure they understand the facts that support both points of view.

5. Role Reversal occurs where each pair in the group argues the opposing pair's position. The goal is to elaborate on what already was said by the other pair.

6. Group members develop a position upon which all members can agree. They summarize the best arguments for both points of view. When a decision is made the group organizes their arguments to present to the entire class. The group needs to be able to defend the validity of their position.

ASSUMPTIONS FOR APPROPRIATE COOPERATIVE CONTROVERSY

Conflict which results when students discuss their personal conceptions of science can increase learning and improve student relationships. The cooperative nature of the activity serves to negate competitive experiences which can close off logical thought and damage relationships. The following assumptions are critical to construct an appropriate cooperative controversy.

1. Conflicts among alternative conceptions of the nature of science are frequent and inevitable in any science classroom. No two students can ever see the nature of science in exactly the same way. The question is not whether there are differences in conceptions, but how different are the conceptions among students. Knowing that there are potential differences in the conceptions of the nature of science, it would make sense to explore all the possible alternative conceptions before trying to decide which is most accurate. This divergence before convergence procedure enhances scientific thinking and is the focus of the cooperative controversy lesson.

2. Cooperative Controversy requires a cooperative context. An important point is that controversy is inevitable when students are put in a situation of disagreement. The most effective controversy exists when students are

allowed to cooperatively strive toward an agreed upon conception of the nature of science. Research supporting cooperative relationships in the classroom have been well documented (Johnson and Johnson, 1983; Johnson, Maruyama, Johnson, Nelson, Skon, 1981; Slavin, 1977; Deutsch, 1962). The cooperative format is essential to the controversy lesson. If students see themselves in a competitive situation they will lose sight of the objective at arriving at a consensus and move into a "lose-win" situation. The most appropriate context for conflict resolution is a cooperative situation where students are engaged in finding the most appropriate conception for the group.

3. Students must have cooperative skills needed to work toward resolution. Students should be encouraged to emphasize the alternative conceptions of science, differentiate alternative conceptions from their own, listen to alternative ideas, negotiate alternative conceptions in consideration of their own ideas, and emphasize rationality in seeking the group's consensus regarding the best conception of the nature of science.

THE TEACHER'S ROLE IN STRUCTURING COOPERATIVE CONTROVERSY

The science teacher has the important job of structuring the cooperative controversy exercise by informing the students about the nature of the controversy. Specific information on structuring cooperation is detailed in Johnson and Johnson (1975) and Johnson, Johnson, Holubec and Roy (1984). The second important role for a science teacher is to facilitate the controversy and to intervene when necessary. The third role for a science teacher is to encourage controversy by questioning viewpoints, playing alternative conceptions, taking alternative sides, encouraging students to take alternative conceptions and monitoring and helping students.

IMPACT OF THIS STRATEGY ON STUDENTS' CONCEPTIONS OF THE NATURE OF SCIENCE

I have found that elementary methods students hold varying conceptions of the nature of science. While reflecting upon their ideas, students noted that it was difficult for them to "figure out why" they held various alternative conceptions. In many cases when asked to think about how they obtained or why they held alternative conceptions, students attributed them to the teaching practices of science teachers when they were students in science classrooms both in the precollege and college environment. From this discussion came the comment that because of their previous experience in science classes, science had not become one of their favorite classes. Further, the students agreed that they wanted to teach science in a manner different from their past experience, using hands-on, discovery, or inquiry oriented approaches.

The cooperative controversy lesson is useful in challenging students' alternative conceptions of the nature of science. Many of the students' conceptions of the nature of science fell somewhere on a continuum between the two sides, science as a "fact" or science as a "belief." The responses were analyzed to reveal trends of students' views of the nature of science.

Before the cooperative controversy lesson, 73% of the students felt that the nature of science was fact based. After participating in the cooperative controversy exercise, 60% of the students felt that the nature of science was a combination of factual information and belief. Table II illustrates the character of the students' conceptions of the nature of science before and after the lesson.

TABLE II
Students' conceptions of the nature of science before and after participation in the cooperative controversy lesson.

Responses before the Cooperative Controversy Exercise			
Fact	Belief	Fact/Belief	No Response
27 (73%)	7 (19%)	2 (5%)	1 (3%)

Responses after the Cooperative Controversy Exercise			
Fact	Belief	Fact/Belief	No Response
8 (21%)	7 (19%)	22 (60%)	0 (0%)

Further analysis of students' responses revealed that their conceptions of the nature of science vary and that they were influenced by the cooperative controversy lesson. For example, one student's conception of the nature of science before participating in the cooperative controversy lesson was that science is fact based, "learning about the world to find out the true answers." This same student's conception of the nature of science after participating in the cooperative controversy lesson changed to viewing science as both belief and fact based, "science is a combination of beliefs and facts . . . science is always changing." There were also students who changed their conceptions of the nature of science from one of belief that "science is a series of beliefs that we construct in our minds," to science as fact based that "science is facts that have been proven." Finally, some students' conceptions changed from one of belief, "science is a group of our beliefs that are ever-changing," to one of both belief and fact, "science as a combination of fact and belief, some science is constant while some science is changing." Of 37 students, only three remained constant in their conception of science after participating in the cooperative controversy lesson.

CONCLUSIONS

Students' conceptions of the nature of science do vary but are influenced by participating in the cooperative controversy lesson. Whether or not the cooperative controversy lesson results in a permanent conceptual change on the part of the students remains to be seen. Logical steps for further research and exploration of students' conceptions of the nature of science are to determine how effective the cooperative controversy lesson is at producing long-lasting conception changes on the part of the students and whether or not students' conceptual changes concerning the nature of science impacts how they will eventually teach science.

The students participating in the cooperative controversy lesson were elementary teacher candidates and they indicated that their conception of the nature of science did influence how they perceived the teaching of science. The majority of students stated before participating in the cooperative controversy lesson that their conception of the nature of science was fact based. The students further indicated that they did not have favorable experiences in science classes due to the manner in which the courses were structured. This experience led the students to state that they wanted to teach science in a manner different from the way they had been taught. Therefore, this lesson provides the opportunity for students to come to an understanding not only of how their conception of the nature of science affects how they learn science content, but also how they perceive the teaching of science. Teacher educators can use the opportunity in the cooperative controversy lesson to assess students' prior and alternative conceptions of the nature of science, which can guide the development of activities around this awareness.

College of Education, Temple University, Philadelphia, Pennsylvania, USA

REFERENCES

Bohm, B. & Peat, F. (1987). *Science, order, & creativity*, New York, Bantam.

Deutsch, M. (1962). 'Cooperation and trust- Some theoretical notes', in M. Jones (ed.), *Nebraska Symposium on Motivation (Vol. 10)*, Lincoln, NE, University of Nebraska Press.

Driver, R., Asoko, H., Leach, J., Mortimere, M. & Scott, P. (1994). 'Constructing scientific knowledge in the classroom', *Educational Researcher*, (23) 5-12.

Johnson, R.T. & Johnson, D.W. (1979, April). *Structuring conflict in science classrooms*, paper presented at the annual meeting of the National Association of Research in Science Teaching, French Link, IN.

Johnson, D.W., & Johnson, R.T. (1975). *Learning together and alone: Cooperation, competition, and individualization*, Englewood Cliffs, NJ, Prentice-Hall.

Johnson, D. W., & Johnson, R.T. (1983). 'The socialization and achievement crisis: Are cooperative learning experiences the solution?', in L. Bickman (ed.), *Applied Social Psychology Annual 4*, Beverly Hills, CA, Sage Publications.

Johnson, D.W., Johnson, R.T., Holubec, E., & Roy, P. (1984). *Circles of learning: Cooperation in the classroom*, Alexandria, Virginia, Association for Supervision and Curriculum Development.

Johnson, D.W., Maruyama, G., Johnson, R., Nelson, D., & Skon, L. (1981). 'Effects of cooperative competitive, and individualistic goal structures on achievement: A meta-analysis', *Psychological Bulletin*, (89), 47-62.

Johnson, D.W., Johnson, R.T., Pierson, W.T. & Lyons, V. (1985). 'Controversy versus concurrence seeking in multi-grade and single-grade learning groups', *Journal of Research in Science Teaching*, (2), 835-848.

Millar, R., Driver, R., Leach, J., & Scott, P. (1993). *Students' understanding of the nature of science: Philosophical and sociological foundations to the study*, working Paper Two from the project The development of understanding the nature of science, Centre for Studies in Science and Mathematics Education, The University of Leeds, United Kingdom.

Scharmann, L.C. (1990). 'Enhancing an understanding of the premises of evolutionary theory: The influence of a diversified instructional strategy', *School Science and Mathematics*, (90), 91-100.

Slavin, R. (1977). 'Classroom reward structure: Analytical and practical review', *Review of Educational Research*, (47), 633-650.

Smith, E. L. (1990, April). *Implications of teachers' conceptions of science teaching and learning*, paper presented at the annual meeting of the National Science Teachers Association

Solomon, J., Duveen, J., Scot, L. & McCarthy, S. (1992). 'Teaching about the nature of science through history: Action research in the classroom', *Journal of Research in Science Teaching*, (29), 409-421.

CATHLEEN C. LOVING

7. NATURE OF SCIENCE ACTIVITIES USING THE SCIENTIFIC PROFILE: FROM THE HAWKING-GOULD DICHOTOMY TO A PHILOSOPHY CHECKLIST

The purpose of this chapter is to provide many strategies for teaching different aspects of the nature of science to enlighten undergraduate or graduate science education students. These strategies include becoming familiar with the variety of positions held by equally distinguished scientists, philosophers and science educators. Also included are activities designed to highlight how science is portrayed in texts, curriculum guides, informal settings and in the media. The important role of scientific theory is dealt with in a way that emphasizes both its uniqueness and value and its limitations. The chapter introduces these aspects of the nature of science through a series of enablers with specific concepts and activities listed — enough for a whole course or for use individually.

The Scientific Theory Profile (STP) (Figure 1) plots positions of twelve well-known philosophers, historians and sociologists of science relative to their written positions about two crucial attributes of theories in science. Theories are the detailed explanations for how phenomena in nature are thought to work. According to Duschl (1990) theories can be viewed as Core, Central or Fringe--depending on how accepted, tentative or wacky they are. The first attribute (on the x-axis) of the STP is how theories are judged for acceptance or rejection. The extremes on the x-axis are judgments by purely rational means versus "natural" means--those involving psychological, social or cultural factors rather than rational (that is, purely intellectual based on logical thinking). The second attribute of theories (on the y-axis) is what scientific theories represent--from an external Truth at one end of the axis (a staunch realist perspective) to an individual human construction at the other end (radical constructivist; anti-realist)--with "what works best for the time being" somewhere in the middle (instrumentalist, for example). Views on these two attributes of scientific theories--judgment and representation--help establish one's position on the nature of science, perhaps more than any other. The reader is encouraged to consult Appendix A where a representation of the primary works used to establish positions on the STP are listed. In addition, Loving (1991) gives a brief description of the essence of the views on theories held by each of the twelve on the profile. This helps clarify comparisons one might wish to make with contemporaries in science, philosophy or science education. Students can be assigned to develop expertise on one or more positions--a tactic used successfully with graduate students.

Not only did Giere provide the idea for the STP, his maverick position on the Profile itself also serves as fodder for discussion among science educators in particular, since he bases his notions about scientific theories on the cognitive activity, what he calls the "constructive realism," of the scientist. A number of colleagues have indicated that they use the STP to introduce their science education students to the world of philosophy of science and the various positions held by well-known scholars. This chapter, however, gives numerous specific ideas for using the STP to show the central role of theory in science; it then expands to other concepts and activities on the nature of science, including suggestions for plotting well-known science educators.

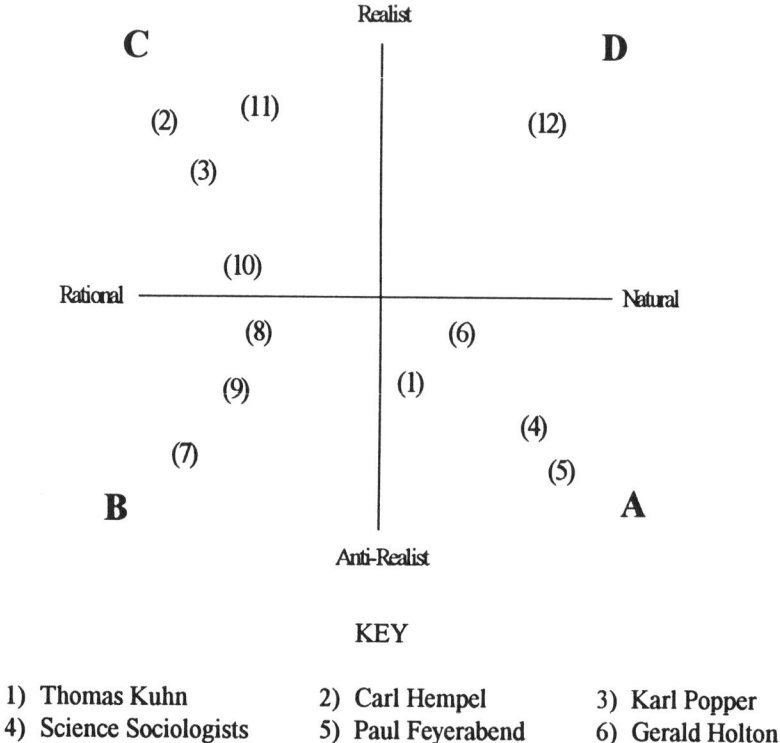

KEY

1) Thomas Kuhn   2) Carl Hempel   3) Karl Popper
4) Science Sociologists 5) Paul Feyerabend   6) Gerald Holton
7) Stephen Toulmin   8) Dudley Shapere   9) Larry Laudan
10) Imre Lakatos   11) Clark Glymour   12) Ronald Giere

Figure 1. The scientific theory profile illustrating the positions of various philosophers and sociologists of science on a vertical axis from realist to anti-realist and horizontal axis from rational to natural. The x-axis is judgment (theory's value) and the y-axis is representation (theory's truth).

"The farther from the origin of the STP, the more radical (Feyerabend) or unique (Giere) is the position on scientific theories. "Balance" is indicated by closeness to the origin of the graph. Those nearest (Holton, Shapere) are perhaps most balanced, acknowledging in their writings the rational, fragile and sometimes stubborn nature of judgment (X-axis) and the moderate ontological position that what works as best explanations *may* be a truth, but not necessarily *the* truth (Y-axis). One of the major thrusts of my research has been to suggest that science teachers, more than bench scientists, need to be familiar with this profile enough to be aware of various positions on science theories and to be able to settle into a position--preferably near the middle. They then need to be able to analyze what they do and say with students in terms of how they are portraying science and its explanations."

Finally, in support of Lederman (1992) who uncovered a wide array of teacher and student conceptions of the nature of science and emphasized the importance of not trying to oversimplify our judgment of these views, this chapter describes activities that will facilitate the development of views by science education students informed by knowledge from all sides and appreciative of the complex nature of this thing called "science," which, in fact, has many different modes.

METHOD OF PRESENTATION

Various nature of science activities are described here as part of a course outline. The activities emerged as a result of the development of a carefully stated mission, course goals, and performance proficiencies one would hope teacher-students would achieve well enough to influence their views and teaching behavior. While space will not permit listing these, some key phrases serve to summarize the essence of the overall purpose of these activities. For the mission, the key phrase might be to "learn how scientific theories are derived, judged and represented . . . leading to more accurate perceptions of what science is and what it is not." While there are nine course goals, summary phrases might be to "develop a philosophy and recognize it in others . . ." " to become . . . aware of the developmental nature of knowledge . . ." "consider . . cultural significance and validity of past scientific explanations" . . . "create bridges between the culture of science and the culture of humanities." The ten proficiencies can be highlighted with a few phrases: "make teaching congruent with dynamic nature of science . . ." ". . . avoid final form science . . ." " . . . avoid treating all knowledge claims equally . . ." " . . . recognize difference between good and bad science, pseudoscience and real science . . ." "facilitate changing flaws in students' intuitive thinking."

Finally, ten "enablers" were developed based on the mission, goals and proficiencies. These enablers are described in detail in the next section. Once conceived, the enablers led this author to investigate current scientists, current controversies, current media interpretations of science and the extent to which they can be explored using the tenets of the Scientific Theory Profile.

The ten enablers, along with the resulting sample concepts and activities described below are not prioritized, but each has been selected because if carried out by students it would support certain course goals and performance proficiencies.

Enablers with Concepts and Activities

1. Distinguish between philosophers' assigned positions on the Scientific Theory Profile and attempt to determine a comfortable position for yourself (both at beginning and end of course).

Sample Concept:

a) How do various experts use the term theory, science, truth, rational, explanation?

Sample Activities:

a) Using the Scientific Theory Profile and suggested readings, students become expert on one philosopher and roleplay or debate some predetermined issue.
b) Students read Horwood (1988) who addresses the differences between scientific explanations, teacher explanations and descriptions.

2. Describe, plot on the STP, and compare philosophical perspectives (stated or implied) of various scientists based on the way they are presented in the popular press. This strategy may also be applied to well-known science educators.

Sample Concepts:

a) Compare the value placed on metaphysical, mystical, beliefs vs. knowledge.
b) Contrast their descriptions of rational, logical, and science or their views on the nature of knowledge (epistemology), of reality (ontology) and that relationship to pedagogy in science teaching.
c) Look for evidence of postmodern vs. Baconian modern science views in descriptions of actual scientific work.

Sample Activities:

a) Compare descriptions by Hawking and Gould of why scientists resist certain theory change; determine the degree to which these writers connect the

emotional significance of certain theories to their power.
b) Have a debate on four views of the value of the mystical--Einstein, Hawking, Medawar, Gould.
c) Determine to what extent consistency and symmetry are driving the scientific fields you read about.
d) Invite local scientists, philosophers, or historians of science to join the class in a discussion of one of more of the books recommended at the end of this monograph or arrange a seminar with a panel of local experts serving to answer questions the class provides ahead of time--based on what they learned from their reading. (This author spoke with a well-respected engineering professor who, when discussing the nature of science said, "You know, biology isn't a real science, don't you?").
e) Invite a group of local scientists in various fields to discuss the degree to which techniques like meta-analysis are anti-theory or "indirect" versus precise observation--like the French electron microscope which "sees" an atom as small as two angstroms--is acceptable in their particular field; this compares methods and aims in different disciplines.
f) Student groups read works on the nature of science from selected science educators (e.g., Matthews, Duschl, Aikenhead, Brickhouse, Lederman, Good) then attempt to plot these scholars on the STP and compare plots.

3. Analyze current scientific debates about quality of rival theories.

Sample Concepts:

a) Compare validation procedures, predictive power; look for language distortion, social pressures, good vs. bad science, and quality of an explanation.

Sample Activities:

a) Read the New York Times and other popular press summaries of debates, then compare to peer reviewed scientific publications on debate issues; report on which if either appears more philosophically valid.

Examples include:

- Freud's theories today--New York Times, 3/6/90, B5; 3/25/90, A18
- Bellak's theories on German aggression--New York Times, 4/25/90, A15; 5/10/90, A18;
- Turco's nuclear winter theory-- New York Times, 1/23/90, B5;
- Nazi Data on hypothermia--New York Times, 5/17/90, A14;

- Kepler's "fudged" data--New York Times, 1/23/90, B5; 2/13/90, A18;
- Freeman's vs. Mead's vs. Geertz's anthropological theories, Chronicle of Higher Education, 8/2/89, A5; New York Times, 5/11/88, Y23;
- Dinosaur Extinction Debate--New York Times, 1/15/88, B5; 1/19/88, B19; Chronicle of Higher Education, 9/1/88, A5; New York Times, 8/15/89, B5;
- Cold Fusion Debate -- Chronicle of Higher Education, 5/24/89, A1; New York Times, 6/10/90;
- Global Warming Theories-- New York Times, 6/24/88, A1;
- Antelman's controversial drugs/brain theory--New York Times, 6/21/88, Y25;
- Masters' and Johnson's AIDS Study--New York Times, 3/6/88, Y14.

4. Critically evaluate philosophical perspectives in texts and curriculum guides in your field.

Sample Concepts:

a) Determine if science is presented as a process or a product, if it is distinguished from technology, if various disciplines of science are compared at all for nature of methods, aims, or theories, or if historical, philosophical or sociological aspects of science are apparent.
b) What is actually mentioned in the text or curriculum guide about the nature of science versus implied or left for the reader to infer.
c) Determine if "final form science" is predominant in order to cover everything.

Sample Activities:

a) Students design a philosophical checklist for use in evaluating texts and curriculum guides with a criteria for judging.
b) Students examine current texts and/or teachers' editions and curriculum guides in their specialty to determine what classroom practice is encouraged explicitly or implicitly.
c) Students read Hodson's (1988) comments on the philosophically valid curriculum and augment his suggestions with personal ideas.

5. Describe several current controversies related to the culture of science, rather than to the formal propositions within particular disciplines.

Sample Concept:

a) Evaluate controversies related to the ability to make something truly random, the issue of publication bias, gender bias in science, journal bias, the militarization of science, and dishonesty among scientists.

Sample Activities:

a) Select from the articles listed below for discussion related to above concepts.

- "The Quest for True Randomness Finally Appears Successful," New York Times, 4/18/88, Y23;
- "They Had Reason to Be Humble" (a review of the book, The Launching of Modern American Science 1846-1876), New York Times Book Review, 5/17/87, 26-27;
- "Science Writing: Too Good to Be True?," (shows how science writing, both popular and professional, probably makes science look "far more precise and reliable that it really is"), New York Times Book Review, 9/27/87, 1, 45-47;
- "Nobel Winner is Caught Up in a Dispute Over Study," (controversy surrounding Nobel Prize winner David Baltimore's raw data versus his published work), New York Times, 4/12/88, Y19, 24;
- "Nova" Television Show on the Public Broadcasting System, "Do Scientists Cheat?";
- "Scientists Must Demonstrate to Would-Be Regulators That They're Capable of Responsible Self-Regulation," The Chronicle of Higher Education, 5/10/89, A44;
- "Publication Bias Said to Skew Information in Medical Journals," The Chronicle of Higher Education (Research Notes Section) 1/4/89;
- "A Flaw in the Research Process: Uncorrected Errors in Journals," New York Times, 5/31/88, Y21;
- "Impossible Idea Published on Purpose," New York Times, 6/30/88, Y9 (Biologists tell of a reaction between something and nothing--this was later shown to be seriously flawed, but "science" fell for it as this article reveals);
- "Two Critics of Science Revel in Their Role," New York Times, 4/20/89, 23, 28 (Regarding two National Institutes of Health scientists whose job is to weed out misconduct and error in science--have since lost their positions and the office has been disbanded)
- "Vast Sums for New Discoveries Pose a Threat to Basic Science," New York Times, 5/27/90, 1,13 (highlights 18 big projects which will cost an estimated $64 billion and the controversies surrounding their effect on the rest of science--since then SSC has been abandoned and Congress has spoken against that Big Science), New York Times Editorial, "Big Science, Little Gain," 6/24/90,20;
- "Military Spending, the SDI, and Government Support of Research and Development: Effects on the Economy and the Health of Science," in Public

Interest Report (Federation of American Scientists), 1987, Holdren & Green.
- "Science's Anti-Female Bias," New York Times, 6/4/90, A19 (using the example of all the reports showing effects of alcohol on pregnancy and what the author terms the "pregnancy police," she shows that our society is "drunk on random bits of information wrested from scientific journals before the ink has dried.").

6. Elaborate on ways that informal science entities present a philosophical perspective about science (planned or unplanned).

Sample Concepts:

a) Determine if science is portrayed as process, product or both at science museum exhibits and in the media.
b) Evaluate media presentations of the role of scientific theories.

Sample Activities:

a) Students are assigned one video or television show related to the nature of science to evaluate in terms of whether a clear philosophical perspective is stated, is crafted but implicit, or is clearly unplanned. Examples are: *Search for Solutions* (Phillips Petroleum), *The Ascent of Man*, (Jacob Bronowski's Public Television (PBS) series), *Newton's Apple* and *Nova* (both PBS TV series), *Cosmos* (Carl Sagan's video series), *Ring of Truth* (Phillip Morrison's PBS series depicting how we know what we know), and the recent series bringing physics alive, *Mechanical Universe*.
b) Students design a "misconceptions" exhibit for younger children (can use the Exploratorium Science Snackbook obtained from Exploratorium Publications, 3601 Lynn, San Francisco, CA 94123 to get the "recipes" for accurate exhibits, then can adjust them).
c) Students evaluate a museum known for its "hands-on" approach.

7. Relate current models of cognition to improved understanding of key concepts, theory formation, and the developmental nature of science.

Sample Concepts:

a) Use concept maps to depict both accepted and erroneous theories (see Novak 1981, 1984, 1987).
b) Compare expert vs. novice methods of arriving at conceptual understanding and depiction of theories.

- c) Compare the history of a theory's development and stages of common misconceptions.
- d) Compare the value of "less is better" related to material covered and notion of situated cognition.
- e) Compare the value of specific procedural knowledge and pedagogical content knowledge in various domains.

Sample Activities:

- a) Student uses Inquire computer program (Hawkins and Pea, 1987) to explore their personal understanding in a weak scientific area; have partner who is more expert in field act as teacher--get "steered" to expert.
- b) Student draws a detailed concept map of a familiar theory, then compares to drawing asked of volunteer experts.
- c) Read Wandersee and Mintzes' (1987) account of how the history of the development of a theory (photosynthesis in this case) parallels stages of cognitive misconceptions students go through; students could then do historical study of another theory's development (including the various people's work that contributed to the understanding) and compare it to concept maps they draw (or have various novices draw).
- d) Have guest scientist provide just enough of the relevant math and science background for students to investigate several stages in the development of a new, important theory in science to see the process involved and the developmental nature of scientific knowledge. Example ideas include chaos theory, string theory, and the punctuated equilibrium theory of evolution.
- e) Students read key ideas regarding how to teach based on how students think. Examples include Reif, 1986; Linn, 1987; Sherwood et al, 1987.

8. Analyze current topics that exemplify the limitations of scientific theories as explanations of the human condition.

Sample Concepts:

- a) Cosmology theories (which affect overall world view of a particular culture) seem to change every few years and have since recorded history--showing limitations of scientific theories.
- b) Explanations of cell aging have provided many possible causes for aging and have resulted in a number of medical breakthroughs related to cancer treatments and other age-related afflictions. But the human species has had the same life span (100-110 maximum) for over 25,000 years and none of this knowledge has changed that.

c) Social science theories about the causes of aggressive behavior in humans have done little to lessen such behavior in our culture.
d) Scientific theories about the effects overpopulation has on so many aspects of life have done little to lessen the problem in most developing countries.
e) Theories such as those describing the causes and effects of acid rain are so politicized that they often lose their effectiveness.

Sample Activities:

a) Compare Hawking's current theories of origin and patterns in the universe with several others in the last twenty years.
b) Compare Ptolemy's world view, Einstein's world view, and the Navajo Indian world view and the effect of a reigning world view on culture, science, and theory choice (See Cobern, 1991).
c) Investigate the different theories on causes of certain life spans in different species (i.e. 20 minutes for bacteria, 2,000 years for a Bristlecone pine tree); determine which species do not "age," the various theories on that process, and try to connect to life span studies.

9. Critically evaluate a scientific idea which was once accepted and then shown to be unsuccessful.

Sample Concept:

a) Compare criteria for acceptance (through time) in various disciplines and sub-disciplines in science (e.g., evolutionary biology versus taxonomy).

Sample Activities:

a) Students choose from a list of "failed" theories and theorists (caloricists, phlogistonists, Lamarck, Ptolemy etc.); try to evaluate the degree of the "failure" of their theory (their method versus their luck, their timing, and the reigning criteria for acceptance).

10. Describe the stages in the development of a successful scientific explanation.

Sample Concepts:

a) Evaluate serendipity vs. logic in science--to what extent are both necessary? Distinguish between the logic of discovery and the logic of justification. Read recent work of Bechtel (1993) describing varied, non-linear search paths leading to discovery in science.

b) Determine possible sociological effects on theory choice, methods, aims.

Sample Activities:

a) Students design an interview questionnaire that would help uncover the working philosophy of a living scientist, her/his theoretical "bedrock," degree of scientific community dependence etc.
b) Interview a living scientist who has developed a "revolutionary" theory (e.g. John Archibald Wheeler, the "father" of black holes, Steven Weinberg, who proposed the Weinberg-Salam unified theory of forces, or Stephen Jay Gould and Nils Eldredge who developed the punctuated equilibrium theory of evolution) to determine what they think the primary reasons for their "theoretical success" were. John Wheeler, reminiscent of Newton's comment about standing on the shoulders of giants, once told my class that he believed a long chain of mentors probably led up to all great scientists, and that someone ought to write a book about it. Students then read a few of the publications related to that scientist's theory over a span of time, looking for evidence of change, process, and re-appraisal. Students compare their results with each other to see if new or common information emerges.

CONCLUSIONS

The goal of this chapter has been to provide activities that can develop better conceptual understanding of the nature of science for prospective and experienced science teachers, as well as graduate students who will become science teacher educators. The ultimate intent is not for science teachers to go into the classroom and teach the Scientific Theory Profile to fourth, seventh, ninth or even twelfth graders. The intent is for teachers to profit from this exposure, to use their texts and curriculum guides more judiciously, and to give their students some opportunities to read about and do real science. In addition, their students are more likely to meet real scientists or historians and philosophers of science, and to spend more time on less material to discover how this knowledge came to be and just how "human" are the people who have constructed this picture of nature.

There are several examples which illustrate how this exposure has helped students. I have used some of these activities while teaching two courses--one an existing graduate course on science curriculum, the other a special problems course on the nature of science and science teaching. One slightly resistant graduate student with a master's degree in geology and twenty years teaching in community college described his learning as "opening doors I never knew existed." Another physics teacher was able to view his summer work at the Los Alamos labs in New Mexico as an opportunity to do a little of what Giere did in a cyclotron--watch and record how scientists do what they do--and then compare his first hand experience with what he had read. Two master science teachers said they understood much better the unique attributes of each others' disciplines--physics and biology--so as to be in a better position to highlight and maintain the uniqueness while encouraging integration of disciplines where appropriate. Creating course work in this area has encouraged me to hold discussions with philosophers and historians of science on our campus, and to explore ways to encourage graduate students from departments such as Educational Psychology as well as the Colleges of Science, Geosciences and Agriculture to come over to the Curriculum and Instruction Department for a few lessons on the nature of science.

In some ways the goals for students involved in these nature of science activities are similar to those of Steven Shapin and Philip Kitcher when they launched a new science studies program at the University of California, San Diego about six years ago. In an interview Shapin described his desire to "make science a glass box instead of a black box..." (Raymond, 1990, A7). Providing students with an opportunity to see science in new and deeper ways is critical to their development as science educators.

Texas A&M University College Station, Texas, USA

ACKNOWLEDGMENTS

I wish to acknowledge a debt to philosopher Ronald Giere (1988) whose suggestions gave me the idea for the development of the Scientific Theory Profile (STP) (Loving, 1990, 1991, 1992), the heuristic device which provides a framework for many of the activities described in this chapter.

REFERENCES

Bechtel, W. (1993). *Discovering complexity: Decomposition and localization as strategies in scientific research*, Princeton, NJ, Princeton University Press.
Cobern, W.W. (1991). *World view theory and science education research*, (National Association for Research in Science Teaching, Monograph 3. Manhattan, KS, National Association for Research in Science Teaching.
Duschl, R.A. (1990). *Restructuring science education: The importance of theories and their development*, New York, Teachers College Press.
Hawkins, J & Pea, R. D. (1987). 'Tools for bridging the cultures of everyday and scientific thinking', *Journal of Research in Science Teaching*, (24), 291-307.
Hodson, D. (1988). 'Toward a philosophically more valid science curriculum', *Science Education*, (72), 19-40.
Horwood, R.H. (1988). 'Explanation and description in science teaching', *Science Education*, (72), 41-49.
Lederman, N.G. (1992). 'Students' and teachers' conceptions of the nature of science: A review of the research', *Journal of Research in Science Teaching*, (29), 331-359.

Linn, M.C. (1987). 'Establishing a research base for science education: Challenges, trends, and recommendations', *Journal of Research in Science Teaching*, 24, 191-216.
Loving, C.C. (1990). 'Current models in philosophy of science: Their application in science teacher education', *Dissertation Abstracts International, 51-10*, 3376A. (University Microfilms No. 91-05, 603).
Loving, C.C. (1991). 'The scientific theory profile: A philosophy of science model for science teachers', *Journal of Research in Science Teaching*, (28), 823-838.
Loving, C.C. (1992). 'From constructive realism to deconstructive anti-realism: Helping science teachers find a balanced philosophy of science', *Proceedings of the Second International Conference on History and Philosophy of Science in Science Education*, Kingston, Ontario, Queens University Press.
Novak, J. (1981). 'Applying learning psychology and philosophy of science to biology teaching', *The American Biology Teacher*, (43), 12-20.
Novak, J. (1984). 'Applications of advances in learning theory and philosophy of science to the improvement of chemistry teaching', *Journal of Chemical Education,* (61), 607-613.
Novak, J. (1987, July). *Human constructivism: Toward a unity of psychological and epistemological meaning making*, paper presented at the Second International Seminar on Misconceptions and Educational Strategies in Science and Mathematics Education, Ithaca, NY.
Raymond, C. (1990, May 9). 'Scholars take a new approach in studying the institution of science', *Chronicle of Higher Education*, A7.
Reif, F. (1986). 'Scientific approaches to science education', *Physics Today*, 49.
Sherwood, R.D., Kinzer, C.K., Bransford, J.D., & Franks, J.J. (1987). 'Some benefits of creating macro-contexts for science instruction: Initial findings', *Journal of Research in Science Teaching*, (24), 417-435.
Wandersee, J.H. & Mintzes J.J. (1987, July). *A bibliography of research on students' conceptual development in the life sciences*, paper presented at the Second International Seminar on Misconceptions and Educational Strategies in Science and Mathematics, Cornell University, Ithaca, New York.

APPENDIX A: SOURCES FOR SCIENTIFIC THEORY PROFILE DEVELOPMENT

Barnes, B., & Edge, D. (1982). *Science in context: Readings in the sociology of science*, Cambridge, MIT Press.
Feyerabend, P.K. (1975). *Against method: Outline of an anarchistic theory of knowledge*, London, NLB.
Giere, R.N. (1988). *Explaining science: A cognitive approach*, Chicago, IL, University of Chicago Press.
Glymour, C. (1980). *Theory and evidence*. Princeton, NJ, Princeton University Press.
Hempel, C. (1965). *Aspects of scientific explanation and other essays in the philosophy of science*, New York, The Free Press.
Hempel, C. (1966). *Philosophy of natural science*, Englewood Cliffs, NJ, Prentice Hall.
Hempel, C. (1974). 'Formulation and formalization of scientific theories', in F. Suppe (ed.), *The structure of scientific theories*), Urbana, IL, University of Illinois Press, 244-265.
Holton, G. (1978). *The scientific imagination: Case studies*, London, Cambridge University Press.
Holton, G. (1986). *The advancement of science and its burdens: The Jefferson lectures and other essays*, London, Cambridge University Press.
Kuhn, T.S. (1970a). 'Logic of discovery or psychology of research', in I. Lakatos & A. Musgrave (eds.), *Criticism and the growth of knowledge*, Cambridge, Cambridge University Press, 1-22.
Kuhn, T.S. (1970b). *The structure of scientific revolutions*. (2nd ed..), Chicago, The University of Chicago.
Kuhn, T.S. (1974). 'Second thoughts on paradigms', in F. Suppe (ed.), *The structure of scientific theories*. Urbana, IL, University of Illinois Press, 459-482.
Kuhn, T.S. (1977). *The essential tension*, Chicago, The University of Chicago Press.
Lakatos, I. (1970). 'Falsification and the methodology of scientific research programs', in I. Lakatos & A. Musgrave (eds.), *Criticism and the growth of knowledge*, Cambridge, Cambridge University Press, 91-196.
Laudan, L. (1977). 'Progress and its problems', Berkeley, The University of California Press.
Laudan, L. (1984). *Science and values*, Berkeley, The University of California Press.
Popper, K. (1959). *The logic of scientific discovery*, London, Hutchinson.

Popper, K. (1965). *Conjectures and refutations: The growth of scientific knowledge* (2nd ed.), New York, Basic Books.
Popper, K. (1983). *Realism and the aim of science,* Totowa, NJ, Rowman and Littlefield.
Shapere, D. (1974). 'Scientific theories and their domains', in F. Suppe (ed.), *The structure of scientific theories* , Urbana, The University of Illinois Press, 518-600.
Shapere, D. (1982, March). 'The concept of observation in science and philosophy', *Philosophy of Science,* 1-23.
Shapere, D. (1984). *Reason and the search for knowledge: Investigations in the philosophy of science,* Dordrecht, Holland, Reidel.
Toulmin, S. (1961). *Foresight and understanding,* Bloomington, IN, Indiana University Press.
Toulmin, S. (1982). 'The construal of reality: Criticism in modern and post-modern science', *Critical Inquiry,* (9), 93-111.

FRED JANSEN AND PETER VOOGT

8. LEARNING BY DESIGNING
A CASE OF HEURISTIC DIRECTED THEORY
DEVELOPMENT IN SCIENCE TEACHING

In teaching, science is often presented as a black box. Students are typically confronted with only the results of scientific research. They are not given any insight into what is actually happening in the black box which is the often tedious process of formulating and testing theories. Science presented in this way, characterized by Joseph Schwab (1962) as the "rhetoric of conclusions," can easily lead students to an incorrect image of both the products and processes of scientific research. Students typically consider theories not as provisional solutions to problems, but rather as facts that are simply gathered from nature by scientists.

OPENING THE BLACK BOX

How can we give students a more adequate and complete picture of science? In order to do this, the black box must be opened to provide students with an insight into how a theory is developed and on what grounds it is accepted or rejected. This can be achieved by telling them about theory development (through autobiographies, scientific papers, etc.), but better still is to allow students to participate themselves in the process of theory development. This assignment is not an easy one for the teacher. In the typical classroom it is the teacher's responsibility simply to explain the theory and subsequently test to see if students understand it. On the other hand, allowing students to participate in theory development, the teacher must first teach students how to formulate problems, how to create possible solutions and how to test these solutions. In this chapter, we will discuss the content and framework of a course in which teachers learn how to help students participate in the process of theory development.

We will first ascertain what we mean by theory development. On the basis of two lesson fragments we will illustrate and describe our view of theory development. The concept of a heuristic (search strategy) is central to this discussion. Such heuristics can generally be derived from the basic principles of a discipline and are thus specific to a particular discipline. We have developed a heuristic for teaching biology. We have called this heuristic a "design heuristic" because pupils use it to develop theories related to various biological systems by designing them from scratch — learning by designing.

Next we will describe how a teacher can prepare and supervise the process of heuristic directed theory development. Then, we consider how teachers can reflect on the premises and method of theory development together with the students. Preparation, supervision and reflection on theory development are illustrated using several examples from biology education. Finally, we focus on the effectiveness of our approach.

THEORY DEVELOPMENT

We will illustrate our conception of theory development by referring to two lesson fragments (Table I). In the first fragment, pupils are taught a theory in a traditional manner. In the second, pupils actually participate in developing the theory. The lessons are in biology with specific focus on the immune system. The pupils are in their final year of a high level, secondary school, biology course. The teacher has given a short introduction on bacteria and viruses before both lessons.

Table I
Lesson fragments illustrating two instructional strategies

Lesson Fragment 1: Typical instructional discourse

Teacher:	White blood cells are responsible for eliminating bacteria and viruses. There are three types of white blood cells. I shall begin with the macrophages. Who thinks he knows what the word macrophage means?
Joost:	Doesn't macro mean big?
Teacher:	Exactly. Macro means big and phage means an eater - a big eater, in other words! When a macrophage encounters a bacterium, it envelops the bacterium and subsequently digests it. We call this process phagocytosis. Where do you think that we can find this macrophage in the body?
Anja:	In the blood - you just said that they are white blood cells!
Cees:	I think near the skin that is where the bacteria enter the body.
Teacher:	You are both correct actually. Macrophages can be found in blood, but also in tissues where bacteria enter the body, such as skin and lungs.

Lesson Fragment 2: Discourse leading to theory development

Teacher:	We now know that bacteria and viruses can be harmful. How can a bacterium or virus be rendered harmless?
Klaas:	With a sort of stabbing cell. This could stab through the bacterium.
Marjon:	But then you still haven't got rid of the bacterium.

Teacher:	What do you mean?
Marjon:	Toxic substances inside the bacterium will be released. You will get sick.
Teacher:	Very good. Has anyone a solution to this problem?
Carla:	You have little gobbling cells in the body and they just gobble up the bacteria whenever they come across them.
Joop:	No, that can't be correct - that type of cell would also gobble up your own body cells.
Teacher:	Exactly. Devouring a bacterium is much better than just bursting it open. There are in fact such "gobbling" cells in the body. They are called macrophages. But we still have the problem that Joop has mentioned. How do you prevent a macrophage ingesting the body's own cells and substances?

In both lessons pupils receive instruction on macrophage function and solve a number of problems. Yet, there are important differences. In lesson fragment one, the pupils are first given information about the macrophage theory, subsequently students are presented with a problem. In order to solve the problem of where the macrophages can be found, students are required to apply the knowledge they have just acquired.

In lesson fragment two, the sequence of events is reversed. The teacher begins by introducing a problem (how can pathogens be rendered harmless?). Pupils devise a trial hypothesis for this problem (such as a stabbing or gobbling cell). The solutions are evaluated by the pupils and any errors in the hypothesis are identified (stabbing cell is not an option because of the release of toxic substances). This error elimination results in an adjustment of the proposed solution (not a stabbing cell but gobbling cell) and in the formulation of a new problem (how do you prevent the gobbling cell eating your own body's cells?). Finally, the teacher presents a part of the scientifically-accepted theory (in this case, the macrophage).

The lesson fragments are represented schematically in the following diagrams:

Lesson Fragment 1: Theory (T) ---> Problem (P)   [T ---> P]
Lesson Fragment 2: Problem (P) ---> Hypothesis (H) ---> Error Elimination (EE) ---> Problem (P)   [P ---> H ---> EE ---> P]

The first lesson represents a structure that figures prominently in traditional natural science education (Kuhn, 1977; Schwab, 1962). The explanatory theory is taught first, and subsequently the pupils apply their newly acquired knowledge to solving problems. The science historian Thomas Kuhn refers to the former not as problems but as puzzles since a solution to the problem (the theory) has already been given to the pupils, albeit in a broad outline. This advance knowledge would affect students' future consideration of the problems. Using this instructional method, pupils are introduced to the theory and its applications but do not learn how the theory originated and on what grounds it was evaluated.

In the second lesson we can recognize several important aspects of hypothetical deductivism, in the form characterized by Karl Popper (1972). Theory development begins with a problem, for which a hypothesis is formulated. This hypothesis is subsequently tested to identify and eliminate inherent errors. This process leads to a provisional acceptance, a modification or a rejection of the hypothesis and often to new problems. The fruitfulness of a hypothesis is tested against a preconceived criterion (in this case, minimal disadvantages for the rest of the organism). In science, a hypothesis will, of course, also be tested empirically. In this case, observable predictions are deduced from the hypotheses, and tested. The major defect of Popper's scheme is that no attention is paid to the generation of hypotheses, a central issue in the model discussed here.

HEURISTIC-DIRECTED THEORY-DEVELOPMENT

During the past twenty-five years, Herbert Simon has developed a scheme of scientific discovery that draws on his own and others' problem-solving research (see Langley, 1987). In his plan, scientists are seen as searching in an enormous problem space consisting of all possible solutions for the problem at hand. There are at least two search methods one can use. One can conduct trial and error searches or one can use an appropriate heuristic. Such heuristics serve as quick and effective ways of generating problem solutions by reducing the size of the set of possible solutions one needs to test.

To illustrate the role of heuristics in problem solving, consider an example put forward by Langley et al. (1987). If a safe has 10 dials numbered 0 to 99 and it will open when all the dials are set correctly, than it will take, on average, 50 billion billion ($\frac{1}{2} \times 100^{10}$) trials to open it. However if one knows that each dial emits a faint click when it is turned to the correct setting one can open the safe with only 500 ($\frac{1}{2} \times 10^{10}$) trials. In this case, the heuristic could be expressed as the rule "if you detect a click when turning a dial, then leave the dial on the setting that produced the click." This example illustrates the need for information about the problem domain for the construction of useful heuristics. The trouble with most scientific problems is that specific information about the problem domain is lacking. In fact, when we propose to investigate a given subject that is to say that we are, in large part, ignorant of it. So what kind of information is available for the construction of heuristics?

The science philosopher Imre Lakatos (1978) has shown that heuristics can be adopted from very general hypotheses about the research object (the so-called hard core). Consider the usefulness of the following hypotheses:

- atoms are very small particles that cannot be further divided or changed (the

atomic programme)
- gasses are composed of moving molecules (the kinetic programme)
- organisms have certain characteristics that function in the maintenance and reproduction of the organism (the functional programme)

Such general hypotheses generate problems and lend direction to the formulation of more specific hypotheses (see Schwab, 1962). Dalton's atom concept, for example, allowed chemists to pose a question on how the realignment of particles takes place following a given reaction. The kinetic programme allowed physicists to explain phenomena such as pressure and temperature in terms of the movement of molecules. We can convert these very broad ideas into powerful heuristics, which in turn can assist students with formulating problems and hypotheses. We have developed such a heuristic for biology education based on the broad idea of function.

Our biology education heuristic is based on the simple idea that we can accrue knowledge of a functioning system (biological or man made), by redesigning the system from scratch (Jansen & Voogt, 1995; Dennett, 1995). In this manner we can discover the function of each part of the system and how this function is carried out (see the final section where we discuss the differences between biological and man made systems). Our design heuristic allows students to design a biological system from scratch (see Table 2).

TABLE II
The design heuristic

1.	What is the function of the system as a whole?
2.	Reformulate this function in a problem.
3.	Devise a simple solution for this problem.
4.	Consider what could be a disadvantage of this solution.
5.	Reformulate this disadvantage in a problem.
6.	Go back to step three.

Table 3 offers an example of the design heuristic in action. A student teacher uses the heuristic to design several components of the immune system. An immunology textbook is used to control the devised solutions. Textbook definitions are printed in italics.

TABLE III
The design heuristic in action

System: The Immune System Function: To destroy invading pathogens		
Problem	Solution	Disadvantage
1. How can pathogens (bacteria and viruses) be rendered harmless?	1. By an "eating cell" that ingests viruses and bacteria (macrophage)	1. It can also ingest its own body cells and material.
2. How can a distinction be made between self (the body's own cells) and non-self (pathogens)?	2a. Recognition of self (own body cells): the host's own cells carry a label on their surface that can be bound by macrophages (via a receptor); once the macrophage has been bound, the cells will not be ingested.	2a. Bacteria can imitate the label, and If the receptor on the macrophage changes, then the cell will fail to recognize self (its own cell and material). This is fatal (X).
	2b. Recognition of the pathogen; the pathogen carries its own specific label (antigen). If the macrophage's receptor binds to this antigen then the pathogen will be ingested.	2b. Bacteria and viruses all carry different antigens for which corresponding receptors must be made.
3. How can a different receptor be made for every antigen?		

PREPARING HEURISTIC-DIRECTED THEORY DEVELOPMENT

The teaching situation which we hope to encourage begins with theory development which must start with the initial situation (prior knowledge and interests) given to the students and via successive problem solving steps (P1 - H1 - EE1 - P2 etc.). Students gradually arrive at the accepted scientific theory. It is clear that when we allow the students to follow their instincts, there is a significant chance that each student will follow different problem solving routes and that many will not arrive at the accepted scientific theory. The teacher must therefore decide, prior to the lesson, which problems and what order for those problems will be presented. We call such a sequence of problems a problem-structure.

The problem structure can be developed in two steps. First, the teacher formulates problems and hypotheses using the design heuristic (see Table II). Designing a biological system in this manner provides us with a starting point for developing a problem structure. The problems that can arise are set out in Table III. Second, this problem structure, based on the teachers' background knowledge and interests, must be adapted to the background knowledge and interests of the students. To do so for every problem we must determine if the students want to solve the problem and if the students able to formulate and evaluate a hypothesis for this problem.

The first question compels us to ensure that the first problem to be solved concur with the global interests of the students. We must offer a problem situation that students cannot immediately answer with their background knowledge but have the willingness to do so. The question as to how pathogens can be rendered harmless is only interesting and relevant for students when they know what damage bacteria and viruses can cause and how difficult it is to keep these pathogens outside our body. When the first problem has been formulated, we can use as a rule of thumb, the notion that the formulation of the next problem must arise from the evaluation of the solutions that the students had devised for solving the aforementioned first problem. In lesson fragment two (Table 1) we see how the evaluation of the "gobbling cell" hypothesis leads to a new problem: how can it be prevented that the "gobbling cell" also destroys the body's own cells and substances. In this way, every problem prepares the student for the following problem and students experience the process of theory development as a process which they themselves had guided by posing relevant questions.

The second question compels us to consider how big the steps are between the problems. We must continually strive to place ourselves in the position of the student and ask: Can students, possessing the background knowledge that they do at that moment, devise a solution by themselves for the given problem, and can they critically evaluate this solution? Let us once again consider the problem: How can we ensure that a "gobbling cell" does not also destroy the body's own cells and substances? A "gobbling cell" solves this problem by having receptors specific for substances foreign to the body. Once the foreign substances have reacted with the receptors, they are subsequently ingested and destroyed. Pupils from the last two years of secondary school education, higher level, cannot formulate such a problem in one step. We need to split the problem into subproblems. For example, we need to ask how a "gobbling cell" can recognize a foreign component? It must be able to recognize what it ingests. Other problems include the question of how a "gobbling cell" can recognize a foreign component by feeling? A "gobbling cell" can "feel" something by using receptors. We can ask which type of receptor does a "gobbling cell" need to ensure it does not ingest its own body components? There are specific receptors for foreign components. We must, therefore, make the steps between the problems big enough to allow students to independently bridge the gap between problem and solution.

ASSISTING STUDENTS WITH HEURISTIC DIRECTED THEORY DEVELOPMENT

When a problem structure has been developed, we can supervise the pupils during the development of the theory. Using the following example, we will demonstrate how this works in practice. In the lesson fragment provided below, the following problem is discussed with pupils in their final year of secondary school: How can it be prevented that a "gobbling cell" (macrophage) ingests the body's own material? The pupils have already discovered that recognition is necessary and that recognition takes place by way of receptors. They are now faced with the problem as to with which set of receptors a distinction can be made between self and foreign or non-self. The following lesson fragment illustrates this point.

TABLE IV
A lesson fragment

Teacher:	So, the question is: How can a macrophage ensure that it ingests foreign material and not material from the body?

The pupils write down their solutions on worksheets.

Teacher:	May I have a look?
Marjon:	I don't really know how to write it down. You can't have receptors for things that you already have in your body, otherwise you'll get them. Therefore, you must have receptors for material foreign to the body. I'm afraid I can't go any deeper than that.
Teacher:	That's O.K. You don't have to.
Mischa:	But is it not so that every time a new bacterium comes along, a new receptor is necessary?
Teacher:	Yes, very good. Write that down as a disadvantage of this solution. Joost, do you have another possible solution?
Joost:	Well, its in fact the same. The white blood cells - macrophages, or whatever they're called, only possess receptors for foreign material, and not for the body's own material. So it's actually exactly the same as what she has written down.
Crista:	Yes, it is.

Mischa: Yes, but what if a new bacterium comes along that you don't know yet?

Teacher: Yes, we will discuss that later. First, I want to propose a simpler solution. You have all said: "Make receptors for foreign material." If a macrophage recognizes a foreign antigen, it will ingest it. Now, I have a more simple solution.

In the lesson described above the following methodology was used:

1. The teacher introduced the problem.

2. Students, individually or in groups, are given the assignment to devise one or more hypotheses and their possible disadvantages. They write these hypotheses in their notebooks. Students are given certain elements from the heuristic, in this case: look for the simplest possible solution to the problem -- the one with the lowest number of disadvantages for the rest of the organism.

3. The teacher then, together with the students, makes an inventory of all hypotheses students devised and their possible disadvantages. This is then evaluated. This evaluation of the hypotheses often leads to new hypotheses. In cases where the students come up with a solution that does not occur with the accepted scientific solution, the teacher can give hints to help them along. The teacher may also introduce incorrect alternative hypotheses, to stimulate students into critical evaluation of their own hypotheses. At this stage of making inventories and evaluation, the teacher must ensure that all pupils remain working on the same problem.

4. The teacher then presents the scientifically-accepted solution to the problem (this step has not been included in the lesson fragment above). Eventually, we can give the pupils some problems to which they can apply their newly acquired knowledge. The teacher then reformulates the disadvantage inherent to this solution into a new problem (See step 1).

REFLECTING ON HEURISTIC DIRECTED THEORY DEVELOPMENT

When students themselves participate in theory development, it offers a good opportunity to pupils to reflect on the nature of science. To illustrate this point we will show here how we can stimulate pupils to reflect on the true nature of biological systems and the consequences for the method of theory development. Using the method of teaching by designing, students regularly formulate a solution that appears to be "better" (i.e., simpler and with fewer disadvantages) than the "correct" (i.e., scientifically accepted) solution. When, in this case, the teacher proceeds with the "correct" solution, the pupils respond by saying, "But I thought my solution was better?" This question allows us to reflect on the aim of theory development in biology. The aim is more fully to understand an existing biological system and not (as in technology) to design the best system.

We can then pose the following question to students: Why shouldn't biological systems always be optimal? This question refers to the historical beginnings of biological systems. We generally expect from a good design (such as a watch) that it has been assembled with a clear purpose, by a designer. However, Charles Darwin has shown that a designer is not strictly necessary for the existence of biological systems (Dawkins, 1986). Biological designs can originate through a process of natural selection. This process of inheritable mutation and selection generally leads to good designs. This process is, however, also responsible for the multitude of less than optimal characteristics of biological systems (Gould and Lewontin, 1979). Think, for example, of the fact that light must first pass four layers of neurons before reaching the photoreceptor in the eye. One of the reasons for this is that evolution, as opposed to technology, does not begin with a clean slate. Variations are always based on existing structures and this imposes restrictions. The inefficient position of our photoreceptor is probably an inheritance from the period when primitive chordates lived in muddy soil. When we have established that biological systems cannot always be considered as an optimal design, this can lead to the following question by students: "If we know that biological systems are not always optimally designed, why is this assumption such a central point in theory development?" Using this question, we can explain the heuristic function of this assumption. By accepting that all components have been optimally adapted to fulfil a certain function, we can trace the components and their associated functions. If we do not work from this assumption, then we cannot even begin our research. If a certain component has not been optimally adapted, we can discover this using the same assumption. In this case, however, we will have to adjust our assumption for this component.

TEACHING ABOUT HEURISTIC-DIRECTED THEORY-DEVELOPMENT

Heuristic-directed theory-development is a teaching-approach demanding much both of teachers and students. The following fragment of a conversation between four teachers who have explored this strategy is illustrative. Patricia and Frank have already applied the learning by designing strategy in their own classrooms, but Peter and Petra have not. Frank and Patricia are enthusiastic about learning by designing. This approach corresponds with their view of teaching. Peter and Petra are still reserved. They still use the traditional approach; they instruct the theory and the students apply it.

TABLE V
A lesson fragment

Patricia:	I've applied learning by designing in teaching immunology. I've also applied it at other occasions. After teaching photosynthesis I asked them: Now we have to find something to prevent the plant from losing too much water. Well, they devised very creative solutions. And not in the first place the stomata.
Petra:	No, really?
Patricia:	In fact, you can use it whenever you talk about form and function.
Peter:	But isn't it time-consuming?
Patricia:	Well, it turned out better than expected. And now I have the feeling that they actually think about the subject.
Peter:	Yes, that could be.
Frank:	And in addition you make it very clear to them. I remember that you (Patricia: F. J.) told me enthusiastically about the lesson in kidney working. Students first invented filtration but then realized the same time that the nutrients were lost. So they have to be pumped back. This is just an example.
Patricia:	While in former lessons they couldn't imagine what resorption meant.
Frank:	No, exactly. And now they do.
Patricia:	They remember it better.
Frank:	They remember it better and put it in a bigger context. And it is, let me put it this way, a pleasant surprise for them. Yes, now I'm exaggerating a little, but they really like playing with ideas.
Petra:	Yes, but Frank, I understand that you already worked a little bit like this. You didn't instruct so much, did you?
Peter:	Therefore it must be easier for you.
Frank:	Well, I was glad to get an impulse in the opposite direction at last. Although I already preferred the attitude: "Keep your mouth shut until they don't know any more."
Patricia:	Yes, but now they have to think more than they used to do. One can say: we have always demanded more of our students. But how often are you instructing anyway, while they don't think about it.

For teachers who are not used to letting students think for themselves, we advise implementing learning by designing gradually. To assist in this process, we offer some suggestions below:

- It is not easy to develop a problem structure for complex systems such as the immune-system. Therefore, it's better to let students design parts of a system first. In order to design stomata we can pose the problem; how can the plant prevent itself from losing water (see Patricia)?

- The teacher has to compare and evaluate alternative solutions in the design-approach. When students come up with a lot of alternative solutions, this is not an easy task. Therefore, it's advisable to keep the number of solutions limited. For instance, the teacher can present a restricted number of solutions to choose from. It is also possible to break the problems into subproblems to limit the possible solutions. Of course, the teacher could also pose problems to the class and decide which the students are to address.

- Finally, many problems can be prevented by preparing the lessons well. Think especially about the solutions which students could give and how they relate to the scientifically accepted solutions.

- The following is a list of topics and problems that could be explored by students using the design strategy. How:

  - to transmit signals to individual target cells (transmission along neurons)
  - to maintain a relatively constant body temperature (thermoregulation)
  - to pump the blood out of the heart (ventricles)
  - plants absorb light (chlorophyll) and/or make food (photosynthesis)
  - mammals exchange oxygen and carbon dioxide (lungs)
  - to ventilate the lungs (negative pressure breathing)

University of Utrecht, The Netherlands

REFERENCES

Dawkins, R. (1986). *The blind watchmaker*, Harlow, Longman.
Dennett, D.C. (1995). *Darwin's dangerous idea*, New York, Simon & Schuster.
Gould, S.J. & Lewontin, R. (1979). 'The spandrels of San Marco and the Panglossian paradigm. A critique of the adaptionist programme', *Proc. R. Soc. London*, 205, 581-598.
Janssen, F. & Voogt, P.A. (1995). 'Learning by designing in biological education', in F. Finley and D. Alchin (eds.), *Proceedings of the Third History, Philosophy and Science Teaching Conference*, University of Minnesota, Minneapolis, 557-563.
Klaassen, C. (1995). *A problem-posing approach to teaching the topic of radioactivity*, Utrecht, Centrum for Science and Mathematics Education Press.
Kuhn, T. (1977). *The essential tension: Tradition and innovation on scientific research*, Chicago, University of Chicago Press.
Lakatos, I. (1978). 'Falsification and the methodology of scientific research programmes', in J. Worrall & G. Currie (eds.), *The methodology of scientific research programmes*. Vol I. Cambridge, Cambridge University Press., 8-93.
Langley, P., Simon, H, Bradshaw, G. & Zytkow, J.M. (1987). *Scientific discovery*, Cambridge, MA, MIT.
Popper, K.R. (1972). *Objective knowledge*, Oxford, Claredon Press.
Schwab, J.J. (1964). 'The teaching of science as enquiry', in *The Teaching of Science*, Cambridge, MA, Harvard University Press.

KAREN R. DAWKINS AND ALLAN A. GLATTHORN

9. USING HISTORICAL CASE STUDIES IN BIOLOGY TO EXPLORE THE NATURE OF SCIENCE: A PROFESSIONAL DEVELOPMENT PROGRAM FOR HIGH SCHOOL TEACHERS

This chapter provides a discussion of a strategy to meet North Carolina's new nature of science objectives. In an attempt to bridge the gap between the state mandate and teachers' resources, an inservice pilot project was designed involving eight biology teachers from three high schools in a predominantly rural county. For purposes of evaluation, eight biology teachers from different high schools in the same district served as a control group in assessing the impact of the program on teachers' understandings of the nature of science.

DESCRIPTION OF THE PLAN

Key Considerations

Two major considerations influenced the design and implementation of the project, including knowledge of the general criteria for effective professional development programs (Fullan & Steigelbauer, 1991; Sparks & Loucks-Horsley, 1990) and appropriate choices for instructional strategies and content. Decisions based on exemplary practices in professional development included recognition of teachers' time limitations in scheduling sessions; provision of opportunities for teachers to develop instructional materials for use in their classrooms using available resources; providing models of appropriate teaching strategies; and personal encouragement through the instructor's classroom visits. The decisions regarding the content and delivery focused on connecting selected issues related to the nature of science with biology content topics through historical cases, a strategy that has been recommended for use with students at both the high school and college level (Arons, 1988; Bentley & Garrison, 1991; Wandersee, 1985).

Project Design and Scheduling Considerations

Beginning with the considerations mentioned above, a detailed plan was developed to focus on a narrow range of issues, conforming to the time restrictions necessarily imposed on an after-school project. The nature of science program consisted of eight two-hour sessions scheduled over a four-month period during the spring semester. Although the instructor determined a starting date and ending date, a timetable for the other sessions was determined in consultation with the participants at the first meeting. The revised schedule included a four-week interval between Session 7 and Session 8 to permit classroom visits by the instructor during implementation efforts. Each session began half an hour after school closed so that participants could travel from outlying schools to the workshop site.

Selection of Topics

The scope of the content focused on the most critical aspects of the nature of science for these particular teachers. Three criteria guided the topic selection task: issues naturally embedded in historical cases in biology; nature of science misconceptions targeted in the literature; and nature of science topics included in North Carolina's statewide curriculum. Historical cases in biology logically lent themselves to considerations of interactions between scientists (both antagonistic and collegial), the progression of scientific thinking related to particular theories, and the influence of personal and community values on scientists' thinking. The literature on misconceptions suggested the inclusion of topics such as the objectivity of scientists (Duschl, 1994; Longino, 1983; Zoller, Donn, Wild, & Beckett, 1991), the nature of scientific methodologies (Gallagher, 1991), the tentativeness of scientific knowledge (Horner and Rubba, 1978), and distinctions between facts, hypotheses, theories, and laws (Horner & Rubba, 1979; Lorsbach, 1992). North Carolina's Standard Course of Study for Science (1994) addressed five topics within its nature of science competency goal: the public nature of science, the historic nature of science, the replicability of scientific work, the tentativeness of scientific knowledge, and the probabilistic nature of scientific data. Except for the subject of probability, these components reflected issues appearing in historical cases and in the misconceptions literature.

A synthesis of the information resulted in the selection of the following topics: scientific methodologies; tentativeness of scientific knowledge, including the public nature of science (the role of the scientific community in confirmation/rejection of ideas) and the historical development of science; and the nature and role of theories, including theories as inventions, the role of theories in limiting objectivity of scientists, and distinctions between facts, hypotheses, theories, and laws.

Since it takes historians and philosophers of science decades of study to derive generalizations about the nature of science from the historical evidence, it is unrealistic to expect teachers or students to develop profound understandings from a necessarily limited exposure to the world of science. For that reason, the project included readings and discussions related to the selected issues before the teachers began their research into historical cases in biology. In essence, the historical studies in biology provided evidence to confirm or refute generalizations from the current literature on the history and philosophy of science, emphasizing that the views on the nature of science are themselves tentative. Ideas about the nature of science have changed through history, and even current philosophers and historians of science are not necessarily in agreement on every issue. Resources used by the instructor to prepare discussion topics are listed in Appendix A. In general, the books were used as a source for the preparation of a background paper by the instructor for the teachers, summarizing major ideas from the philosophy of science. The articles were copied and distributed to the teachers. Assigned as out of class readings, the background paper and the articles provided a basis for discussion as well as resources for the teachers in preparation of their instructional units.

Model lesson

Addressing the nature of science through historical cases was a new approach for the teachers; therefore, the instructor presented a model lesson to illustrate one way to connect biology content with nature of science issues. With cell theory as the biology content focus, the lesson used the strategy of a dramatization with a narrator and characters from history whose work contributed to our modern ideas about cells as the structural and functional basis of organisms. In their assigned roles, the teachers read from a script written by the instructor (see Appendix B) which recounted major historical events from the perspective of the scientists. The narrator's script introduced the characters and provided connections between the historical vignettes and ideas about the nature of science embedded implicitly in the accounts.

Developing instructional units

An important component of the project design was the development of instructional units by the teachers. Applying Little's (1993) principle of offering meaningful intellectual engagement with ideas, with materials, and with colleagues, this task served two key purposes--to reinforce the teachers' learning and to increase the likelihood that ideas from the workshop would move to the classrooms. The group of eight teachers formed two groups of four, each of which focused on a single biology content topic. Although the teachers were free to choose any topic in biology (such as cells, genetics, ecology), in reality they were restricted by their curriculum and by the resource books and articles available to them. Textbooks provided little or no relevant information about historical cases or about the nature of science. Although it was originally thought that each group could develop two units, the process of selecting appropriate topics was very time-consuming; thus, the research and development process for one unit consumed all the time available. One group chose "evolution theories" and the other group selected "circulatory/respiratory systems."

During the research and development process, the teachers identified key players (both early philosopher-scientists and modern experimental scientists), noting their methodologies, their findings, their interactions with each other or with the written records of other scientists, the influence of society's reaction, and the influence of the prevailing scientific paradigms. Using nature of science issues as a framework, participants connected the historical accounts as evidence to support or refute contemporary views of the nature of science. Once the teachers determined which issues could appropriately be addressed in the context of history, they developed instructional strategies to involve their students in making connections. As an example, one teacher developed a subunit focusing on the nature of science to incorporate into a larger traditional unit on circulatory and respiratory systems in animals. She modified the research and development model used in the professional development project, identifying historical cases and library resources in advance and assigning student groups to research one case each. A summary of the teacher's unit is included as Appendix C.

Resource materials

Since teachers were unlikely to have appropriate resources related to history and philosophy of science or to historical cases in biology, the instructor provided copies of selected readings and a collection of books from the university library to the teachers for their historical case study research. The only helpful resources found in the high school libraries were histories of medicine which addressed studies in human anatomy and physiology. The resources used by the teachers in the development of instructional materials are listed in Appendix A.

COURSE SYLLABUS

The sessions included in the syllabus occurred at two to three week intervals.

Session 1: As an introduction, the instructor briefly presented excerpts from the National Science Education Standards (1996) and North Carolina's Standard Course of Study in Science (1994) to show the presence of nature of science goals in both national and state reform documents. Teachers informally assessed their personal ideas about the nature of science by using Cobern's card exchange game. This strategy is available in an enhanced version in this volume or as published earlier (Cobern, 1991). The instructor contrasted traditional views of science with contemporary views, providing teachers with a written background summary, articles focused on misconceptions, and a review of the literature to be read before the second session. The assigned readings are listed in Appendix A.

Session 2: The second session consisted of three elements: a conversation about misconceptions, focusing on the assigned readings from Session 1; introduction to the use of historical cases to build understandings about the nature of science, using a model lesson to illustrate; and a discussion about teacher logs. In reference to logs, the instructor gave the teachers stenographer pads, asking them to begin making brief entries each time they addressed a nature of science topic in their classes (including the date, the biology context, the nature of science issue, and other details which they chose to include) and to turn in their logs at the last session of the project.

Sessions 3 and 4: Working in groups of four, the teachers chose a biology topic and traced its historical development, using the resources in history of biology listed in Appendix A.

Sessions 5 and 6: The teachers completed their research and developed instructional plans appropriate for implementation in their biology classes. Although the research element was a collaborative effort, teachers chose to develop individual units, sharing some elements with each other but creating unique products for their own use.

Session 7: The teachers presented their units to the group for critique, discussed their plans for implementation, and scheduled classroom visits by the instructor to be completed before the final session.

Session 8: Focusing primarily on evaluation of the project, the final meeting included brief reports by the teachers on their implementation efforts, a focus group interview conducted by the instructor, and the administration of a questionnaire addressing teacher understandings of the nature of science. The instructor collected the teachers' logs, which documented their formal and informal attention to nature of science topics in their classrooms.

MEASURES OF EFFECTIVENESS

We gauged the effectiveness of the project on improving teachers' understandings of the nature of science and on supporting their incorporation of these understandings into their planning and their teaching. Methods of evaluation included a focus group interview, analysis of unit plans, analysis of teacher logs, classroom observations, and a questionnaire administered to both the participants and a control group composed of biology teachers from different high schools in the same district. The data from the questionnaires indicated that teachers participating in the project increased their understandings of the issues and provided stronger rationales for their views than did the control group teachers. Furthermore, as indicated by the log entries, participating teachers incorporated nature of science topics not only in the units prepared during the course of the project but also into other biology units presented during the semester.

East Carolina University, Greenville, North Carolina, USA

REFERENCES

Arons, A.B.(1988). 'Historical and philosophical perspectives attainable in introductory physics courses', *Educational Philosophy and Theory, 20*, 13-23.
Bentley, M.L., & Garrison, J. W. (1991). 'The role of philosophy of science in science teacher education', *Journal of Science Teacher Education, 2*, 67-71.
Cobern, W. W. (1991). 'Introducing teachers to the philosophy of science: The card exchange', *Journal of Science Teacher Education, 2*, 45-47.
Duschl, R.A. (1994). 'Research on the history and philosophy of science', in D.L. Gabel (ed.), *Handbook of research on science teaching and learning* (pp. 443-465), New York, Macmillan.
Fullan, M.G., & Steigelbauer, S. (1991). *The new meaning of educational change*, New York, Teachers College Press.
Gallagher, J. J. (1991). 'Prospective and practicing secondary school science teachers' knowledge and beliefs about the philosophy of science', *Science Education, 75*, 121-134.
Horner, J., & Rubba, P. (1978). 'The myth of absolute truth', *The Science Teacher, 45*, 29-30.
Horner, J., & Rubba, P. (1979). 'The laws are mature theories fable', *The Science Teacher, 45*, 31.
Little, J. W. (1993). 'Teachers' professional development in a climate of educational reform, *Educational Evaluation and Policy Analysis, 15*, 129-151.
Longino, H. (1983). 'Beyond "bad science": Skeptical reflections on the value-freedom of scientific inquiry', *Science, Technology, and Human Values, 8*, 7-17.
Lorsbach, A. (1992, March). *An interpretive study of prospective teachers' beliefs about the nature of science*, paper presented at the meeting of the National Association for Research in Science Teaching, Boston, MA.
National Research Council (1996). *National science education standards*, Washington, D.C., National Academy Press.
North Carolina Department of Public Instruction (1994). *Standard course of study for science*, Raleigh..
Sparks, D., & Loucks-Horsley, S.(1990). 'Models of staff development', in W.R. Houston (ed.), *Handbook of research on teacher education* (pp. 234-250), New York, Macmillan.
Wandersee, J. H. (1985). 'Can the history of science help science educators anticipate students' misconceptions?', *Journal of Research in Science Teaching, 23*, 581-597.
Zoller, U., Donn, S., Wild, R., & Beckett, P. (1991). 'Students' versus their teachers' beliefs and positions on science/technology/society-oriented issues', *International Journal of Science Education, 13*, 25-36.

APPENDIX A:
INSTRUCTIONAL RESOURCES

Resources Used by Instructor to Prepare Background Paper on Philosophy of Science

Brown, H.I. (1977). *Perception, theory, and commitment*, Chicago, IL, Precedent Publishing Inc.
Chalmers, A.F. (1982). *What is this thing called science?* (2nd ed.), St. Lucia, Queensland, University of Queensland Press.
Kuhn, T.S. (1970). *The structure of scientific revolutions* (2nd ed.), Chicago, IL, University of Chicago Press.
Matthews, M. R. (1994). *Science teaching: The role of history and philosophy of science*, New York, Routledge.

Articles Provided to Teachers as Reading Assignments

Horner, J., & Rubba, P. (1978). 'The myth of absolute truth', *The Science Teacher, 45*, 29-30.
Horner, J., & Rubba, P. (1979). 'The laws are mature theories fable', *The Science Teacher, 45*, 31.
Lederman, N.G. (1992). 'Students' and teachers' conceptions of the nature of science: A review of the research, *Journal of Research in Science Teaching, 29*, 331-359.
Storey, R. D., & Carter, J. (1992). 'Why THE scientific method?', *The Science Teacher, 59*, 18-21.

Resources Used by Teachers to Explore Historical Cases in Biology

Bender, G.A. & Thom, R.A. (1966). *Great moments in medicine*, Detroit, MI, Northwood Institute Press.
Buck, A.H. (1978). *The dawn of modern medicine*, New York, AMS Press.
Canguilhem, G. (1988). *Ideology and rationality in the history of the life sciences*, Cambridge, MA, MIT Press.
Gardner, E. J. (1972). *History of biology*, Minneapolis, Burgess.
Lanham, U. (1968). *Origins of modern biology*, New York, Columbia University Press.
Magner, L. N. (1994). *A history of the life sciences*, New York, M. Dekker.
Nitecki, M. H., & Nitecki, D.V. (1992). *History and evolution*, Albany, NY SUNY Press.
Oliver, R. P. (1963). *History and biology*, Oxnard, CA, Griff Press.
Rogers, F.B. (1962). *A syllabus of medical history*, Boston, MA, Little, Brown.
Shippen, K. B. (1957). *Men of medicine*, New York, Viking.
Silverberg, R. (1967). *The dawn of medicine*, New York, Putnam.
Singer, C. J. (1950). *A history of biology*, New York, H. Schuman.
Singer, C. J. (1959). *A history of biology to about the year 1900*, New York, Abelard-Schuman.
Walker, K. M. (1955). *The story of medicine*, New York, Oxford University Press.
Wilder, A. (1901). *History of medicine*, New Sharon, ME, New England Eclectic Publishing.

APPENDIX B:
THE CELL THEORY AS A MODEL LESSON

Objective: Through historical cases, students will infer the following about the nature of science:

1) Science is public: the scientific community plays an essential role in the confirmation and rejection of theories, so publication of research and presentations at scientific meetings is the means for judging reliability and value of research findings

2) Science shows historical development, but is not necessarily cumulative in that some ideas are "erased" as better replacements occur

3) Theories play a role in providing the context for a scientist's work--scientists do not start with a theory-free mind

4) Scientists are not individually objective because they all come with a certain mental framework which means a degree of subjectivity, but the scientific community helps to keep scientific knowledge from being totally at the mercy of individual biases

5) All scientific knowledge remains tentative, subject to revision as new findings emerge--it is never authoritative

6) All serious scientific findings are valuable, even those which may eventually be replaced, because they may stimulate productive research in a field of study

7) Although there are occasional revolutionary breakthroughs such as the early cell theory of Schleiden, Schwann, and Virchow, the work of most scientists is focused on supporting, enlarging, or refining theories that already exist

8) If there are two competing theories in a particular area, the better of the two is, in the final analysis, a matter of consensus among scientists arising from critical scrutiny

9) Replication of scientific studies is an important part of the role of the scientific community -- studies conducted by only one scientist are generally viewed with skepticism until validated by replication

10) Some people think that theories have not been proved; therefore they are not very useful. In reality, it is not important to "prove" a theory, just to support it with evidence, in which case it plays a fundamental role in the work of most scientists

11) Theories are broad explanations which are made up of related hypotheses that have gained substantial support.

A Dramatization of the Cell Theory

Narrator: Occasionally new theories or explanations of natural phenomena create a scientific revolution, dispelling all previous attempts to provide reasons for what we observe in the world. That was the case with the explanation of structure and function of organisms based on cell theory which developed in the nineteenth century. Previous thinking explained the functions of plants and animals on all kinds of "vital forces" such as humors or vapors, none of which were supported by experimental evidence.

Remember, a theory is an explanation of some natural phenomenon. So cell theory is an explanation of the structure and function of living things in terms of the individual units we have come to call cells. This idea, which we accept without question, was not always so obvious--as we shall see from our version of "The History of Cell Theory."

The first character I will introduce to you is Marie Francois Xavier Bichat, a French doctor and anatomist who performed at least 600 autopsies per year during his short career, which ended when he died at the age of 31. (Probably overworked.)

Please meet Dr. Bichat:

Dr. Bichat: I must say that I owe a debt to one of my predecessors in France, Phillipe Pinel, who contended that diseases must be understood by tracing them back to organs in the body, which he thought to be composed of different elements. I decided, however, that there must be sub-components smaller than we had previously imagined--and I found those sub-components to be tissues. In fact, through my dissections of cadavers, I discovered 21 kinds of tissues which I categorized into systems such as respiratory, nervous, and digestive.

I cannot for the life of me understand why some scientists insist on studying human anatomy through those confounded microscopes. The way I see it, "When one gazes into the darkness, everyone sees in his own way and is affected accordingly."

Narrator: Did you hear what Bichat just said? He implied that scientists may not be entirely objective in their observations. Some may see differently from others. Interesting point of view!

Another historical note of interest: Even after cell theory had been well grounded for plants and animals, some reactionary French physicians still considered tissue as the natural unit of structure and function for organisms. They were highly influenced by Bichat and tended to disregard new evidence based on microscope studies, even after microscopes had been improved to the point of providing almost indisputable observations of cells. It just shows that even the best supported theories may be accepted very reluctantly by the "establishment."

It could be argued that Bichat had some justification for his disdain of the microscope. In fact, the sight of "cells" (actually plant cell walls) by Robert Hooke much earlier in the 17th century still influenced many in the 19th century and misled them into thinking of cells as walls surrounding empty spaces. Even though Hooke and his colleague Nehemiah Grew also described the appearance of fluid inside the cells of younger plant growth, there did not seem to be any comparable structures in bodies of animals. Listen to our old microscope maker friend, Leeuwenhoek.

Leeuwenhoek: I made the most beautiful microscopes in the world. I even saw corpuscles in animal blood. They certainly didn't look anything like those cells Mr. Hooke saw in slices of cork! I'm afraid I did not make the connection between plant cells and animal cells. In fact, I wasn't really interested in all that scientific theorizing. I just wanted to create beautiful lenses and look at all kinds of strange things through them. Nobody ever said I was an Einstein!

Narrator: So--even though the primitive microscopes provided the technology for seeing cells in both plants and animals, the scientific players had conceptual limits which did not make a connection between the two. Perhaps there must be a historical "readiness" to accept new ideas in a particular field of study.

You might be surprised to know that one of the contributors to the development of cell theory is a fellow most noted for his work in physics--Robert Brown (the Brownian effect).

Robert Brown: Actually, I was interested in studying pollen grains in the context of reproduction in orchid plants. As I viewed them under the microscope, I saw tiny little specks that seemed to be in constant motion. I thought maybe the specks might also be in nonliving materials such as coal dust, powdered glass, and minerals. I did not observe the phenomena in these materials.

In 1832, I published a scientific paper on reproduction in orchids, and described my observation of a "circular areola" in each cell, a dark area that appeared to be more opaque than the cell membrane. I saw these spots in the cells of other plants as well.

Narrator: Note that Brown published his findings--otherwise, his observation of what we now know to be cell nuclei would not have been used by other scientists and added to the collection of knowledge that was beginning to form the cell theory.

Other minor, but essential, players in the evolution of cell theory include Lorenz Oken, Johannes Evangelista Purkinje, and Johannes Peter Muller. I'll let them describe their work.

Lorenz Oken: I was a German naturalist heavily influenced by the Romanticists including, the poet, Johann Goethe. My speculations on the nature of living things was more philosophical than scientific. I did u' e microscopic observations, but I used a lot of creativity to come up with the idea of a jelly-like materials from which spherical vesicles arose to produce the simplest living things, which I called "infusoria." All animals and plants are just colonies of infusoria which have given up their independence to subordinate their life to the organism as a whole. Isn't that a beautiful, poetic idea!

Narrator: Oken's ideas lacked rigorous experimental verification, but perhaps his imaginative speculations stimulated more serious scientists to undertake the necessary investigations.

Here is Purkinje.

Johannes Evangelista Purkinje: My area of interest was physiology, how organisms function, as well as histology, the study of tissues. Unfortunately, my university did not always support my progressive research and provide adequate space and equipment, I had to do a lot of my work at home. That's sometimes what happens when you have a bunch of old fuddy duddies in charge of the budget. I finally persuaded them to buy one of the new achromatic microscopes, the compound ones without the distortions they had at first. That really helped me with my research. Although I published many papers in the areas of physiology and histology, those most influential in the development of cell theory were my descriptions of nerve cells, their nuclei and dendrites.

Narrator: Remember that observations of animal cells came along later than plants--so Purkinje's work with animal tissues was particularly valuable.

Let me introduce Johannes Peter Muller.

Johannes Peter Muller: I am probably best known in the realm of cell theory as the professor of Rudolf Virchow and Theodor Schwann, whom you will meet momentarily. I was a professor in human and comparative anatomy, embryology, physiology, and pathological anatomy at the University of Berlin. I was Rector of the University when the revolution of 1848 broke out, and I had to deal with violent disturbances among the students and staff. The mental and physical stress ruined my health. It gives me great pleasure to tell you, however, that when I died three people were appointed to replace me.

Narrator: Now what you have all been waiting for--the main event--the big dogs of cell theory--Matthias Jacob Schleiden and Theodor Schwann. I will let them tell their own stories. Here's Herr Schleiden.

Matthias Jacob Schleiden: Some people say I was a mental and emotional cripple. I studied law at Heidelberg and practiced as a barrister in Hamburg, but I was such a failure that I tried to commit suicide. I couldn't even do that very well. I aimed at my forehead and just managed to skim my scalp enough to cause a lot of bleeding and pain but no permanent damage. How embarrassing!. After I recovered, I decided to study science. I enrolled in botany and medicine at the Universities of Gottingen and Berlin and earned doctorates in both medicine and philosophy. SO--you may say I'm crazy, but I sure am not stupid. I worked as a professor of botany for a while, but I got bored and decided to wander around Germany. It was good for my frazzled nerves.

A fortuitous event was my meeting with Herr Schwann in Berlin--which would provide a very productive collaboration. You'll hear about that from him a little later.

I published a paper in 1838, referring to the work by Robert Brown on the nucleus and elaborated on its special relationship to cell development. From my observations of plants, I came to understand that all plants of any complexity were aggregates of "fully individualized, independent, separate beings, namely the cells themselves." Furthermore, I conjectured that all aspects of plant physiology were fundamentally manifestations of the vital activity of cells. In other words, the cell runs the whole plant.

As far as how cells are formed, I emphasized the theory of "free-cell formation" which likened cell growth to the simple physical process of crystallization--a new cell crystallized inside an old cell. I realize that the idea was soon to be dropped like a hot potato, but give me some credit. I certainly never thought that cells could arise from spontaneous generation--from nonliving things. Some of my predecessors did.

Narrator: Oops. Did you hear that version of cell reproduction? Should we ignore everything that Schleiden said just because he messed up on that one? Heavens no. Over time, the helpful ideas will last and the less useful ones will be replaced. That is a very natural part of historical development of theories. Ideas that seem really goofy to us today might have been very reasonable at the time and might have been valuable in paving the way to more useful knowledge.

Now you will hear from Theodor Schwann, who was quite a contrast with the abrasive Schleiden. In fact, Schwann was viewed by his peers as a timid, introspective, and excessively pious person. But here--let him tell you himself.

Theodor Schwann: I studied medicine at three universities, graduating from the University of Berlin in 1834, where I studied with Professor Muller, whom you met earlier. Dr. Muller provided much encouragement for me and great opportunities to make new observations in the areas of histology, physiology, and microbiology--not to mention the work that made me famous: my contributions to cell theory. I know that violent quarrels are a natural part of the academic tradition, but I just could not take the criticism which one always had to endure within the German scientific establishment. I essentially retreated into scientific exile.

I had observed nuclei in the process of making slides of animal tissue for Professor Muller as a student, but I really did not think much about the meaning until I met Herr Schleiden in Berlin on that momentous day in 1838. We were having lunch together one day, and he pointed out to me the important role the nucleus plays in the development of plant cells. I at once recalled having seen a similar organ in the cells of the notochord, and in the same instant I grasped the extreme importance that my discovery would have if I succeeded in showing that this nucleus plays the same role in the cells of animals as does the nucleus of plants in the development of plant cells.

My job was difficult because animal cells are so diverse, and since they are generally quite transparent, they are hard to see, even with the new and improved compound microscopes. One thing was obvious to me, however. All the cells had a nucleus. I published my findings and my hypotheses in a work that I must modestly admit became a classic: Microscopical Researches. I observed that, like plant tissues, animal tissues contained cells, cell membranes, cell contents, nuclei, and nucleoli. I set up an analogy between plant and animal tissue. But it was quite an ambitious task to provide convincing evidence that most or all animals tissues developed from cells. I did, however, put forth a theory that there is universal principle of development for all organisms, both plants and animals and this principle is the formation of cells.

Narrator: A sad note in Schwann's career which I will tell since it would be painful for him to recount. It concerns his description of yeast cells. From what we now know, he was quite correct in describing fermentation granules as fungi and fungal cells similar to all other cells. His idea was accepted only after a bitter controversy between a German chemist, Liebig, and the famous French scientist, Louis Pasteur. Liebig published an anonymous cartoon satire of Schwann's theory which really embarrassed the timid scientist. He was destroyed psychologically and left the field to be defended and conquered by the more self-confidant Pasteur, who continued research into fermentation and established that living organisms produced the carbon dioxide and alcohol products and were not themselves produced by the chemicals as some opponents thought. Ultimately, it was the scientific community which decided which theory was correct--Pasteur's or Liebig's, and they chose Pasteur's which was essentially the same conclusion that Schwann had made.

Back to Schleiden and Schwann: As important as their work was in the development of cell theory, there were some corrections that were yet to come. Specifically, two ideas were replaced: free-cell formation and the notion of the cytoblastema. They thought that cells arose from a mucous-like material called the cytoblastema and went on to change into different kinds of cells dictated by "the blind laws of necessity." (Remember: nobody knew about DNA back then.)

A number of scientists added to an identification of protoplasm and to division of the nucleus-- but I am going to hurry along to introduce a man who put two and two together to get four: Rudolf Ludwig Carl Virchow.

Herr Virchow.

Rudolf Virchow: I am very grateful for the influence of my professor, Dr. Muller. I worked as a pathologist with a special interest in infectious diseases and ended up at the University of Berlin where I became involved in politics as well as academics. I was elected to the City Council and to the Prussian Legislature. Although I respected my colleagues, Schleiden and Schwann, I just could not accept their idea about free cell formation from the cytoblastema. I replicated many of their studies, using the best microscopic evidence I could attain, but my observations led me to a different conclusion from theirs. I came to understand that "there is no life but through direct succession." I was certain enough of my hypothesis that I published it in 1855--stating that all cells come from pre-existing cells. I was very adamant to convince other scientists of the importance of using the microscope in medicine and of recognizing the cell as the locus of disease. In fact, I saw cells as "the last constant link in the great chain of mutually subordinated formations that form tissues, organs, systems, the individual."

Narrator: Since Virchow's publications in the mid-1800s, many refinements of cell theory have been made--new observations of cells, and parts of cells, discoveries of entities only observable after the development of the electron microscope. But after about 1875, the scattered diverse observations of cell division and nuclear inclusions were finally organized and cell theory came to include two more generalizations: cells in animals and plants are formed by equal division of existing cells and division of the nucleus precedes division of the cell. At this point, the cell theory was a strong unifying idea in our understanding of the structure and function of all living organisms. But, as with all theories, it has continued to evolve as we constantly learn new things about cells. Just think of the research into DNA science, all of which adds bits and pieces to the cell theory. No scientist in our modern day world would attempt to start from scratch in understanding the function of a plant or an animal. They would start with the preconceived notion that everything they observe in living things originates with activity in cells. Every serious biologist working today does research in the context of cell theory--it guides their thinking and helps direct the focus of their research.

One other thing I would point out as we think back over our historical analysis. Probably the most important thing we can learn about the nature of science from looking at its history: There is no idea which is absolutely proved forever and ever. All scientific findings are considered tentative, even those on which we depend and in which we have great confidence. It is part of the mind-set of scientists to leave open the possibility for change as new evidence comes to light.

Questions to Guide Discussion. What does the history of cell theory illustrate? :

1. the importance of making scientific findings public through publications and through presentations

2. the idea that scientific knowledge shows historical development, though it is not necessarily cumulative

3. that theories play a role in guiding the work of a scientist

4. that scientists are not necessarily objective when they begin their research

5. the tentativeness of scientific knowledge

6. that all scientific findings have value, even those which may eventually be replaced

7. much scientific work is focused not on revolutionary new theories but on enlarging, supporting, or refining existing theories

8. that a consensus among the scientific community is the criterion for judging how good a theory is

9. the importance of replication of studies by other scientists

10. that theories don't have to be "proved" to be useful

11. that theories are broad explanations which are made up of related hypotheses that have gained substantial support?

APPENDIX C
SAMPLE UNIT PLAN PREPARED BY TEACHER:
CIRCULATORY AND RESPIRATORY SYSTEMS

Objectives

Biology Content and Nature of Science Objectives: To

- Conduct activities relative to major anatomical features and systems of organisms.
- Illustrate correlation between structure and function.
- Document with examples how scientific knowledge is tentative.
- Discuss how scientific findings of the past provide a basis for modern science.

Daily Lesson Plans

Day 1: Describe the assignment and state that science is not just about test tubes, lab coats, and stuffy old facts that never change. On the contrary, this discipline is quite a human endeavor and a dynamic endeavor. For too long, we have neglected this human, historical component. Therefore, you will help unearth the long forgotten roots of science, learning who the early scientists were, what they did, how they did it, and how they learned from each other. Specifically, we will look at scientists' work in the circulatory and respiratory systems of animals. Through our research and presentations, we will clearly see how science is an ever-changing realm of knowledge that both influences and is influenced by society.

Each group will pick one scientist and will conduct library research on that scientist. After gathering information, a narrative script will be written. Each member of the group must have a speaking part. One member will be the scientist and the other members will play roles such as mother, spouse, news reporters, etc. Each group will make a class presentation in a creative, dramatic, and interesting manner. Group members should dress in costume of the scientists' era. Answer as many of the following questions as possible to ensure that your group has the information necessary for a good presentation.

1. When did the scientist live?
2. What was his nationality?
3. What was his contribution to scientific knowledge?
4. What contribution (if any) did he make to knowledge about the circulatory and respiratory systems?
5. Were there any ideas concerning circulation and respiration that he was able to refute?
6. Explain the idea of tentativeness in science.
7. What were some interesting circumstances that led to your scientist's discoveries?
8. Which part of his work was replication of studies done by other scientists? Is this important? Why?
9. Identify any facts that this scientist learned and communicated.
10. Identify any hypotheses that this scientist put forth.
11. Identify any theories that this scientist supported with evidence. Do scientists prove theories?
12. Did this scientist work to develop a new theory, or was he trying to support or enlarge a pre-existing theory? (Discuss the pre-existing theory if that is the case.)
13. How did the public receive the new information?
14. Should scientific knowledge be shared with the public? Why or why not?
15. What part of this scientist's work has been replaced by a more modern idea?

Students will pick from this list of people who researched respiratory and/or circulatory systems and can be found in the school library. An approximate time is included with each scientist.

3000 B.C.E.	Sekhetenach	1516	Columbo
2980 B.C.E.	Imhotep	1578	William Harvey
700 B.C.E.	Asclepios	1628	Malpighi
500 B.C.E.	Pythagoras	1665	Leeuwenhoek
460 B.C.E.	Hippocrates	1722	Leopold Auenbugger
129	Galen	1781	Lannec
1514	Versalius	1900	Landsteiner

The remainder of the class period will be spent in the library using reference books previously pulled from the collection.

Days 2-4: Student groups will work in the library, collecting research and writing their scripts.

Day 5: Each group will give a five minute presentation. The audience members will take notes.

Day 6: Each student will choose a scientist with whom to compare and contrast their scientist in an essay.

NAHUM KIPNIS

10. A HISTORY OF SCIENCE APPROACH TO THE NATURE OF SCIENCE: LEARNING SCIENCE BY REDISCOVERING IT

My interest in teaching the nature of science came from attempts to integrate history into physics courses for teachers at the Bakken Library and Museum beginning in 1985. My background in the history of science and in teaching physics appeared ideal for helping teachers transform the science of facts and equations they typically teach into science as a human activity. Michael Matthews labels this a distinction between "technical" and "liberal" science. The trouble was that many course participants had degrees in biology or chemistry, and were unfamiliar even with "technical physics." Having no foundation to build on, I choose to teach both, and the question was how to do it.

By that time, I had already been convinced that physics could be taught to all students by shifting the emphasis from the memorization of facts to developing skills of thinking, reasoning, and systematic purposeful work by students. I decided to try investigative laboratory experiments as one of the main vehicles for achieving this purpose. Teachers conducted experiments in groups and individually. When they brought the new labs to their schools, they appealed to the majority of students and raised their interest in learning science. The peculiar feature of these laboratory activities was that many of them recreated historical experiments.

The idea of reproducing historical experiments in the classroom came from Devons and Hartmann (1970). I could have achieved my goal with other experiments as well, but I wanted to use history as much as possible, and the historical experiments did have an advantage over the "real-life investigations" practiced by some teachers, such as finding the cause of the clogged classroom sink or determining which paper towel is most absorbent. First, with the historical approach, the result is always known to the instructor, which is not the case with the sink and towels. Second, a historical experiment can be chosen so as to help teach a narrow scientific topic, while most real-life phenomena are too complex and complicated for students' study. Third, for the first two reasons, the students' chances to succeed are higher with the historical experiments than with the other ones. Finally, a success in repeating a historical scientific discovery may boost students' self-confidence much more than in fixing the plumbing. It is not that technological problems are unusable; if carefully chosen, they are. However, it is easier to learn the necessary investigative skills in historical scientific experiments and then use them to tackle the frequently more complicated problems in technology.

Soon, it became clear that with all its advantages, the historical approach brings to the forefront very difficult questions of science usually suppressed in textbooks, such as the changeability of theories, the meaning of experimental support and its application in selecting one of several competing theories, and others. Some teachers came up with such questions on their own, while others recalled similar challenges from their brighter students, who having found out from popular science books and television programs that theories come and go concluded that scientific knowledge is of no real value. Teachers realized a necessity in answering those questions, but they did not know how to do so.

Thus, I realized that in the environment that emphasizes thinking and usage of history, teachers had no choice but to address basic issues of the nature of science, and they must be taught how to do it. Teachers certainly could have benefitted from proper philosophy of science courses (Matthews, 1990; McComas, 1995), but I expected them to obtain such training elsewhere. On my part, I tried a more practical approach that could be complementary to theoretical courses: to bring up the issues of the nature of science from within the science, in particular when engaging teachers into "doing science." This chapter will give an example of such an activity and the way it addressed the questions related to the scientific method[1].

THE NATURE OF SCIENCE: A HISTORICAL APPROACH

The historical approach in science teaching has had several periods of popularity. In the late 1950s, James Conant suggested that all necessary knowledge of the nature of science could be gained from studying a few historical cases. The cases must come from the "old" science (usually, this means scientific ideas from the 17-19th centuries), for only the old science can allow a nonspecialist a sufficiently good grasp of the subject matter (Conant, 1959).

While Conant's cases were written for universities, several authors incorporated his idea into the secondary school curriculum, some explicitly and others implicitly. At one extreme, the nature of science was advocated as a separate instructional element, the understanding of which is emphasized more strongly than that of the science content involved (Klopfer, 1961, 1992). In Klopfer's scheme, student booklets, prepared for each case, contain excerpts from primary sources alternated with the author's narrative, and also questions directing students' attention to various aspects of science as described in the text. Blank spaces are provided for student to write answers to these questions. Some of the questions refer to various aspects of the nature of science (Klopfer, 1964). The idea of using original historical materials

---

[1]The term "scientific method" used here means the *common* features of research strategies applied by scientists in modern times that we want students to learn.

together with questions to answer was appealing, however, I found its practical realization unsatisfactory. While some historical experiments were included, their purpose was largely illustrative. Most important, I did not believe that "nature of science" can survive in the secondary school as a separate subject.

At the other extreme, history was used solely to aid in learning scientific concepts, and if students were expected to draw conclusions about science and scientists, they had to do this on their own, using historical examples incorporated in the text (Rutherford et al., 1964). There were also attempts to provide a middle ground between these two extremes by balancing the study of nature of science with that of scientific concepts, or using inquiry as a vehicle for learning science (Schwab, 1964). The general approach is appealing, but the role of a historical element in it is rather limited. In general, Schwab's approach appears sound, but the historical element in it plays only a casual role, if any. For instance, although Galileo's name is mentioned, students' experiments with pendulums have nothing to do with the historical ones.

After a period of neglect, historical approaches reappeared in the 1980s, with the emphasis on evaluating case studies, role playing and doing historical reading. Some issues of the nature of science come up when students discuss historical accounts in the class (Lochhead & Dufresne, 1989). Unfortunately, while the role-playing activities will be of some interest to all students, only those few who are actively involved as panelists or actors will get the full intellectual benefit from these activities, because only they had done all the reading. For the majority of students, a better way to integrate the historical element would be to recreate history in action rather than on stage. Some historical experiments have already found their way into teachers' education (Devons & Hartmann, 1970; Teichmann, 1986). My intention was to do it on a more comprehensive basis and to make history an organic component of science.

Learning the Nature of Science While Learning Science

Although teachers report that they like learning about the history of science, they fell that it is unlikely that they will pass much of that knowledge on their students, because they have no time for any "extra" units, such as those devoted specifically to the history and nature of science. For this reason, I subordinated the discussion of historical and philosophical issues of science to learning scientific concepts superimposing them so as to make them inseparable. The topics of units are the same as in regular science courses, such as "electrical conductors and nonconductors," and the goal is the same: to formulate the laws of phenomena. The difference is in the ways the unit is taught.

History and the nature of science are involved in three areas. The first concerns the way a new scientific concept is introduced. I have found that understanding of a concept improves if it is "rediscovered" with active participation on the part of the learner. Usually, it begins with the instructor's demonstration of a relevant historical experiment, for instance, the first experiment of Stephen Gray[2]. After a short discussion, students begin repeating and modifying this experiment following a certain plan (see Table I). The instructor stops students every 10-15 minutes to discuss the intermediate results. If some groups differ in their results, this particular experiment is repeated until a consensus is established. After the final conclusion is formed, the instructor informs students of Gray's subsequent procedures and results, and the participants can compare how closely they came to the scientist. Although technically the experiments are simple, it is not easy to draw a conclusion from them, because it involves introducing a new concept of conductivity (Kipnis, 1996). Thus, students get a lesson about a possibility of various interpretations of experimental results (including the erroneous ones) and a difficulty of making the best general conclusion (it was not Gray who introduced the general concept of "conductors and "nonconductors").

If pedagogically advisable, the chain of "rediscoveries" follows historical events. This is the case with static electricity, for instance, which led some teachers to replace their whole static electricity unit with the one developed at The Bakken (Bakken, 1995). However, occasionally, history defies what we call "common sense," as with the case of interference of water waves being discovered after interference of light. In such cases, I prefer to sequence instruction in ways that students will find easier to grasp. Later, we discuss that scientific developments are not always "logical" and may be very convoluted.

The second area concerns the type of experiment used. I emphasize investigative experiments which require students to imitate scientists rather than use artificial or contrived experiments. In another lesson from history we see that since much of physics remained a qualitative science up to the middle of the nineteenth century, an obsession with measurements in high-school physics appears unjustified. Thus, I began to emphasize qualitative experiments where students study a certain phenomenon to find a qualitative empirical law to describe it (Kipnis, 1995). The result is considered "true" if obtained by all groups. These experiments are preceded and followed with theoretical discussions in such a way as to "recreate" an introduction into science of a new physical concept (potential, for instance) or a law (such as the relationship between potential and capacitance) (Kipnis, 1992, 1996). The investigation is, to a certain extent, guided by the teacher. However, since students have some freedom of action, we can consider this imitation of the work of

---

[2] In 1731, the English physicist Stephen Gray observed, while rubbing a glass tube corked at the ends, that not only glass attracted a feather, but so did the cork. Gray supposed that electricity produced in glass by rubbing somehow moved to the cork. When he inserted a nail in the cork, electricity reached the nail too. Through his work with long "communicating lines," Gray discovered the concept of "conductors" and "non-conductors."

scientists sufficient to give students a taste of the problems scientists face. By making students "discoverers" we achieve two objectives. First, we remove some of the "mystery" from science by showing it to be a professional activity that requires certain skills, and that everyone with necessary skills and motivation can discover something new and useful. Involving history has additional advantages. Seeing that they are capable of repeating on their own certain important steps of famous scientists or deciding who was right in a scientific dispute (Lawrenz & Kipnis, 1990) gives students a tremendous boost to their self-confidence. Second, they will see how easy it is to choose an incorrect direction of research, how difficult it is to invent a good procedure, and how easy it is to reach a false conclusion. Students will realize that while not everyone can become Newton or Volta, everyone can discover something. This will teach them a proper appreciation of great discoveries of the past, on the one hand, and a habit of critically analyzing information about modern discoveries, on the other.

The third area deals with the strategy of experimenting. The one discussed here is distilled from scientific treatises of the past (Table I). The strategy consists of two stages: Preliminary and Main Parts. The Preliminary Part is the one with which teachers are least familiar. It comprises the origin of the problem and initial experiments, which do not follow any plan but aim at discovering a few plausible variables and a reliable experimental procedure. The problem with the widely spread method of doing open- ended experiments is that usually a teacher solicits students' suggestions about the variables to study after they observe a single demonstration. Having received a considerable number of ideas, the teacher asks each group to investigate a different variable, with the idea that all these partial results can be combined at the end into a general one. One problem with this approach is that bringing variables "out of the blue" is inefficient. In our case, students do a sufficient number of modifications of the original experiments until they can say which changes appear to affect the result, and select the corresponding parameters as variables. Also, only the most obvious two or three variables are selected for the further examination in the Main Part, which allows all groups to study all variables eliminating a possibility of false results due to poor statistics.

The Main Part is the planned stage of an investigation (Table I). Each variable is examined while keeping the rest constant, and this examination follows a certain plan identical for all variables. We start with preliminary experiments which result in a certain hypothesis about the way this variable affects the phenomenon. Although students may have some idea of this hypothesis from the Preliminary Part, this knowledge is not definite, for in some initial experiments several variables are changed at a time. Unlike those, the preliminary experiments are "clean," for everything but one variable remains constant. To advance a hypothesis even two experiments may be enough. But to test it more experiments are needed, and they must be different from the preliminary ones. If the test shows that the hypothesis is false, we do additional experiments and suggest another hypothesis. Finally, we summarize all conclusions about separate variables and formulate a law or a rule. While studying each subsequent variable we use the results about the variables previously examined. Thus, it makes sense to insist that all groups investigate the variables in the same order, at least, during the period when students learn how to investigate and need more assistance.

It must be noted that teachers cannot have too many full-scale investigations with 2-3 variables, similar to the one in the example, because they take from one to three hours and require a careful preparation on the teachers' part. For this reason, teachers supplement them with a greater number of shorter experiments (10-20 minutes each), which deal only with a single variable.

TABLE I
A strategy for investigation

I. PRELIMINARY PART

    1. Background (origin of the problem)
    2. Initial observations and experiments
    3. Formulating a problem
    4. Selecting variables
    5. Selecting a procedure

II. MAIN PART

    Variable 1
        a. Preliminary experiments
        b. Hypothesis
        c. Test
        d. Conclusion

    Variable 2
        a. Preliminary experiments
        b. Hypothesis
        c. Test
        d. Conclusion

    Additional Variables

    General Conclusion Formation

An Example of the Strategy in Action: Does Light Travel Along a Straight Line?

Some features of this technique are illustrated in the following example. In this context, "student" can refer to any learner, while "teacher" can be either a university instructor or a secondary school teacher who knows this strategy. The description of the experiment is a hypothetical student report which follows the plan illustrated in Table I, in which several possible answers imitate results from different groups. In an actual experimental report, a student can use the same format writing in the the blank space under relevant headings.

The law of rectilinear propagation of light is one of many that may be "rediscovered" during a unit on optics and vision. This example is based on the book for teachers *Rediscovering Optics* (Kipnis, 1992), and the following discussion includes excerpts from it. This study is recommended for secondary students (12-18 years of age), although its modified version is quite applicable for those in upper primary grades (9-12 years of age). If students have not had geometry, the teacher should familiarize them with the concept of similar figures using models and drawings.

The teacher begins the unit on light with the question: "what conditions are necessary to see a certain object?" Students provide a variety of answers, including: "a healthy eye," "a sufficient amount of light," "the requirement that the line of vision should not be blocked," etc. Subsequently these answers serve to introduce a variety of topics to study, beginning with the eye and the meaning of "vision." During this first lesson the teacher describes some early theories of vision using a combination of lecture and discussion. In particular, in the "extromission" theory, the eye sends out a "fire" which touches an object and receives information of its shape, size, and color (something similar to a radar). According to the "intromission" theory, something (a "mask" or an "image" of the object) detaches itself from its surface and moves toward the eye. When students start smiling at such "naive" ideas, it is useful to challenge them to refute any of these notions. Students will quickly realize that the matter is not as obvious as it may have appeared at first, and the teacher may use this case as an example of an important maxim of science that "obvious is something you have never thought about."

The teacher summarizes that ancient philosophers finally realized they could not resolve the problem of vision in full, and began focusing on some part of it that could be resolved. That was done by Euclid (ca. 300 BC) who ignored what agent is moving between the eye and the object, which direction, and how it interacts with the eye, and paying attention only to the trajectory of the agent's motion. That is how the concept of a *ray* was born. While Euclid was thinking of a *visual ray* coming from the eye, to Ibn al-Haytham (ca. 965-1039 AD) it was a *light ray* coming from the object. Whatever the case, they agreed that the trajectory must be a straight line. The teacher challenges students to prove this experimentally. As illustrated by the following interaction, students will probably come up with the following idea that many teachers use to study reflection and refraction.

> Mary. I would use three pins and a ruler. First, draw a straight line on paper using a ruler and place two pins at its ends. Second, move a third pin between the two until all three appear to coincide. Finally, check whether the third pin is on the same straight line with the others.
>
> Teacher. This sounds like an easy experiment. Let us do it following Mary's instructions. ... we'll repeat the experiment three times. What are the results?
>
> John. It's true, and all the pins make the same straight line.
>
> Teacher. Incidentally, do you take it for granted that a ruler's edge is straight?
>
> John. No, I can prove it. For instance, if I look along the edge of a ruler and see all its points on a straight line, the edge is straight.
>
> David. Wait a second! We've just proved experimentally that light moves rectilinearly on the basis of the coincidence of the line of vision with the line drawn with a ruler. Now, you are saying that the ruler's edge is straight because it coincides with the line of vision. This is circular reasoning!

Indeed, it appears that the proof is impossible. For this reason, Euclid simply *postulated* that a ray is a straight line. However, to al-Haytham that was not good enough, and he tried to find a physical demonstration.

To test whether the light of a flame spreads rectilinearly, Ibn al-Haytham employed a tube with a pinhole. To extend his proof to the light scattered in the atmosphere, he let light into a room through a one-foot hole into the outer wall and two identical holes in the inner wall. He found that a luminous spot C (Figure 1) on the floor appeared only when this spot and the holes B and A on both walls were on the same straight line as verified by a stretched thread.

Students may say that because the holes are big, points A and B have a large range of movement within each hole, which implies many different lines connecting them and producing different projections C on the floor. Then the teacher suggests another test based on comparing the shape of an object and of its shadow. The idea came from Aristotle (384-322 BC), who claimed to deduce from it that light rays cannot be straight lines. Students may look surprised at this conclusion, and the teacher suggests to investigate whether Aristotle was right. The following excerpt from *Rediscovering Optics* (Kipnis, 1992) describes a hypothetical student's investigation. The discussion is dramatized to present different possible views or results achieved by different groups.

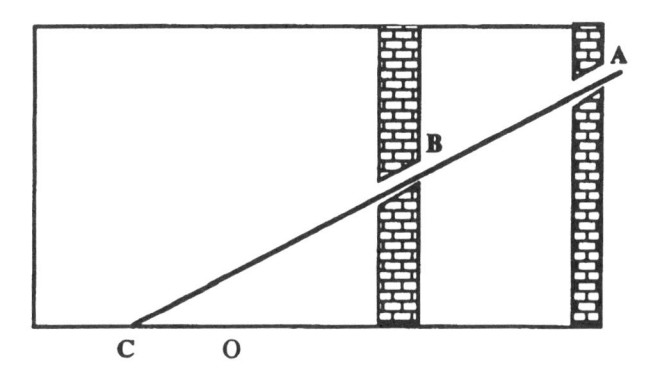

Figure 1. Alhazen's demonstration of the rectilinear propagation of light

As discussed with reference to Table I, the actual investigation consists of two elements, the preliminary part and the main part. Each of these elements has subsections. The lesson begins with a short overview of the previous lesson on vision and a short demonstration by which the teacher introduces the subject for an investigation connecting it with Aristotle's problem -- what should the shadow resemble: the object or the sun?

> Teacher. Here I have two holes cut in an index card, one circular and the other rectangular. I will shine light on them from a desk lamp, and you watch their images on a white screen. What do you see?
>
> John: Both images are round. But this is impossible!
>
> Teacher: What would you expect?
>
> John: A circle and a rectangle.
>
> Teacher: That was what Aristotle thought too. He asked:

Why is it that when the sun passes through quadrilaterals [a rectangular grid], as for instance in wickerwork, it does not produce a rectangularly-shaped figure, but one that is circular? Is it because the sun's rays fall in the form of a cone and the base of a cone is a circle, so that no matter what object they fall upon, the rays of the sun must appear circular? For if the rays were straight the figure formed by the sun would necessarily be bounded by straight lines. For when the rays fall straight onto a straight line they do produce a rectilinear figure (Aristotle, *Problems*, 334-5).

Here Aristotle is referring to the geometrical theorem that if the straight lines originating from a single point touch the extremities of a figure, its projection on a parallel plane must be similar to the original (Figure 2).

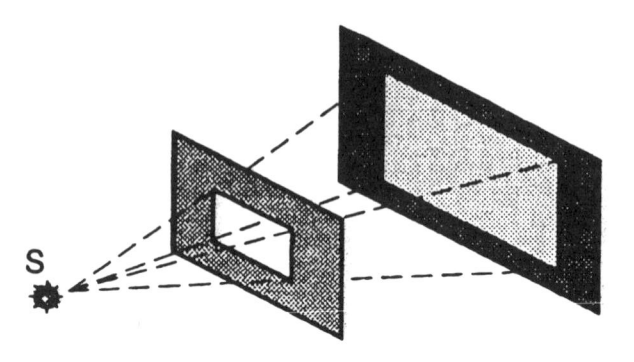

Figure 2. An illustration of Aristotle's argument.

However, having seen a round image instead of a rectangular one made him suggest that perhaps the sun's rays are cones with their apexes on the sun rather than straight lines. While observing a solar eclipse, Aristotle discovered again that the sun's image resembled the sun rather than the opening in a screen:

Why is it that in an eclipse of the sun, if one looks at it through a sieve, or through the leaves of a broad-leaved tree, or if one joins the fingers of one hand over the fingers of the other, the rays are crescent-shaped where they reach the earth? Is it for the same reason as that when light shines through a rectangular peephole that it appears circular and cone-like? (Aristotle, *Problems*, 341).

The problem raised by Aristotle of why the image made by sunlight passing through an opening is similar to the luminous body and not to the opening, had baffled scientists for almost two thousand years.

Let us investigate this problem with the following simple supplies: index cards, masking tape, razor blade or scissors, paper or cardboard white screen. In each index card make two round holes with a paper-punch and several square holes of different size using a razor blade or by cutting pieces with scissors and pasting them together.

The teacher should explain the significance of each step of this procedure. The Preliminary Part (Table I) begins with background, which shows the origin of the problem, in this case a chance observation by Aristotle. The teacher should emphasize that although scientists do not rely on chance, sometimes they benefit from it. An accidental observation becomes a discovery only when a scientist recognizes something new and unusual.

TABLE II
An illustration of the strategy in action

PRELIMINARY PART OF THE INVESTIGATION

Background

Teacher. We begin this investigation with a problem suggested by Aristotle. The problem is that a rectangular hole produces a round image of the sun? We can repeat Aristotle's experiment by observing an image produced by sunlight passing through a square hole in a cardboard sheet. Incidentally, some 16th century astronomers also became interested in this problem, but their curiosity came not from reading Aristotle but from the needs of their science.

Initial experiments

Ruth. Our group discovered that Aristotle's conclusion is correct only when the screen is far away from the index card. He probably never tried to bring the screen close to the hole, for if he did, he would have seen a square image instead of a round one.

John. Distance is not the only thing that matters. We've found that at the same distance between the screen and the card a large square opening produced a square image, and a small square hole projected a circle.

Formulating a problem

Teacher. We see that Aristotle was only partially right. This finding suggests that we examine a more general problem than his: why does the image produced by sunlight coming through the same opening at some circumstances resemble the sun and at others, the opening?

Selecting variables

Mary. We found that two factors affect the shape of the image: 1) the distance from the hole to the screen, and 2) the size of the hole. These could be our variables.

The next stage (initial experiments) consists of reproducing the phenomenon a number of times to be sure it is real and not spurious. Also, by modifying the experimental conditions, the scientist tries to determine which factors affect the phenomena: these will be selected for further study as "variables." In addition to this, during the initial experiments, the scientist develops a satisfactory experimental procedure. Unlike the Main Part, in the Preliminary Part the experiments are conducted without any plan, and several factors can be changed simultaneously. For this reason, all conclusions made, for instance, about the choice of relevant variables, are tentative. In the Main Part, each variable is studied with other variables kept constant. The study of each variable begins with preliminary experiments, from which students deduce a hypothesis about how this variable changes. This hypothesis is tested by additional experiments. If different groups disagree on the conclusion about a specific variable, they conduct additional experiments until they reach a consensus.

TABLE III
The example continued

MAIN PART OF THE INVESTIGATION

Variable 1: Distance between an Opening and a Screen

Preliminary experiments

David. I would suggest placing the screen in the shadow and moving the index card to and from it.

Ruth. It looks as if all square holes produce square images when close to the screen.

Hypothesis

John. The closer the hole is to the screen, the more the image resembles the opening.

Mary. Why do you call this conclusion a hypothesis? Haven't we just proved it?

Teacher. We compared only two holes of a specific shape (square). Are we sure that holes of other shapes behave in the same way? Do we know that the result will not change if we make the hole either very large or very small? In both cases, the answer is "No." That is why we need additional experiments. The preliminary experiments are necessary to advance a hypothesis, but to prove it one has to perform more experiments and of a somewhat different kind.

Test

Ruth. Let us see how holes of different shape but about the same size behave at different distances. Instead of cutting additional holes we changed the shape of the original openings by covering them partially with tape. In our experiments the images of triangles and rectangles closely resembled the apertures when the screen was no further than 15 cm from the card.

John. We tried parallelograms and trapezoids. In our view, the similarity was preserved for distances up to 20 cm.

Conclusion

The image resembles a middle size opening only when the latter is close to the screen.

Variable 2: The Size of the Opening

Preliminary experiments

John. Let us watch the images of two different squares, first, when the card is close to the screen, and then when the card is far from it. Apparently, in both cases the larger square hole makes a sharper square image.

Hypothesis

Mary. The larger the opening, the closer its picture resembles the original, whatever its distance from the screen.

Test

John. We can check this in two ways: 1) by making the hole much larger and much smaller; 2) by studying openings of other shape (for instance, triangles of different size).

Conclusion

The hypothesis is true, and the conclusion of the previous part is confirmed again. Note: if there is a disagreement among different groups, all groups repeat the experiment at the conditions that produced the "different" result.

General Conclusion Formation

Sunlight reproduces the shape of any given opening only when the screen is close to it. On the other hand, at a sufficiently large distance any opening produces a round image. Perhaps at large distances we obtain the image of the luminous body. This hypothesis can be tested by experimenting with light sources of different shapes.

Again, as in the Preliminary Part, the teacher explains to students the necessity of various steps in the Main Part (see Table 1). In particular, students may argue that they have already obtained the necessary Hypotheses in the Preliminary Part, and thus no more Preliminary Experiments are necessary. The teacher explains that any hypothesis from the Preliminary Part is tentative, because the experiments leading to it were conducted without keeping everything but one variable constant. While the Preliminary Experiments are necessary for formulating a hypothesis, the Test aims at confirming it, which requires experiments different from those used to suggest the hypothesis.

To extend the importance of solving Aristotle's problem, the teacher notes that the rounding-off of the shadow's corners at large distances was not understood until the work of Johannes Kepler (1571-1630). The teacher can provide a geometrical drawing representing Kepler's ideas, which high-school students can follow. Since this drawing is based on straight lines representing light rays, the conclusion is made that Aristotle was mistaken about his phenomenon contradicting the concept of light rays being geometrical straight lines.

Then the teacher discusses with students several points this unit offers about the nature of science. First, we see that even famous scientists can err. If students wonder why they were able to get better results than Aristotle, the teacher may explain that the scientific method of Greek scientists somewhat differed from the modern one. The ancient Greeks were keen observers of phenomena in nature, but their experimentation was sporadic. If instead of limiting his observations to the "natural objects" (the shadow of a fence), Aristotle recreated this phenomenon artificially the way we did, he would have discovered the correct result. Second, we see that solving some problems require a tremendous amount of time (in this case about 2000 years!). Third, the case provides an opportunity to talk about the meaning of an experiment supporting a theory. Perhaps, some students will challenge the teacher's final conclusion that Kepler demonstrated the rectilinearity of light. The teacher answers that a theory is confirmed when none of the experiments contradicts it. To decide whether there is an agreement or a contradiction, a scientist must make certain assumptions about the phenomenon, which may be true or not. For instance, Aristotle based his conclusion on the idea of a point source of light, while the sun is not one because its angular size is quite large.

There may be other more generic questions as well, such as: 1) How many experiments are enough to test a hypothesis? 2) What would have happened if we picked a false variable? The teacher explains that repeating an experiment involves varying a parameter to be sure that its quantitative change alters the result. For instance, if an experiment is conducted with only two different sizes of the opening, one would never be certain of the results at much larger or smaller dimensions. If a false variable (the shape of an opening, for one) is selected, changing it does not affect the result. Since the time invested by a scientist in a given investigation is limited, an unlucky choice of a variable may lead to failure. Failure may also occur because of an inadequate experimental procedure. Aristotle, for instance, could not see a rectangular shadow if he never changed the distance between the aperture and the "screen." There are other difficulties as well. That is why, although all scientists use a similar research strategy, some of them make great discoveries, while others only minor ones.

It is worth noting that the plan of investigation described here was tested with such subjects as electricity and bioelectricity, optics and vision, acoustics and hearing, waves and vibrations, and graduates of our programs have applied it to topics in mechanics, chemistry, biology, astronomy, and car troubleshooting. By consulting an appropriate work on the history of science (see Appendix A for suggestions), the general strategy provided here could be used in any science discipline.

CONCLUSIONS

Teaching aspects of the "nature of science" through an historical approach appeals to teachers if it is intertwined with teaching science content so that it makes the latter more stimulating without taking additional classroom time.

Historical reading combined with investigative experimentation, especially of a historical nature, appears to be a promising way for students to learn the basics of the scientific method and understand some other issues of the nature of science even when this subject is presented unobtrusively. While reading and a discussion with a teacher may be sufficient for a few motivated and curious students, this will not work for the majority of them. Teachers may try to appeal to their practical sense, by explaining that their future job opportunities depend on their ability of systematic and critical thinking, but this is a too far-fetched perspective to many. On the other hand, involving any student on a daily basis in investigative experimentation where they can show their creativity without fear of being punished for errors, can gradually change their attitude toward science and learning it.

However, to teach students how to do investigative experiments, teachers must be trained in a similar way. This requires an investment of time and effort, but the benefits are considerable. This skill is applicable for life and may be used with any science subject and beyond science. Reading without experimentation, naturally, takes less preparation on the teacher's part, but its benefits are more limited; an experiment described in a book does not provide as much information as the real one, nor does it create a sense of participation.

Taking into account that few educators have an interest in experimentation and probably even fewer in the history of science, the value of my recommendations seems to be very limited. However, it is possible to split a course on the nature of science into two parts, theoretical and experimental, taught by two people. Moreover, the experiments must not necessarily be historical: one can find other experiments for investigations, the results of which are known and comprehensible to teachers (two important advantages of historical ones!). And it is very possible that some educators will decide to try the fruit of history and find it interesting, if not delicious.

Bakken Library and Museum, Minneapolis, Minnesota, USA

NOTE

The Bakken Library and Museum, founded by Earl E. Bakken, inventor of the first transistorized cardiac pacemaker, is a center for education and learning that furthers the understanding of the history, cultural context, and applications of electricity and magnetism in the life sciences and medicine. It holds a vast collection of rare books and scientific instruments relating to the historical role of "electricity in life." In addition to serving as a research center, The Bakken is a pioneer in using the history of science to enhance K-12 science education, through in-service training programs for science teachers, workshops for students in grades 4-12, publications and kits.

REFERENCES

Aristotle XV, *Problems* I (1970). In the Loeb Classical Library. Cambridge, MA, Harvard University Press.
Bakken Library and Museum (1995). *Sparks and shocks: experiments from the golden age of static electricity*, Dubuque, IA, Kendall/Hunt.
Conant, J. B. (Ed.). (1959). *Harvard case histories in experimental science*, Cambridge, MA., Harvard University Press.
Devons, S. & Hartmann, L. (1970). 'A history-of-physics laboratory', *Physics today*, (2), 44-49.
Kipnis, N. (1992). *Rediscovering optics*, Minneapolis, MN, BENA Press.

Kipnis, N. (1995). 'Qualitative Physics in High School: A Lesson from History', *Proceedings of the Third International History, Philosophy, and Science Teaching Conference*, Minneapolis, MN, (pp. 624-635).
Kipnis, N. (1996). 'The 'Historical-investigative' approach to teaching science', *Science and Education*, (5), 277-292.
Klopfer, L. & Cooley, W. (1961). *Use of case histories in the development of student understanding of science and scientists*, Unpublished manuscript, Harvard University, Cambridge, MA.
Klopfer, L. (1964-66). *History of science cases*, Chicago, IL, Science Research Associates.
Klopfer, L. (1992). Historical perspective on the history and nature of science on school science programs in BSCS/SSEC, *Teaching about the history and nature of science and technology: Background papers*, Colorado Springs, CO, The Biological Sciences Curriculum Study, 105-129.
Lawrenz, F. & Kipnis, N. (1990). 'Hands-on history of physics', *Journal of Science Teacher Education*, (1), 54-59.
Lochhead, J. & Dufresne, R. (1989). 'Helping students understand difficult science concepts through the use of dialogues with history. History and philosophy of science in science education (*Proceedings of the First International Conference*), 221-229
Matthews, M. R. (1990). 'History, philosophy and science teaching. What can be done in an undergraduate course?', *Studies in Philosophy and Education*, (10), 93-97.
McComas, W. (1995). 'A thematic introduction to the nature of science', *Proceedings of the Third International History, Philosophy, and Science Teaching Conference*, Minneapolis, MN, 726-737.
Rutherford, J. et al. (1970). *Harvard project physics*, New York, Holt, Rinehart & Winston.
Schwab, J. (1964) 'The teaching of science as enquiry', in J. Schwab & P. Brandwein (eds.), *The Teaching Of Science*, Cambridge, MA, Harvard University Press, 31-102.
Teichmann, J. (1986). 'The historical experiments in physics education: theoretical observations and practical examples. Science Education and the History of Physics', (Proceedings of the multinational teacher and teacher-trainer conference at the Deutsches Museum, Munich), 189-221.

APPENDIX A

This list of references, added by editor, addresses episodes from the history of science that may be useful in constructing scenarios for use with the strategy described in this chapter. This list was abridged from one originally developed by Robert Lovely and updated by HsingChi Wang. The editor sincerely appreciates the willingness of these two individuals to allow their work to be used here.

Teaching the History of Science

Conant, J. B. (ed.). (1952). *Case studies in experimental science*, 1-2, Cambridge, MA, Harvard University Press.
Matthews, M. R. (1994). *Science teaching: The role of history and philosophy of science*, New York, Routledge.
Shortland, M., & Warwick, A. (eds.), (1989). *Teaching the history of science*, Oxford, Blackwell.

General Studies of the History of Science

Boorstin, D. J. (1983). *The discoverers*, New York, Random House.
Bronowski, J. (1978). *The origins of knowledge and imagination*, London, Yale University Press.
Cohen, I. B. (1985). *Revolution in science*, Cambridge, MA, Harvard University Press.
Conant, J. B. (1951). *On understanding science: An historical approach*, New York, Yale University Press.
Kuhn, T. S. (1977). *The essential tension: Selected studies in tradition and change*, Chicago, IL, University of Chicago Press.
Harré, R. (1981). Great scientific experiments, Oxford, Phaidon.
Lloyd, G. E. R. (1996). *Advisaries and authorities: Investigation into ancient Greek and Chinese science*, New York, Cambridge University Press.

Ronan, C. A. (1982). Science: Its history and development among the world's cultures, New York, Facts on File.
Sarton, G. (1936). *The study of the history of science*. Cambridge, MA, Harvard University Press.
Sarton, G. (1952). *Horus: A guide to the history of science*, Waltham, MA, Chronica Botanica.
Thagard, P. (1992). Conceptual revolutions, Princeton, NJ, Princeton University Press.

History of Science: Antiquity through the Renaissance

Goldstein, T. (1995). *Dawn of modern science: From the ancient Greeks to the renaissance*, New York, Da Capo Press.
Hall, A. R. (1984). *The revolution in science 1500-1750*, New York, Longman Inc.
Hall, M. B. (1994). *The scientific renaissance: 1450-1630*, New York, Dover.
Lindberg, D. C. (1978). *Science in the middle ages*, Chicago, IL, University of Chicago Press.
Lindberg, D. C. (1992). *The beginning of western science*, Chicago, IL, University of Chicago Press.
Montgomery, S. L. (1996). *The scientific voice*, New York, Guilford.
Sobel, D. (1995). *Longitude: The true story of a lone genius who solved the greatest scientific problem of his time*, New York, Penguin Book Inc.
Schrodinger, E. (1996). *Nature and the Greeks and science and humanism*, New York, Cambridge University Press.
Toulmin, S., Bush, D., Ackerman, J. S., & Palisca, C. V. (1961). *Seventeenth century science and arts*. Rhys, H.H. (ed.), Princeton, New Jersey, Princeton University Press.

The Scientific Revolution

Boas, M. (1966). The scientific renaissance, 1450-1630, New York, Harper & Row.
Butterfield, H. (1966). *The origins of modern science*, New York, Free Press.
Bylebyl, J. (ed.), (1979). *William Harvey and his age*, Baltimore, Johns Hopkins Press.
Cohen, I. B. (1985). *Revolution in science*, Cambridge, MA, Harvard University Press.
Drake, S. (1957). *Discoveries and opinions of Galileo*, Garden City, NJ, Doubleday.
Hall, A. R. (1984). *The revolution in science 1500-1750*, New York, Longman Inc.
Koyré, A. (1978). *Galileo studies*, (trans. John Mepham), Atlantic Highlands, NJ, Humanities Press.

History of Science: Nineteenth and Twentieth Centuries

Brockman, J. (1995). *The third culture*, New York, Touchstone.
Buchwald, J. Z. (1997). *Archimedes: 1996 New studies in the history and philosophy of science and technology: Scientific credibility and technical standards in 19th and early 20th century Germany and Britain*, Boston, MA, Kluwer Academic Publishers.
LaFollette, M. C. (1990). *Making science our own: Public Images of science, 1910-1955*, Chicago, IL, University of Chicago Press.

Environmental Science

Carson, R. (1962, 1987). *Silent spring*, Boston, MA, Houghton Mifflin.
Clark, J. G. (1987). *Energy and the federal government: Fossil fuel politics, 1900-1946*, Campaign.
Graham Jr., F. (1970). *Since silent spring*, Boston, MA, Houghton Mifflin.
Leopold, A. (1949, 1984). *A Sand County almanac*, New York, Ballantine Books.
Levine, A. (1987). *Love Canal: Science, politics, and people*, Lexington, MA, Lexington Books.
McIntosh, R. P. (1985). *The background of ecology, concept and theory*, Cambridge, Cambridge University Press.
Pinchot, G. (1947). *Breaking new ground*, New York, Harcourt Brace.

Astronomy and Cosmology

Hawking, S. W. (1988). A brief history of time: From the big bang to black holes, New York, Bantam Books.
Kane, G. (1995). The particle garden: Our universe as understood by particle, New York, Addison-Wesley.
Kuhn, T. S. (1985). The Copernican revolution: Planetary astronomy in the development of western thought, New York, MJF Books.
Ferris, T. (1988). Coming of age in the Milky Way, New York, William Morrow and Company, Inc.
Kuhn, T. S. (1957). The Copernican revolution, Cambridge, MA, Harvard University Press.
Toulmin, S., and Goodfield, J. (1965). The fabric of the heavens: The development of astronomy and dynamics, New York, Harper & Row.
Westfall, R. S. (1983). The construction of modern science: Mechanisms and mechanics, Cambridge, Cambridge University Press.

Biology/ Medicine/Physiology/Evolution/Genetics

Allen., G. (1978). Thomas Hunt Morgan: The man and his science. Princeton, Princeton University Press.
Bowler, P. J. (1989). Evolution: The history of an idea, Berkeley, CA, University of California Press.
Bowler, P. J. (1989). The Mendelian revolution: The emergence of hereditarian concepts in modern science and society, Baltimore, Johns Hopkins University Press.
Coleman, W. (1987). Biology in the nineteenth century: Problems of forms, function, & transformation, Cambridge, MA, Cambridge University Press.
Crick, F. (1988). What mad pursuit: A personal view of scientific discovery. New York, Basic Books Inc.
Darwin, C. (1859). The origin of species, London, Penguin Books.
Darwin, C. (1988). The voyage of the Beagle, New York, Mentor.
Desmond, A., & Moore, J. (1991). Darwin: The life of a tormented evolutionist, New York, W. W. Norton.
Dyson, F. (1985). Origins of life, New York, Cambridge University Press.
Eiseley, L. (1958, 1961). Darwin's century: Evolution and the men who discovered it. Garden City, NJ, Anchor books Doubleday & Company, Inc.
Gardner, E. J. (1972). History of biology. New York: Macmillan.
Jacob, F. (1988). An Autobiography: The Statue Within. New York: Basic Books Inc.
Keller, E. F. (1983). A feeling for the organism: The life and work of Barbara McClintock. New York: W.H. Freeman and Company.
Mayr, E. (1991). One long argument: Charles Darwin and the genesis of modern evolutionary thought, Cambridge, MA, Harvard University Press.
Olby, R. (1974). The path to the double helix, Seattle, WA, University of Washington Press.
Ruse, M. (1979). The Darwinian revolution, Chicago, IL, University of Chicago Press.
Stern, C., & Sherwood, E. (1996). The origin of genetics: A manual source book. San Francisco, CA: Freeman.
Watson, J. (1980). The double helix: A personal account of the discovery of the structure of DNA: Text, commentary, reviews, original papers, (Gunther S. Stent, ed.). New York, W. W. Norton.
Young, D. (1992). The discovery of evolution, Cambridge, MA, Cambridge University Press.

Chemistry

Brock, W. H. (1993). The Norton history of chemistry. New York, W. W. Norton and Company.
Hall, M. B. (1958). Robert Boyle and seventeenth century chemistry. Cambridge University Press.
Hartley, H. (1971). Studies in the history of chemistry. Oxford, Clarndon Press.
Ihde, A. (1964, 1984). The development of modern chemistry. New York: Dover.
Russell, C.A. (Ed) (1985). Recent developments in the history of chemistry. London: Royal Society of Chemistry.

Geology and the Earth Sciences

Dalrymple, G. B. (1991). The age of the earth. Stanford, CA, Stanford University Press.

Glen, W. (1982). The road to Jaramillo: Critical years of the revolution in earth science. Stanford, CA: Stanford University Press.
Gohau, G. (1990). A history of geology. New Brunswick, Rutgers University Press.
Gould, S. J. (1987). Time's arrow time's cycle: Myth and metaphor in the discovery of geological time. Cambridge, MA, Harvard University Press.
Hallam, A. (1989). Great geological controversies. Oxford, Oxford University Press.
Hsu, K. J. (1986). The great dying. New York: Harcourt Brace Jovanovich.
Hsu, K. J. (1983). The Mediterranean was a desert. Princeton: Princeton University Press.
Rudwick, M. J. (1972). The meaning of fossils, Chicago, IL, University of Chicago Press.
Sargeant, W. A. S. (1980). Geologists and the history of geology: An international bibliography from the origins to 1978. London: Macmillan.

Physics

Brackenridge, J. B. (1996). The key to Newton's Dynamics: The Kepler problem and the Principia. Los Angeles, CA: University of California Press.
Boyer, P. (1985). By the bomb's early light. New York: Pantheon Books.
Brush, S. G. (1972). Resources for the history of physics. Hanover: University Press of New England.
Brush, S. G., & King, A. L. (1972). History in the teaching of Physics. Hanover, New Hampshire: New England University Press.
Dahl, P. F. (1997). Flash of the cathode rays: A history of J. J. Thomson's electron. Bristol, PA: Institute of Physics.
Darrigol, O. (1992). From c-number to q-number: The classical analogy in the history of quantum theory. Los Angeles, CA: University of California Press.
Davis, N. P. (1968). Lawrence and Oppenheimer, New York: Simon and Shuster.
Davis, E. A., & Falconer, I. J. (1997). J. J. Thomson and the discovery of the electron. Bristol, PA: Taylor and Francis.
Einstein, A. (1961). Relativity: The special and the general theory. New York: Crown Publishers, Inc.
Fermi, L. (1982). Atoms in the Family, Albuquerque, NM, University of New Mexico Press.
Hankins, T. L. (1985, 1987). Science and the enlightenment, Cambridge, Cambridge University Press.
Hofmann, J. R. (1995). Andre-Marie Ampere: Enlightenment and Electrodynamics. New York: Cambridge University Press.
Holton, G. (1973). Thematic origins of scientific thoughts: Kepler to Einstein, Cambridge, MA: Harvard University Press.
Holton, G. (1995). Einstein, history, and other passions: The rebellion against science at the end of the twentieth century. New York: Addison-Wesley.
Holton, G. (1978). The scientific imagination: Case studies. New York: Cambridge University Press.
Kelves, D. (1987). The physicists. Cambridge, MA: Harvard University Press.
Rayleigh, L. (1942). The life of Sir J. J. Thomson. Cambridge: Cambridge University Press.
Reston, J. (1995). Galileo: A life. New York: Harper Perennial.
Rhodes, R. (1988). The making of atomic bomb. New York: Simon and Schuster, Inc.
Shamos, M. (ed.). (1987). Great experiments in physics: Firsthand accounts from Galileo to Einstein. New York: Dover.
Spielberg, N., & Anderson, B. D. (1987). Seven ideas that shock the universe. New York: Wiley.
Westfall, R. (1983). The Construction of modern science: Mechanisms and mechanics. Cambridge: Cambridge University Press.

MICHAEL P. CLOUGH

11. INTEGRATING THE NATURE OF SCIENCE WITH STUDENT TEACHING: RATIONALE AND STRATEGIES

Preservice science teacher education programs typically culminate with student teaching. Because science teacher education programs vary widely in their quality of instruction concerning the nature of science, and because individual student teachers have unique strengths and weaknesses, the cooperating teacher's role in promoting science teaching that accurately portrays the nature of science is critical.

Those entering the teaching profession have markedly different experiences and impressions about the nature of science. Some preservice teachers have had courses in the history and philosophy of science taught by a science educator who tailors the experiences to the needs of classroom science teachers. In spite of such a thorough introduction when these individuals become student teachers they still need guidance concerning how to incorporate accurate portrayals of the nature of science when teaching science content. Other student teachers have taken a course in the social studies of science taught in the history or philosophy departments. Such courses typically lack explicit connections to science teaching. While some of these student teachers do show an interest in the nature of science, they are not always aware of its value for science education, nor do they possess strategies to incorporate it into instruction. Finally, the majority of student teachers come from universities requiring no formal study in the nature of science nor is it integrated within their science methods courses. These student teachers typically are not aware of their misconceptions concerning the nature of science, and seem genuinely surprised when introduced to contemporary views on the subject. Obviously, these new teachers have not developed strategies to accurately portray the nature of science. A recent study (Bell, Lederman and Abd-Khalick, 1997) provides a glimpse of just how difficult incorporating the nature of science is for student teachers. This chapter is a description of my strategies I have developed as a former cooperating teacher to address these varying backgrounds of expertise.

DEVELOPING GOALS FOR SCIENCE INSTRUCTION

The student teachers working in my high school science classroom begin their experience by reading several articles addressing a desired state of science education

Included in this list are readings addressing the importance of accurately portraying the nature of science when teaching science content (AAAS, 1993; Clough, 1994; Gibbs and Lawson, 1992; Lederman, 1992; Matthews, 1989; NRC, 1996; NSTA, 1982). After several meetings during which we talk about these and other readings, my student teachers create a list of general student goals (i.e., what he or she would like to see students doing as a result of science instruction). Because these goals provide a foundation for the entire student teaching experience, prolonged discussion and refinement of the goals ensues. Invariably, among all the other student goals, having students demonstrate an understanding of the nature of science emerges in some format. A typical list of student goals is presented in Table I.

TABLE I
Student goals consistent with the desired state of science education

Students will:

- Use critical thinking skills.
- Convey an understanding of the nature(s) of science (i.e., social studies of science).
- Identify and solve problems effectively.
- Use communication and cooperative skills effectively.
- Work toward solutions of local, national and global problems.
- Be creative and curious.
- Set goals, make decisions, and self-evaluate.
- Convey a positive attitude about science.
- Access, retrieve and use the existing body of scientific knowledge in the process of investigating phenomena.
- Demonstrate deep robust understanding of science concepts rather than mastery of many insignificant/isolated facts.
- Demonstrate an awareness of the importance of science in many careers.

These discussions also serve as vehicles which enable me to probe what student teachers think the nature of science means and its place in science instruction. The development and discussion of student goals are critical because student teachers must come to see each goal, including an understanding of the nature of science, as important in a student's science education experience. Science teachers too often see the nature of science as an "add on" rather than a central theme to be integrated with other student goals throughout the course. Goodlad (1983, p. 468), for instance,

NATURE OF SCIENCE AND STUDENT TEACHING 199

lamented how teachers often placed emphasis only on content acquisition at the expense of all other student goals.

At this point, my student teachers begin the process of building a research-based framework for teaching science using pedagogical research to ensure that decisions in the classroom are directed toward eliciting student actions consistent with the stated goals of instruction. The fundamental components of a research-based framework illustrated in Figure 1 also provide a common ground for discussions concerning how preservice teachers will facilitate their goals for students, including an understanding of the nature of science.

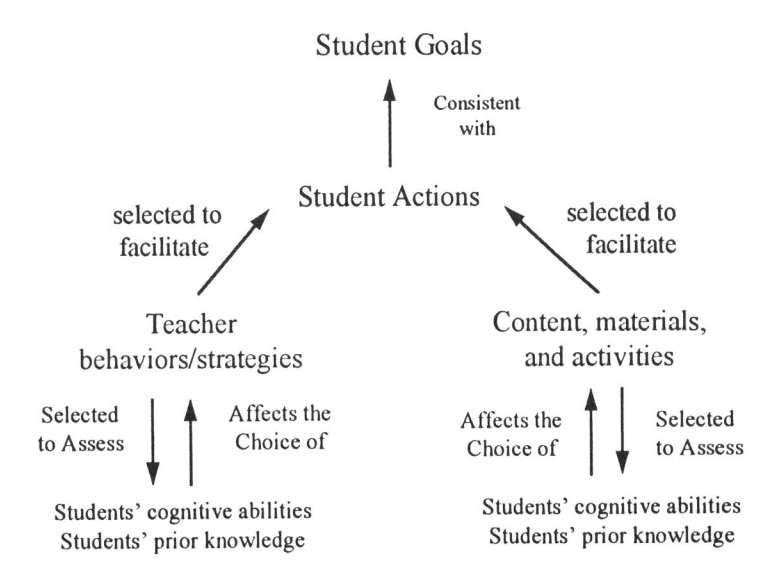

```
                        Student Goals
                             ↑
                         Consistent
                           with

                       Student Actions
       selected to    ↗              ↖    selected to
       facilitate                          facilitate

        Teacher                      Content, materials,
    behaviors/strategies               and activities

  Selected    ↑  Affects the      Affects the  ↑  Selected
  to Assess   ↓  Choice of        Choice of    ↓  to Assess

 Students' cognitive abilities   Students' cognitive abilities
 Students' prior knowledge       Students' prior knowledge
```

Figure 1. Components of a research-based framework for teaching

After having settled on an initial list of student goals, student teachers then describe student actions that are consistent and inconsistent with each goal, including those actions related to an understanding of the nature of science. These lists of student actions are not static and as the student teachers gain experience through reading nature of science materials and teaching, the initial list is often modified. Table II contains a compilation of student actions consistent with an understanding of the nature of science.

TABLE II
Student actions consistent with an understanding of the nature of science

Students should:

- Accurately describe what is meant by the "nature of science."
- List questions that those who study the nature of science might ask.
- Describe the differences and interactions between basic science, applied science, and technology. Provide examples of each.
- Describe how laboratory work and puzzle-solving are alike.
- Articulate why scientific explanations are couched in naturalistic terms with no recourse to the supernatural.
- Describe how science affects society through the technological innovations that flow from it. Provide several examples illustrating this.
- Describe how science affects society through the philosophical issues that its ideas raise. Provide several examples illustrating this.
- Provide examples of how society affects science and technology.
- Provide arguments against a universal step-by-step scientific method.
- Explain the value of imagination and creativity in doing science.
- Provide examples of how scientific ideas are not exact descriptions of laboratory data.
- Justify that science ideas, while durable, are not absolutely certain.
- Properly use significant nature of science language (e.g., "law" and "theory").
- Relate each of the above to laboratory experiences throughout the year.
- Question readings that distort the nature of science.

With these objectives in mind, teaching can now become the purposeful activity of selecting behaviors, strategies, materials and activities to assess students' views and to facilitate desired changes. However, this may be problematic if, as is often the case, preservice teachers themselves have misconceptions concerning the nature of science.

MODELING NATURE OF SCIENCE INSTRUCTION

Early in the school year I teach a unit on the nature of science to high school students, and the student teachers participate alongside those students, taking part in the pretest, activities, discussions, note-taking, and a posttest. In a previous manuscript (Clough, 1997), I set out in greater detail what transpires during this two-week unit and how the nature of science is portrayed for the remainder of the school year. To ensure that this experience occurs for student teachers working with me in the spring semester, for their benefit I explicitly revisit the nature of science with my high school students at this time in the year. Two weeks of instruction are not likely to fundamentally change high school students' or preservice teachers' notions about the nature of science, but the unit raises interest, introduces several fundamental issues that we revisit throughout the course, and serves notice of the importance of the nature of science. The unit also serves as one model for assessing and addressing students' conceptions of the nature of science.

During the two weeks of explicit nature of science instruction, the student teachers and I meet often to discuss the lessons. For those student teachers with an extensive and appropriate nature of science background, these discussions deal primarily with the strategies I use to engage my high school students in nature of science instruction. Discussions often involve: 1) the value of assessing students' preconceptions about the nature of science, 2) the rationale for selecting the readings and activities, 3) the types of questions asked during the lesson and why these, as opposed to other possible questions, were asked, 4) strategies to ensure that students were mentally active during the lesson, and 5) why students' views may not change during this two-week period despite the extensive instruction. However, most student teachers ask additional questions concerning the content of required readings and other basic issues in the nature of science. These discussions help the student teacher grapple with nature of science content and instruction, and help me visualize what the student teacher is thinking, as well as the cognitive conflict taking place as they wrestle with teaching the nature of science. After this two-week experience, the student teachers resume responsibility for teaching one or two classes. At this point, the focus of both the modeling and the character of our discussions shifts toward the incorporation of the nature of science into daily science content instruction.

Nature of Science in the Context of Laboratory Activities

Student teachers need explicit ideas for incorporating the nature of science into the context of daily lessons. One powerful way of doing so is to use laboratory experiences to illustrate important nature of science ideas. Clough and Clark (1994a, 1994b) have suggested modifying chemistry laboratory activities to facilitate student learning of science content and the nature of science. In these investigations, students working in small research teams are responsible for developing experiments to investigate a particular question posed by the instructor. Students must figure out which equipment to use and what type of result might support their investigative approach. Although some groups begin by using trial-and-error, most quickly begin applying what they know to the research question. After having collected what they determine are worthwhile data (students are often asked to justify the use of particular data), each group must then deal with the ambiguity of data when interpreting their results. This experience, sometimes the first for many students (and preservice teachers), forces students back into the lab for additional testing and rechecking of data. At the end of the lab experience, a classroom discussion occurs where each group presents their evidence and interpretation. After much debate and an occasional return to the laboratory, a group consensus is typically, but not inevitably, reached. In these types of laboratory experiences, students are not told whether or not their work is correct. Rather, student skepticism is directed back to their laboratory procedures, evidence accumulated, and interpretations made. In these sorts of activities, my student teachers observe students doing science, and experiencing the frustration, enjoyment, and uncertainty inherent in this distinctly human endeavor!

Such laboratory activities also provide an opportunity to model for the student teachers how explicit instruction in the nature of science can be made a normal part of everyday science instruction. To facilitate reflection on the nature of science as a result of the lab activity, students are often required to keep a journal discussing not only why they performed each experiment and what they learned, but also their thoughts and feelings about the experience. Each student group then reports their findings as if their paper were to be published in a scientific journal. This comparison between personal accounts of science and what appears in the final formal report provides an excellent experience both for students and student teachers to reflect on how poorly scientific papers tell the entire story of the laboratory experience. This is later recalled when students read Medawar's (1963) "Is the Scientific Paper a Fraud?" which argues that scientific papers distort how science research actually occurs. Students and student teachers are asked to reflect on how theory guided the experimental set-up, what criteria were used to decide what observations were relevant, and how the results were interpreted. Laboratory activities such as these force students to *think* about the science content and how content and process are intricately tied together. Post-lab discussions often involve considerable time discussing significant issues in the nature of science such as whether a universal scientific method exists; the role of theories in influencing questions, experimental set-up, observations and interpretations and the truth-value of scientific conclusions.

Nature of Science in the Context of Content Lectures and Reading Assignments

Student teachers also need to see how the nature of science may be effectively integrated into lectures and content readings. Historical case studies, stories about science and scientists, and having students critically analyze their science textbook's portrayal of the nature of science are effective ways to merge science content and the nature of science. Incorporating portions of *The Double Helix* (Stent, 1980) and

Jurassic Park (Crichton, 1990) while teaching genetics, or *The Big Splash* (Frank, 1990) when studying the origin of our planet's water, make excellent science reading while also facilitating reflection on how science works. These sources are far more interesting than dry science textbooks, and if used properly they can show science in the making rather than science ready-made (Latour, 1987). In chemistry, articles such as *Seeing Atoms* (Trefil, 1990) and *For the First Time, You Can See Atoms* (Hoffmann, 1993) can make for a lively discussion on science content and the nature of science. The key here is that student teachers see instruction about the nature of science spiraling through my students' entire science experience.

The Role of Teacher Behaviors in Accurately Portraying the Nature of Science

Cooperating teachers must ensure that student teachers do not adopt the mistaken notion that a teacher's understanding of the nature of science and choice of appropriate activities are sufficient to facilitate student understanding of the nature of science. Both these conditions are critical, but what a teacher does with activities is more important than the activities themselves. The activities suggested in this chapter will enhance students' long-term understanding of the nature of science only if appropriate teacher behaviors and strategies are utilized. Lederman (1992, pp. 346 & 351) writes,

... some method of influence must exist; naturally the influence must be mediated by teacher behaviors and classroom ecology. ... it appears that the most important variables that influence students' beliefs about the nature of science are those specific instructional behaviors, activities, and decisions implemented within the context of a lesson.

Students cannot be rushed through conceptual change experiences, nor can teachers transfer their personal understanding directly to students. Science educators are increasingly calling for less breadth and more depth of science content. Approaches similar to those discussed above to improve students' conceptions of the nature of science require more time and effort than most student teachers are accustomed. Second, transforming existing activities into ones that more accurately portray the nature of science also require thoughtful consideration of pedagogical research. Students must be held responsible for seriously reflecting on and struggling through a number of abstract notions. Eliciting students' conceptions and reasoning is often more difficult than student teachers realize, and facilitating movement from naive to more informed conceptions even more so. Essential teacher behaviors that facilitate desired student reflection include the use of extended-answer questions that build on students' earlier ideas, questions requiring students to elaborate on their ideas, the application of wait-time, responding behaviors that accept rather than judge student ideas, encouragement rather than praise, and a great deal of teacher observation and listening. The teacher's use of language in the classroom is also critical. Zeidler and Lederman (1987, pp. 6-7) conclude that, "the ordinary language teachers use to communicate science content does provide the context in which students formulate their own conceptions of the nature of science." Hence, cooperating teachers must carefully consider what they say about science and how they use significant language in the social studies of science both in communicating with students and modeling for student teachers.

COOPERATING TEACHER/STUDENT TEACHER INTERACTIONS

During the initial three to four-week period when my student teacher and I share responsibility for teaching, extensive discussions occur before and after lessons. While reviewing lessons that student teachers are planning, I typically ask the following questions:

- How is the nature of science valuable for sequencing science content to be taught?
- How does this material, activity, or content portray the nature of science?
- How can this lesson be changed so it better illustrates the nature of science?
- What questions will you ask to facilitate student understanding of NOS?
- What will your post-lab discussion entail? How does this address the nature of science?
- How will your language accurately portray the nature of science?

Then I often create scenarios to force the student teacher to consider how to respond to particular situations. For instance, I might play the role of a frustrated student who wants simple answers concerning how science works, or I might play the role of a student who disagrees with what my student teachers promote.

After a lesson I might ask my student teacher questions such as:

- What did you observe happening that is relevant to the nature of science?
- In what ways was the nature of science accurately and inaccurately portrayed?
- What would you do differently next time?
- How would these changes better facilitate student understanding of the nature of science?
- From observing/listening to students, what conceptions of the nature of science do they have?
- How will you structure future instruction to address these conceptions?

Periodically the student teachers' lessons are videotaped and they examine their teaching for a number of instructional issues regarding the nature of science. Sometimes we discuss their individual analysis and at other times we analyze portions of their videotaped lessons together. By analyzing my own teaching in the same fashion, and by requiring student teachers to do the same, I model the reflection-on-action that all effective teachers should exhibit.

Recognizing that conceptual change in both students and teachers requires a great deal of time and effort, at an appropriate point during the semester I provide the student teachers with a lengthy list of resources concerning the nature of science content and teaching strategies to pursue during and after our experience together. Appendix A contains a partial listing of valuable readings for teachers to pursue when time permits. Throughout the student teaching semester, student teachers are encouraged to reflect on teaching behaviors and activities to determine what they illustrate about the nature of science. Emphasis is placed on how teachers can "keep the heat on" students' misconceptions concerning the nature of science while teaching science content. Activities, strategies, and reflection are designed to confront students' misconceptions and suggest more plausible/accurate alternatives concerning the nature of science. With respect to nature of science instruction, the cooperating teacher should design experiences with student teachers so that they begin to adopt some of the objectives set forth in Table III.

TABLE III
Desired student teacher outcomes regarding the nature of science

Student teachers should:

- Use the precise meaning(s) of significant language in the social studies of science and explain possible consequences of the precise or imprecise use of language on students' perceptions of the nature of science.
- Express potential explicit and implicit view(s) of the nature of science portrayed in several activities (lab or otherwise).
- Evaluate textbooks, audiovisual materials, and other curriculum materials for their accuracy in portraying the nature of science.
- Modify existing materials and activities so they more adequately portray the nature of science.
- Demarcate (using consensus views from the social studies of science) science from non-science and pseudo-science.
- Implement correct historical examples (where appropriate) that effectively convey a more accurate portrayal of the nature of science.

- Convey a research-based rationale for science teaching which includes the necessity of an accurate understanding of the nature of science.
- Self-evaluate classroom performance as it pertains to an accurate portrayal of the nature of science. Demonstrate both reflection-on-action and reflection-in-action.
- Use the nature of science in articulating what learning and teaching science encompasses.
- Evaluate materials designed to assess students' understanding of the social studies of science.

MEASURES OF EFFECTIVENESS

Not surprisingly, the approaches described here significantly impact the manner in which my student teachers teach, but even more meaningful is how former student teachers have incorporated the nature of science into their present teaching and professional development. Many of my former student teachers have developed novel ways of incorporating nature of science instruction appropriate for the grade level and courses they currently teach. This indicates that they are not simply copying what they experienced with me, but are applying what they learned to their own unique situations. In addition, while I continue to suggest readings to them and share ideas for teaching the nature of science, I am pleased how often they now suggest useful readings and activities.

All of my former student teachers have presented papers and activities at science teacher conferences and often incorporate the nature of science into these presentations. One such former student teacher recently wrote an article describing how he modified a chemistry activity so that it better engages students while more accurately portraying the nature of science. Another former student teacher has taken additional courses in the nature of science. Empirical evidence (Clough, 1995) and anecdotal incidents gathered in my class show my student teachers that the strategies incorporated in my class significantly improve students' views concerning the nature of science. Integrating the nature of science throughout student teaching provides preservice teachers with the confidence, strategies, and models necessary to implement such instruction as they begin their own careers.

University of Iowa, Iowa City, Iowa, USA

REFERENCES

American Association for the Advancement of Science (1993). *Benchmarks for science literacy*, New York Oxford.
Bell, R., Lederman, N. G., Abd-El-Khalick, F. (1997). *Developing and acting upon one's conception of science: The reality of teacher preparation*, paper presented at the AETS international meeting, Cincinnati, OH, January 9-12.
Clough, M.P. (1994). 'Diminish students' resistance to evolution education', *The American Biology Teacher* (56), 409-415.
Clough, M.P. (1995). 'Longitudinal Understanding of the Nature of Science as Facilitated by an Introductory High School Biology Course', *Proceedings of the Third International History, Philosophy, and Science Teaching Conference*, pp. 212-221, University of Minnesota, Minneapolis, MN.
Clough, M.P. (1997). Strategies and activities for initiating and maintaining pressure on students' naive nviews concerning the nature of science, *Interchange*, 28(1-2), 191-204.
Clough, M.P. and Clark, R. L. (1994a). 'Cookbooks and constructivism: A better approach to laboratory activities', *The Science Teacher*, (61), 34-37.
Clough, M.P. and Clark, R. L. (1994b). 'Creative constructivism: challenge your students with an authentic science experience', *The Science Teacher*, (61),46-49.
Crichton, M. (1990). *Jurassic park*, New York, Ballantine.
Frank, L.A. (1990). *The big splash*, New York, Birch Lane.
Gibbs, A. and Lawson, A.E. (1992). The nature of scientific thinking as reflected by the works of biologists and by biology textbooks, *The American Biology Teacher*, (54),137-151.
Goodlad, J. I. (1983). 'A study of schooling: some findings and hypotheses', *Phi Delta Kappan*, (64),465-470.
Hoffmann, R. (1993). 'For the first time, you can see atoms', *American Scientist*, January-February, 11-12.
Latour, B. (1987). *Science in action*, Cambridge, MA, Harvard University.
Lederman, N.G. (1992). 'Students' and teachers' conceptions of the nature of science: a review of the research', *Journal of Research in Science Teaching*, (29), 331-359.
Matthews, M. R. (1989). 'A role for history and philosophy in science teaching', *Interchange*, (20), 3-15.
Medawar, P.B. (1963). 'Is the scientific paper a fraud?', in Medawar, P.B. (1990), *The threat and the glory: reflections on science and scientists*, New York, HarperCollins.
National Research Council (1996). *National science education standards*, Washington, D. C., National Academy Press.
National Science Teachers Association (1982). Science-Technology-Society: Science Education for the 1980's, an NSTA Position Statement, Washington, D.C.
Stent, G. S. (Ed.) (1980). *The double helix: text, commentary, reviews, and original papers*, New York, Norton.
Trefil, James (1990). 'Seeing atoms', *Discover*, June, pp. 55-60.
Zeidler, D.L. and Lederman, N.G. (1987). *The Effect of Teachers' Language on Students' Conceptions of the Nature of Science*, paper presented at the 60th Annual Meeting for the National Association for Research in Science Teaching, Washington, D.C., ERIC Document Reproduction Service No. ED286734.

APPENDIX A
ADDITIONAL NATURE OF SCIENCE READINGS
FOR STUDENT TEACHERS

Aikenhead, G. S. (1988). 'An analysis of four ways of assessing student beliefs about STS topics', *Journal of Research in Science Teaching*, 25,607-629.

Aikenhead, G. S. and Ryan, A.G. (1992). 'The development of a new instrument: "Views on science-technology-society" (VOSTS)', *Science Education*, (76),477-491.

Bednarek, D. I. (1993). 'Friction: heat-loving bacterium roils two worlds. Business, academia spar when product of basic research proves practical', *The Milwaukee Journal*, May 9.

Begley, S. (1992). 'Math has on its face: The foundations of mathematics have serious flaws. And that may imperil all the sciences', *Newsweek*, November 30.

Bridgman, P. W. (1960). 'Explanation', in *The Logic of Modern Physics*, pp. 37-39. New York, Macmillan.

Campbell, N.A. (1993). 'A conversation with . . . David Suzuki', *The American Biology Teacher*, (55), 31-36.

Campbell, N.A. (1992). 'A conversation with . . . Michael Bishop & Harold Varmus', *The American Biology Teacher*, (54),476-481.

Einstein, A. and Leopold, I. (1938). *The evolution of physics*, New York, Simon and Schuster.

Elmer-Dewitt, P. (1994). 'Don't Tread on my Lab, *Time*, (143), 44-45.

Feynman, R. P. (1985). 'The value of science', in *What do you care what other people think?* New York, Norton.

Johnson, R. L. and Peeples, E.E. (1987). 'The role of scientific understanding in college: Student acceptance of evolution', *The American Biology Teacher*, (49), 93-96.

Kuhn, T. S. (1970). *The structure of scientific revolutions*, Chicago, IL, University of Chicago.

Latour, B. and Woolgar, S. (1986). *Laboratory life: The construction of scientific facts*, New Jersey, Princeton.

Matthews, M. R. (1994). *Science teaching: The role of history and philosophy of science*, New York, Routledge.

Medawar, P. B. (1973a). 'The cost-benefit analysis of pure research', in Medawar, P.B. (1990), *The Threat and the Glory: Reflections on Science and Scientists*, New York, Harper Collins.

Medawar, P. B. (1973b). 'The pure science', in Medawar, P.B. (1990), *The Threat and the Glory: Reflections on Science and Scientists*, New York, Harper Collins.

Moore, J. A. (1983). 'Evolution, education, and the nature of science and scientific inquiry', in Zetterberg, J. Peter (ed.), *Evolution versus Creationism: The Public Education Controversy*, Phoenix, AZ, Oryx Press.

Munby, H. (1976). 'Some implications of language in science education', *Science Education*, (60), 115-124.

National Institutes of Health (1984). *Why do basic research?* NIH Publication Number 84-660. Bethesda, Maryland.

Overbye, D. (1993, April 26). Who's afraid of the big bad bang? *Time*.

Ryan, A. G. and Aikenhead, G. S. (1992). 'Students' preconceptions about the epistemology of science' *Science Education*, (76),559-580.

Shamos, M. H. (1995). *The myth of scientific literacy*, New Brunswick, Rutgers.

Thomas, L. (1983). *Late night thoughts on listening to Mahler's Ninth Symphony*, New York, Viking Press.

Van Allen, J. A. (1964, February 3-7). 'Science as a human enterprise', *Current Science*.

Westfall, R. S. (1977). *The construction of modern science*, New York, Cambridge.

SECTION III

COMMUNICATING THE NATURE OF SCIENCE:
COURSES AND COURSE ELEMENTS

WILLIAM F. McCOMAS

12. A THEMATIC INTRODUCTION TO THE NATURE OF SCIENCE: THE RATIONALE AND CONTENT OF A COURSE FOR SCIENCE EDUCATORS

This chapter describes the content elements and pedagogical orientation of a semester-long nature of science course designed for science teachers. The central learning tool is discussion based on an extensive set of selections from the works of and commentaries on philosophy, history, sociology and psychology of science. The course content is arranged around a number of overarching nature of science ideas that are both accessible and useful to those involved in science teaching. Major course ideas include knowledge production in science, research programs and revolutions, the development and status of laws and theories, the interrelationships of science and society, and philosophically valid science teaching models.

Through its standards documents, the science education community worldwide has generally defined the content of the nature of science that should be part of the knowledge base of secondary school graduates. By inference, science teachers should also have acquired both the nature of science content background and the strategies for communicating this content if they are accurately to represent what scientists do, how scientific knowledge is generated and the goals and limits of science.

THE COURSE OUTLINE

The *Thematic Introduction to the Nature of Science* is a three-credit course designed for students interested in science and/or science teaching who would like to know more about the "game" of science, the role and work of scientists, the generation of knowledge in science, and the place of science in society. The primary learning tool is discussion with reference to an extensive set of selections from the works of major philosophers of science and commentaries on the relationship between science education and the social studies of science. The course content is drawn from the foundation of scholarship in the nature of science provided by Matthews (1990, 1994), Martin (1972) and Robinson (1968) and from new descriptions of what students should know about how science operates (National Research Council, 1996; AAAS, 1994).

To assist students in what is likely their first encounter with the nature of science, readings have been selected that directly or indirectly address these big ideas. These readings are grouped into themes (see Table I). The themes and related sub-themes contain a number of overarching ideas within the nature of science that are both accessible to students and useful to those with interests in science teaching and learning. In most instances, fundamental notions within the discipline are addressed not only within their thematic unit, but throughout the course. The nature of the selections, the sequence of the readings, and full citations is included in Table I.

TABLE I

Titles, descriptions and readings included in each of the eight thematic units

Reading Set #1   AN INTRODUCTION TO THE SOCIAL STUDIES OF SCIENCE

These articles provide background perspectives on the varied definitions of science from a historical perspective. In addition, students become familiar with the domain of the nature of science with descriptions of the scientific enterprise from the history, sociology, psychology and the philosophy of science. The Gould (1990) selection discusses the unusual fossils of the Burgess shale is the beginning of a recurring theme about the evaluation of new ideas by the scientific community developed later in the course.

1. Lindberg, D. C. (1992). 'What is Science?', in *The Beginnings of Western Science*, Chicago, IL, The University of Chicago Press, 1-4.
2. Definitions of Science and Other Statements about its Nature (See Appendix A).
3. Gjertsen, D. (1989). 'Science and Philosophy: Past and Present', in *Science and Philosophy: Past and Present*, London, Penguin Press, 1-7.
4. Klemke, E. D., Hollinger, R. and Kline, A. D. (1988). 'What is Philosophy of Science?', in *Introductory Reading in the Philosophy of Science, Revised Edition*, Buffalo, NY, Prometheus Books, 1-5.
5. Rhodes, R. (1986). 'Atoms and Void', in *Making of the Atomic Bomb*, New York, Simon and Schuster, 29-39.
6. Gould, S. J. (1990). 'Selections', in *Wonderful Life*, New York, W. Norton and Company, 277-79 and 282-291.
7. Casti, J. L. (1989)., 'Hope and Asperity', in *Paradigms Lost*, New York, William Morrow and Company, Inc., 1-15.

Reading Set #2   THEORIES & LAWS: PRODUCTS AND TOOLS OF SCIENCE

These articles provide an opportunity to discuss the role of facts, hypotheses, theories and laws so students can both use these terms correctly and anticipate sources of confusion when these words are misapplied in textbooks and by the general public.

8. Dilworth, C. (1994). 'On the Nature of Scientific Laws and Theories', in *Scientific Progress, Third Edition*, Boston, MA, Kluwer Academic Publishers, 174-194.
9. Trusted, J. (1979). 'Theories and Laws', in *The Logic of Scientific Inference*, New York, Macmillan, 70-77.
10. Rhodes, G. and Schaible, R. (1989). 'Fact, Law and Theory: Ways of Thinking in Science and Literature', *Journal of College Science Teaching* (19), 228-232, 288.
11. Sonleitner, F. J. (1989). 'Theories, Laws and All That', *National Center for Science Education. Newsletter*, 9,3.
12. Fleisher, P. (1987). 'What is a Natural Law', in *Secrets of the Universe: Discovering the Universal Laws of Science*, New York, Athenaeum, 1-4.
13. Crick, F. (1988). 'Selection', in *What Mad Pursuit*, New York, Basic Books, Inc., 137-142.

Reading Set #3 THE LOGIC AND METHODS OF SCIENCE

This section features a discussion of the knowledge generating aspects of science with particular emphasis on the strengths and limitations of induction and deduction. The second set of readings in this theme examine the issue of what constitutes scientific method from varying perspectives.

14. Wallace, W. L. (1971). 'Selections', in *The Logic of Science*, New York, Aldine Publishing Company, 15-19, 24-29 and 34-39.
15. Anderson, O. R. (1976). 'Contemporary Perspectives', in *The Experience of Science*, New York, Teachers College Press, 9-13.
16. Richards, S. (1983). 'Scientific Argument: The Role of Logic', in *Philosophy and Sociology of Science; An Introduction*, Oxford, Basic Blackwell, 14-27.
17. Gjertsen, D. (1989). 'Is There A Scientific Method?', in *Science and Philosophy: Past and Present*, New York, Penguin Books, 87-113.
18. Richards, S. (1983). 'Philosophies of Scientific Methods and Theories of Science (Part of Chapter 4)', in *Philosophy and Sociology of Science: An Introduction*, Oxford, Basic Blackwell, 44-59.
19. Hempel, C. G. (1966). *A Philosopher Gives His Account of the Scientific Method from Philosophy of Natural Science*, Englewood Cliffs, NJ, Prentice-Hall, Inc., 3-18.
20. Mayr, E. (1991). 'Darwin's Scientific Method', in *One Long Argument*, Cambridge, MA, Harvard University Press, 9-11.
21. Chamers, A. (1990). 'Against Universal Method', in *Science and Its Fabrication*, Minneapolis, MN, University of Minnesota Press, 11-23.
22. Millar, R. (1988). 'What is the 'Scientific Method' and Can It be Taught?', in *Skills and Processes in Science Education: A Critical Analysis*, Wellington, J. J. (ed.), London, Routledge, 47-62.
23. Horgan, J. (May, 1993). 'Paul K. Feyerabend: The Worst Enemy of Science', Scientific American, 36-37.

Reading Set #4 INDUCTIVIST-EMPIRICISM, HYPOTHETICO-DEDUCTIVISM
AND THE ROLE OF CREATIVITY IN SCIENCE

These articles continue the discussion of knowledge generation with added emphasis on the creative element central to science. The science education applications of hypothetico-deductivism are discussed along with the descriptions of how Einstein and others viewed the knowledge generation process.

24. Joshua, S. and Dupin, J. (1986) 'Is Systematization of Hypothetico-Deductive Reasoning Possible in a Class Situation?', *European Journal of Science Education.* 8, 381- 388.
25. Holton, G. (1975). 'Selection from Mainsprings of Scientific Discovery', in *The Nature of Scientific Discovery,* Gigerich, O. (ed.), Washington, D.C., Smithsonian Institution Press, 203-208.
26. Medawar, P. (1982). 'Two Conceptions of Science', in *Pluto's Republic,* New York, Oxford University Press, 28-34.
27. Trefil, J. (1989). 'Science in Context', in *Reading the Mind of God: In Search of the Principle of Universality,* New York, Anchor Books, 31-44.
28. Rachelson, S. (1977). 'A Question of Balance: A Wholistic View of Scientific Inquiry', *Science Education,* 61, 109-117.
29. Atkin, J. M. and Karplus, R. (1962). 'Discovery or Invention?', *The Science Teacher,* 29, 45-51.
30. Medawar, P. B. (1963). 'Is the Scientific Paper a Fraud?', in P. B. Medawar (1990), *The Threat and the Glory,* New York, Harper.

Reading Set #5 IMAGES OF SCIENCE AND ITS METHODS

Kuhn's interpretation of the progress of science is featured in this readings set along with discussion of elements of the "new" post-positivist philosophy of science. Included are the first articles making the connection between the nature of science and science education.

31. Richards, S. (1983). 'Philosophies of Scientific Method: Theories of Science', in *Philosophy and Sociology of Science: An Introduction* (concluding part of chapter 4), Oxford, Basil Blackwell, 60-70.
32. Nadeau, R. and Desautel, J. (1984). *The Kuhnian Development in Epistemology and the Teaching of Science,* Toronto, Canada, Guidance Center of the University of Toronto, 11-21.
33. Wallace, B. A. (1989). 'Views of Science and Reality Through History', in *Choosing Reality: A Contemplative View of Physics and the Mind,* Boston, MA, New Science Library, 24-33.
34. Barrow, J. D. (1988). 'The Different Views of Science', in *The World Within the World,* Oxford, Clarendon Press, 10-12.

35. Hodson, D. (1982). 'Selection from Science – The Pursuit of Truth? Part I', *School Science Review*, (63), 643-652.
36. Gould, S. J. (1993). 'Selection from The First Unmasking of Nature', *Natural History*, (102), 14-21.
37. Elkana, Y. (1970). 'Science, Philosophy of Science and Science Teaching', *Educational Philosophy and Theory*, (2), 15-35.
38. Margetson, D. (1982). 'Some Educational Implications of the Uncertain Identity of Science', European Journal of Science Education, (4), 357-365.

Reading Set #6 VISIONS OF REALITY AND THE ROLE OF OBSERVATION IN SCIENCE

The initial focus in this set is on the distinction between realism and instrumentalism as an important element of the nature of science. A second element examines the special problem and role played by observation in science and in science learning. Students are encouraged to think of all observation as theory-based, an issue that has important implications in both science and science learning.

39. Chalmers, A. F. (1982). 'Realism, Instrumentation and Truth', in *What is This Thing Called Science?* 2nd ed., St Lucia, Australia, University of Queensland Press. 146-160.
40. Munby, A. H. (1976). 'Some Implications of Language in Science Education', *Science Education*, (60), 115-124.
41. Casti, J. L. (1989). 'Faith, Hope and Asperity', in *Paradigms Lost*, New York, William Morrow and Company, Inc., 15-55.
42. Hodson, D (1986). 'The nature of scientific observation', *School Science Review* (68), 17-29.
43. Hainsworth, M. D. (1956). 'The effect of previous knowledge on observation', *School Science Review*, (37), 234-242.

Reading Set #7 CONCEPTUAL CHANGE IN SCIENCE AND IN THE CLASSROOM

Here we reexamine the controversy over the interpretation of the Burgess Shale to make the point that most learners and scientists are resistant to change in their thinking. The science education implications of this issue and a discussion of the parallels between change in science and conceptual change in students are explored.

44. Barber, B. (1961). 'Resistance by scientists to scientific discovery', *Science*, (134), 596-602.

45. Richards, S. (1983). *'Scientific Argument: The Role of Logic'*, in *Philosophy and Sociology of Science; An Introduction*, Oxford, Basic Blackwell.
46. Lessem, D. (1993). 'Weird Wonders Fuel the Battle over Evolution's Path', *Smithsonian*, (23), 107-115.
47. Lewin, R. (1992). 'Whose View of Life?', *Discover*, (15) 18-19.
48. Strike, K. A. and Posner, G. J. (1982). 'Conceptual Change and Science Teaching', *European Journal of Science Education*, (4), 231-240.
49. Gauld (1989). 'A Study of Pupil's Responses to Empirical Evidence', in *Doing Science*, R. Millar (ed.), Philadelphia, PA, The Falmer Press/Taylor and Francis, 62-82.
50. Millar, R. (1989). *Bending the Evidence: The Relationship Between Theory and Experiment in Science Education*, Philadelphia, PA, The Falmer Press/Taylor and Francis, 38-61.
51. Finley, F. (1983). 'Science Processes', *Journal of Research in Science Teaching*, (20), 47-54.
52. Driver, R. (1983). 'Invention and Imagination', in *The Pupil as Scientist?*, Philadelphia, PA, Open University Press, 41-49.

Reading Set #8   THE SOCIETY OF SCIENCE, SCIENCE, SOCIETY & SCIENCE EDUCATION

In this final set of readings, there are several elements of interest. First, students discuss the ethical aspects of science which flows nicely into the social production of scientific knowledge. The final selections examine, in more detail, the link between the nature of science and science education with reference to selections on the nature of science from the national standards projects.

53. Kiefer, G. F. (1979). 'Science and Society', in *Bioethics: A Textbook of Ideas*, Reading, MA, Addison-Wesley, Inc., 413-442.
54. Thompson, D. (November 23, 1992). 'Science's Big Shift', *Time*, 34-35.
55. Collins, H. M. (1983). 'The Sociology of Scientific Knowledge: Studies of Contemporary Science',. *Annual Review of Sociology*, (9), 265-285.
56. Mendelsohn, E. (1977). 'The Social Construction of Scientific Knowledge', E. Mendelsohn, P. Weingart and R. Whitley (eds.), 'The Social Production of Scientific Knowledge', Boston, MA, D. Reidel Publishing Company, 3-26.
57. Abimbola, I. O. (1983). 'The Relevance of the 'New' Philosophy of Science for the Science Curriculum', *School Science and Mathematics*, (83), 182-190.
58. Duschl, R. A. (1985). 'Science Education and Philosophy: Twenty-five Years of Mutually Exclusive Development', *School Science and Mathematics*, (85), 541-555.
59. Matthews, M. (1989). 'A Role for History and Philosophy in Science Teaching', *Interchange*, (20), 3-15.
60. Hodson, D. (1988). 'Toward a Philosophically More Valid Science Curriculum', *Science Education*, (72), 19-40.
61. Matthews, M. (1992). 'History, Philosophy and Science Teaching: The Present Rapprochement', *Science and Education*, (1), 11-47.

62. American Association for the Advancement of Science (1994). *The Nature of Science: Benchmarks for Science Literacy,* New York, Oxford University Press.
63. The National Research Council (1996). 'The Nature and History of Science' from *The National Education Standards,* Washington, D.C., National Academy Press.

Instructional Orientation

There is a constructivist teaching philosophy underlying the course. The general direction of the discussions is, of course, directed by the nature of the readings provided. However, the authors of the readings are permitted to speak for themselves. Students find meaning by reading several views of the same unifying idea frequently from different and occasionally contradictory perspectives. In support of this goal, students form discussion groups to read, discuss, critique and cooperatively evaluate the ideas exhibited by the authors. The instructor guides the small group discussions by acting as a facilitator and occasionally presenting mini-lectures on the more difficult aspects of a readings set. In addition, class time is also devoted to relevant video presentations (such as selections from the *Day the Universe Changed* by Burke to support the historical perspective) or to small group activities such as black box investigations (in support of the discussions of laws, theories and logic). Generally each of the eight themes is discussed over a two-week period covering six hours of class time. Throughout the entire course, students are engaged in a number of formal assessment plans discussed in the next section.

ASSESSMENT

Students are evaluated with reference to a multifaceted scheme which includes a theory/law comparison paper, a theory/law application paper, brief written overviews of the most important ideas in each readings set and a final examination. Each of these elements will be discussed here in some detail.

Theory/Law Assignments

Comparison Paper

The distinction between theories and laws is established in several ways. First, students are asked to consult science content and philosophy of science texts and produce a list of ten definitions of the term "theory" and of the term "law." Following this elucidation of definitions, students are instructed to synthesize what they have found in terms of similarities and differences between these definitions. Invariably, students discover that there is much confusion in the ways these important terms are defined. This is particularly true in science textbooks where the term "law" is used but rarely defined.

Application Paper

The theory/law paper, due late in the course, provides an opportunity for each student to select a specific law or theory (such as the theory of plate tectonics, the first law of motion, the theory of evolution by natural selection or the kinetic-molecular theory of matter) and accomplish several tasks with respect to that law or theory. Students describe what the law or theory states, review what a law or theory is, examine several textbooks and compare the way(s) in which the words law and theory are defined and actually used in the body of the text and analyze the way in which the specific law or theory is presented in those texts.

Points of Most Significance (POMS)

For each of the eight themes, students are asked to write no more than four short overviews (a maximum of 35 words) of the essential ideas communicated in that theme. These short reviews are called "points of most significance" or POMS. The technique of having students write brief review statements serves to increase student opportunities to make connections between important ideas while encouraging them to extract the big course concepts from the readings. Students receive differential credit for this assignment related to the complexity of their POMS. Three points are awarded for appropriate POMS related only to a single set of readings. Four points are awarded for POMS that relate the content of one readings set to an earlier one. Students may receive up to five points if their POMS discusses a significant application for teaching the nature of science not stated explicitly by the author(s).

Final Examination

The final examination is an opportunity for students to synthesize what they have learned by drawing on the readings to evaluate statements such as, "Analyze the statement that scientific insights are in part imposed upon phenomena by the scientists who created these insights." Another part of the examination requires students to present and defend the eight major ideas from the entire set of readings. Not only does this requirement focus students' attention throughout the course, but provides feedback to the instructor about what ideas seem most significant.

TABLE II
Percent of students naming a particular idea as one of the eight most important in the course.

% of naming a particular issue	Rank	Issue
90	1	The role and distinction of laws and theories
75	2	Realism vs. instrumentalism
74	3	The issue of the scientific method
67	4	The role of creativity in science
61	5	Paradigms, revolutions and the progress of science
57	6	Modes of knowledge generation (deduction vs. induction)
49	7	The role of theory-driven (laden) observations
48	8	Conceptual change in science and science learning
39	9	Definitions of the social studies of science
30	10	Philosophy of science in science teaching
26	11	Human aspects of science
25	12	The role of history in understanding science

N = 87 students surveyed during four semesters of the course.

COURSE IMPACT

It is always difficult to know how much any course impacts the professional lives and practices of the students involved, but some measure of what course elements make the most powerful impression has been established. Table II presents a list of the kinds of ideas that students feel are the most important in the course as extracted from the final examinations. Each student is asked to name eight such issues. There has been some variation in this list from semester to semester since the readings list is constantly being modified and the accompanying discussion is always unique to a particular class. However, over time, the list has stabilized considerably.

CONCLUSIONS

It is clear from a review of the new science education standards that elements of the nature of science must play a role in the education of the next generation of science learners. This course was developed to provide teachers a strong background of relevant content knowledge regarding the nature of science. Even if this mode of content delivery is effective for university learners, this knowledge must now be translated through appropriate curriculum models and through the skills of individual educators into an appropriate form for the classroom.

ACKNOWLEDGMENT

The author is pleased to acknowledge that the course described here was inspired by a pair of classes -- the *Nature of Science* and the *Meaning of Science* -- developed by Dr. George W. Cossman of the University of Iowa. Although there are now significant differences between this course and those he developed, the experience described here would not exist without Dr. Cossman's fundamental contributions and inspiration.

University of Southern California, Los Angeles, California, USA

REFERENCES

American Association for the Advancement of Science (AAAS) (1994). *Benchmarks for science literacy*, New York, Oxford University Press.
Burke, J., Lynch, J. (Producer). *The day the universe changed*. [videotape], Los Angeles, CA, Churchill Films (6901 Woodley Avenue, Van Nuys, CA 91406).
Martin, M. (1972). *Concepts of science education: A philosophical analysis*, Glenview, IL, Scott, Foreman.
Matthews, M. R. (1994). *Science teaching: The role of history and philosophy of science*, New York, Routledge.
Matthews, M. R. (1990). 'History, philosophy and science teaching. What can be done in an undergraduate course?', *Studies in Philosophy and Education*, (10), 93-97.
National Research Council. (1996). *National science education standards*, Washington, D.C., National Academy Press.
Robinson, T. (1968). *The nature of science and science teaching*, Belmont, CA, Wadsworth.

APPENDIX A
DEFINITIONS OF SCIENCE AND OTHER STATEMENTS ABOUT ITS NATURE

1. A branch of study which is concerned with a connected body of demonstrated truths, or with observed facts systematically classified and more or less colligated by being brought under general laws, and which includes trustworthy methods for the discovery of new truth within its own domain. (*Oxford English Dictionary*)

2. Science is built up with facts as a house is with stones. but a collection of facts is no more a science than a heap of stones is a house. (*J. H. Poincaré*)

3. It cannot be that the axioms established by argumentation can suffice for the discovery of new works, since the subtlety of Nature is greater many times over than the subtlety of argument. (*F. Bacon*)

4. Scientists are concerned with the generalized, the ideal. Artists are usually concerned with the particular, the individual. The artist is nearer to the scientist in his capacity as problem poser, theorizer, hypothesizer, than he is to the scientist as problem solver, or checker. This latter function is usually the sole prerogative of science. (*A. H. Maslow*)

5. The belief that science proceeds from observation to theory is still so widely held that my denial of it is often met with incredulity. (*K. Popper*)

6. The aim of science is the recording and rational correlation of those parts of our experience which are actually or potentially common to all normal people. (*H. Dingle*)

7. Science is an interconnected series of concepts, and conceptual schemes that have developed as a result of experimentation and observation. (*J. B. Conant*)

8. Science searches for relations which are thought to exist independently of the searching individual. (*A. Einstein*)

9. It is a capital mistake to theorize before one has data. Insensibly one begins to twist facts to suit theories, instead of theories to suit facts. (*A. Doyle as "Sherlock Holmes"*)

10. Surveying the experimental literature... makes one suspect that something like a paradigm (a theory) is prerequisite to perception itself. (*T. S. Kuhn*)

11. Science is organized knowledge. (*H. Spencer*)

12. Science is not for the common public, but only for those spiritually prepared. Otherwise it is like pouring pure water into a muddy well; you only stir up the mud and lose the water. (*Thorn (1873) in his preface to a translation of "De Revolutionibus."*)

13. We need very much a name to describe a cultivator of science in general. I should incline to call him a scientist. (*W. Whewell*)

14. Science is what scientists do. (*J. D. Bernal*)

15. Science is the study of those judgements concerning which universal agreement can be obtained. (*N. Campbell*)

16. The whole of science is nothing more than a refinement of everyday thinking. *(A. Einstein)*

17. Science is a method for testing claims about the natural world, not an immutable compendium of absolute truths. *(S. J. Gould)*

18. The true scientific attitude is "utterly different from the dogmatic attitude which constantly claimed to find "verification" for its favourite theories. Thus I arrived . . . at the conclusion that the scientific attitude was the critical attitude, which did not look for verifications but for crucial tests; tests which could refute the theory tested, though they could never establish it." *(K. Popper)*

19. Science is the attempt to make the chaotic diversity of our sense experience correspond to a logically uniform system of thought. *(A. Einstein)*

20. Science is the interpretation of nature and man is the interpreter. *(C. Gore)*

21. Science is concerned with the general conditions which are observed to regulate physical phenomenon; whereas religion is wholly wrapped up in the contemplation of moral and aesthetic values. *(A.N. Whitehead)*

22. Every grand advance in science has issued from a new audacity of imagination. *(J. Dewey)*

23. A scientist experiences no failures, only unprofitable avenues to pursue. *(J. Salk)*

24. Science contributes to our culture in many ways, as a creative intellectual activity in its own right, as a light which has served to illuminate man's place in the universe, and as the source of understanding man's own nature. *(J. F. Kennedy)*

25. Science is a little bit like the air you breathe -- it is everywhere. *(D. Eisenhower)*

26. Science is a great game. It is inspiring and refreshing. The playing field is the universe itself. *(I. I. Rabi)*

27. Man loves to wonder, and that is the seed of our science. *(R. W. Emerson)*

28. It is a good morning exercise for a research scientist to discard a pet hypothesis every morning before breakfast. *(K. Lorenz)*

29. . . . science is the quest for knowledge, not the knowledge itself. *(H. D. Roller)*

30. . . . no human inquiry can be called science unless it pursues its path through mathematical exposition and demonstration. *(L. da Vinci)*

31. . . . science is the collection of scientific subjects taught in universities. *(D. Flanagan.)*

32.. It is one of the hopes of science that, by careful analysis of past discovery, we shall find a way of separating the effects of good organization from those of pure luck, and enabling us to operate on calculated risks rather than blind chance. *(J.D. Bernal)*

33. Science tends of be created as it is needed. *(R. Feynman)*

34. The theorists . . . often repeated the truism that progress in science comes when experiments contradict theory. *(R. Feynman)*

35. Science when well digested is nothing but good sense and reason. *(Stanislaus, Maxims (No. 43))*

JOHN O. MATSON AND SHARON PARSONS

13. THE NATURE OF SCIENCE: ACHIEVING SCIENTIFIC LITERACY BY DOING SCIENCE

Inquiry teaching is once again advocated as a central element in science teaching (NRC 1990, 1995; AAAS 1989, 1993). This requirement makes it vital for science teachers to know and understand the basic processes and philosophies of science. Unfortunately, we rarely provide useful experiences in this domain to students participating in teacher education programs. The science education program at our university incorporates into its curriculum several strategies and courses where the processes and nature of science are stressed. This chapter describes the means by which the nature of science is blended with a science education program for both preservice and inservice teachers. First, we briefly describe the preservice programs leading to both the elementary credential (grades K-8) and the secondary credential in science (grades K-12); second, we describe how we incorporate the nature of science into the inservice program.

RATIONALE

It has been our experience that few science majors and even fewer teachers are prepared to take on the role of scientists simply by earning a bachelor's degree in science. Most undergraduate course work is content rather than process oriented. The laboratory experiences of undergraduate students tend to be verification experiments, with known results, or are designed to teach techniques rather than investigate processes. Our experience suggests that many practicing teachers are inadequately prepared to satisfactorily teach science via inquiry methods.

An understanding of the processes and nature of science and an ability to do scientific inquiry is a requirement for effective science teaching. It does not suffice to simply teach facts and have students perform "cookbook" laboratory experiments; therefore, science teachers must understand the thinking and behavior of scientists and develop methods to communicate this understanding to students. All those engaged in science teaching and learning must be able to carry out research projects by asking pertinent questions, construct hypotheses, predict outcomes, design experiments, analyze data, and reach conclusions. In brief, science teachers must be able to *do* science. However, experiencing the processes of science by itself is not sufficient. A teacher of science must also bring to the classroom the attitude and world view of scientists. To achieve this, a basic understanding of the philosophies of science is necessary. With a basic science content background and the ability to carry-out the process of science, science teachers can teach science as a conceptually oriented, hands-on/minds-on, problem solving, critical thinking activity which will promote science literacy among students.

One approach that emphasizes the importance of a constructivist epistemology in education was presented by von Glasersfeld (1989). Essentially, in constructivism, knowledge is not considered an absolute. Knowledge is the result of the social, cultural, and historical milieu. In this sense, knowledge is constructed individually based upon a person's socio-cultural background (Saunders 1992). In our program we have noticed that while many science faculty members profess a constructivist philosophy of knowledge, in reality they act as logical positivists, or objectivists in their teaching. In essence, they teach a set of "truths." It would appear that there exists a vestigial transcendentalism among many logical positivists who wish to maintain an idealized concept of 'truth' in nature rather than to embrace the notion of multiple truths dependent upon the perceptions of the viewer. While constructivism would appear to be a powerful epistemological philosophy to bring about conceptual change in the classroom (Nussbaum 1989, von Glasersfeld 1989), few teachers actually use such an approach (Tobin et al. 1991).

Most science educators and philosophers of science view scientific knowledge as tentative (Hempel 1966, Lawson 1995 and others; with Lewis (1993) for an alternate view). Because scientific knowledge is constantly being questioned, re-evaluated, and tested, changes in what we teach will inevitably occur. The conditional nature of scientific knowledge is distressing to some people. They feel that knowledge should be stable; you can add to it but unless it was wrong to begin with it should not change. Interestingly, professional scientists frequently make the same comments. These scientists may have been involved with doing "normal science" in the Kuhnian (1970) sense. I have had similar comments from both preservice and practicing science teachers. Science itself is not static nor should science teaching be unchanging. We need to insure that teachers are aware of this aspect of science.

The faculty of the science education program at our university essentially view knowledge as temporary constructs by which we attempt to make sense of the world. Therefore, our preservice and inservice classes and programs emphasize the nature and processes of science. We have also encouraged change among our university colleagues and their teaching, especially, of basic introductory science courses.

The basic philosophies of science, whether directly taught or modeled within the classroom and laboratory, are integral to our program. In the research courses described below, professors actively discuss with preservice teachers the nature of science, the temporary nature of scientific knowledge, and problems involved in interpreting and analyzing data. Many courses within the sciences also do this but in the science methods courses (both elementary and secondary), the nature of science is a major theme.

NATURE OF SCIENCE STRATEGIES IN PRESERVICE SCIENCE EDUCATION

The Character of our Science Education Program

The preservice education program discussed here is a fifth year program that requires and builds upon a bachelor's degree. For those students pursuing the elementary (K-8) credential, Natural Science is the only science academic major available. The secondary credential builds on a student's bachelors degree in one of the sciences by providing a teaching emphasis (or equivalent course work or experience). Both credentials include, in addition to an array of education courses, a science teaching methods course and two semesters of student teaching.

The major in Natural Science for the elementary credential has relied on courses that historically emphasize a survey of subject content. The Natural Science major is currently being revised to require students to take the core introductory courses required of science majors. An overview of these courses and the inclusion of the nature of science strategies follows those given below for the secondary credential.

Students who seek a Single Subject Credential and earn their science degree at our university are introduced early to the nature of science because we have encouraged the faculty in the various science departments to make drastic changes in their conceptual approaches to teaching by increasing coverage of the processes and nature of science. All science teaching majors are required to take at least one year in each of the four majors (biology, chemistry, geology, and physics) sciences.

Integrating Nature of Science with Science Education

We have changed our conceptual approach most dramatically in the biological sciences. An introductory course "Computers in Biology" introduces students to the basic idea of scientific inquiry and statistics. This causes students early in their academic careers to think quantitatively and ask questions about how we know what we know. Major revisions have occurred in the three-semester introductory core courses for biology majors (Plant Biology, Animal Biology, and Cell and Molecular Biology). These courses now include more guided and open inquiry-based laboratory activities. More emphasis has been placed on concepts rather than the traditional content orientation that is so prevalent in introductory courses. Similar changes and modifications are taking place in the other disciplines. Table I summarizes the similarities and differences between the elementary and secondary programs.

In biology, for instance, "Hypothesis Testing" is an upper division course required of all majors that has had a tremendous effect on the ability of students to conduct real science investigations (Appendix A). Students learn basic univariate and multivariate statistical techniques and use a set of computer generated data to design experiments to test hypotheses. Since this course was initiated, students entering the credential program have shown a much greater understanding of the process of science.

TABLE I
Summary of the nature of science concepts introduced to prospective science teachers in the elementary and secondary programs discussed here.

	Elementary Program	Secondary Program
Epistemological Approach	Constructivism	Constructivism
Approach	Socio-cultural	Socio-cultural
Inquiry-based Science	Process of Science: Focuses on directed inquiry	Process of Science: Focuses on directed inquiry
Majors Non-science Majors	Natural Science Several majors that meet the multiple subjects criteria	Major in one of the science such as biology, chemistry, geology, meteorology or physics
Science Concepts	Limited to general education science courses	All majors in science teaching take one year in each of the four major science areas (biology, chemistry, geology/ meteorology and physics) in addition to specialized courses in their major
Science Methods Courses	Elementary Science Teaching (2 hrs/wk)	Secondary School Science (6 hours per week)
Nature of Science Instruction	No special course, the nature of science is embedded in the methods course	Biology majors: Hypothesis Testing (3 hours) All majors: Research and Investigation Techniques in Science (3 hrs) and NOS content embedded in the secondary methods course

Biology majors who enter the teaching emphasis concentration are required to complete an experiment-based Senior Thesis. This thesis must include defining a problem, forming a hypothesis, designing an experiment to test the hypothesis, and analyzing data using accepted statistical procedures.

Prior to admission to the science credential program, we require that all science credential candidates demonstrate the ability to engage in the processes of science. First, students in our undergraduate program are required to complete both the Hypothesis Testing course and prepare a Senior Thesis. Students who come to us from other institutions must show evidence of similar work as demonstrated by a research paper from a class project or industrial research of which they are the primary or sole author.

If these conditions are not met, we offer a course "Research and Investigation Techniques in Science" (see Appendix B). This course requires students to complete a preliminary research paper. Then, using the results of that first paper the student generates another related hypothesis; evaluates that second hypothesis and writes a second paper that incorporates both research questions. We believe that these experiences, at least, give preservice teachers an authentic view of what science is. It is a basic introduction to the process of science. These requirements are in line with new guidelines and standards for teacher preparation at both the state (Commission on Teacher Credentialing, 1992) and national (NRC 1995) levels.

We have recently designed and implemented a capstone course for senior science and mathematics majors, and available to inservice teachers, that emphasizes integration of science and mathematics. Using the *Science Framework for the State of California* (California Department of Education, 1990) as a guide, the course includes a curriculum based upon the theme of Evolution. The major thrust of this course is to demonstrate to preservice mathematics and science teachers how the various disciplines interact, rely upon each other, and form the basis for scientific knowledge. It emphasizes the nature of scientific inquiry in bringing about or modifying knowledge. In the course, we are more interested in how scientific knowledge is constructed rather than on the end product itself.

During a fifth year, credential students are in one of the two (elementary or secondary) programs. During this year, they take a course in science teaching methods and student teach. These courses emphasize the nature of science as it applies directly to teaching science.

An "Elementary Science Teaching" course (two hours per week) is taken by all elementary teaching credential candidates. Few of these students have a rich science background. While the course places priority on the nature of science, via inquiry, it is not sufficient to prepare students to teach science. Students who major in Natural Science take a separate science methods course taught by science faculty. This course emphasizes the nature of science and builds upon the science background of its students. In these methods courses there is an emphasis on science from a socio-cultural perspective (Barba, 1998).

Essentially, faculty members share a constructivist view of knowledge (Saunders 1992) from a socio-cultural position. A constructivist view of knowledge fits in well with the "new philosophy of science" described by Strahler (1992) who contrasts previous philosophies of science (e.g., logical positivism) with our current understanding of how scientific knowledge is acquired. While scientific knowledge may be viewed through a number of different lenses, constructivism is a more appropriate approach when the multicultural nature of our students is taken into account.

Recently, a science emphasis student teaching section (Parsons and Reynolds 1995) was initiated in the elementary teaching credential program to assist student teachers interested in science teaching regardless of their science background. This section tends to attract students who have had science or technology related work experiences, or who have recently discovered that they have an interest in science. This special section places student teachers with science teacher leaders during their field experience. The science teacher leaders help to facilitate science teaching by the student teachers. While this is still an experimental student teaching experience, the idea has received support from local school districts, science teachers, and science educators. The majority of students in this section plan to complete a supplemental authorization in science which will enable them to teach science through grade 8..

In the secondary science methods course (six hours a week), we include discussions and activities dealing with the limits of scientific inquiry, the tentative nature of scientific knowledge, the differences between science and technology, controversial topics in science, and how science interfaces with society. Students identify major concepts in their field and prepare a teaching unit that addresses one of those concepts. In the preparation of their teaching unit, students address the nature of science by preparing a series of inquiry-based lessons using a variety of teaching techniques.

During the student teaching phases of the secondary credential program, students are teamed with science teacher leaders (described below in the inservice program ISISS). These lead teachers have been chosen because of their science expertise and understanding of the importance of the nature of science in science education.

It is important for future teachers to understand that science is but one way to know and that they have a basic understanding of the process and nature of science. Teachers thus prepared are better able to address the educational concerns and needs of the majority of our population who will not become scientists, nor go on to college.

INSERVICE SCIENCE EDUCATION AND THE NATURE OF SCIENCE

We maintain and cultivate an active and positive relationship with local science teachers. Two of our recent inservice programs, Improving Science Instruction in Secondary Schools (ISISS) and Evolution and The Nature of Science (ENSI), both funded by the National Science Foundation, were directed at improving high school science teaching by improving the basic understanding of the nature of science.

The primary objective of the ENSI project is founded on three basic theses:

> ...first, that a clear understanding of the tentative and self-revising nature of science is an especially powerful vehicle for fostering higher order critical thinking; second, that evolution can only be appropriately understood and taught within a framework emphasizing the nature of science and third, that the teaching of both evolution

and the nature of science can be enhanced considerably by helping teachers make curricular and lesson revision ongoing, self-correcting, and experimental enterprises.

The Evolution and Nature of Science Institute is not limited to local science teachers, but it brings together educators from throughout the country to act as catalysts for change in science teaching. (See Nelson, et al. in this volume for a description).

The ISISS project has had a more local effect. It is designed primarily to update science teachers' content knowledge, give them experience in the process of science, and to develop a cadre of lead teachers that will be able to bring about change in science teaching. During two recent summers the participating science teachers were involved in experimental field investigations and subsequent curriculum development based upon the results of these field studies. The idea was to have the teachers conduct 'real science' investigations of their own design and use the results to generate inquiry-based lessons for the classroom. A series of meetings in the spring and early summer allowed the teachers time to pose questions, develop hypotheses, and design field experiments to test their hypotheses. We then spent two to three weeks in the field to collect data. During the following fall and winter the results were used to generate curricula for use in the classroom. One such unit that was developed involves the population ecology of small mammals implemented this year. Another involves a study of changes in the water chemistry along the entire drainage of the American River system.

San Jose State University, San Jose, California, USA

REFERENCES

AAAS (1993). *Benchmarks for science literacy*, New York, Oxford University Press.

AAAS (1989). *Project 2061: Science for all Americans*, Washington, D.C., American Association for the Advancement of Science.

Barba, R. (1998). *Science in the multicultural classroom*, 2nd ed, Boston, MA, Allyn and Bacon.

California Department of Education (1990). *Science framework*, Sacramento, CA, California State Department of Education.

Cobern, W. W. (1993). 'College Students' Conceptualizations of Nature: An Interpretive World View Analysis', *Journal of Research in Science Teaching*, (30), 935-951.

Commission on Teacher Credentialing (1992). *Science teacher preparation in California: Standards of quality and effectiveness for subject matter programs*, Sacramento, CA, California State Department of Education

Hempel, C. G. (1966). *Philosophy of Natural Science*, Englewood Cliffs, NJ, Prentice Hall, Inc.

Kuhn, T. S. (1970). *The structure of scientific revolutions*, 2nd ed., Chicago, IL, The University of Chicago Press.

Lawson, A. E. (1995). *Science teaching and the development of thinking*, Belmont, CA, Wadsworth Publishing Co.

Lewis, R. W. (1993). 'The nature of scientific knowledge debated, Letter to the Editor', *The American Biology Teacher*, (55), 262.

National Research Council (1990). *Fulfilling the promise: Biology education in the nation's schools*, Washington, D. C., National Academy Press.

National Research Council (1995). *National Science Education Standards*, Washington, D.C, National Academy Press.

Nussbaum, J. (1989). 'Classroom conceptual change: Philosophical perspectives', in D. E. Herget (ed.), *The History and Philosophy of Science in Science Teaching*, Tallahassee, FL, Florida State University, 278-91.

Parsons, S. and Reynolds, K. (1995). *Establishing an action research agenda for preservice and inservice teachers collaboration on self-empowerment in science*, Paper presented at the Annual General Meeting of the National Association of Research in Science Teaching, San Francisco, CA.

Saunders, W. L. (1992). 'The constructivist perspective: Implications and teaching strategies for science', *School Science and Mathematics*, (92),136-141.

Strahler, A. N. (1992). *Understanding science: An introduction to concepts and issues*, Buffalo, NY, Prometheus Books.

Tobin, K., Davis, N., Shaw, K., and Jakubowski, E. (1991). 'Enhancing science and mathematics teaching', *Journal of Science Teacher Education*, (2), 85-89.

von Glasersfeld, E. (1989). 'Cognition, construction of knowledge, and teaching, *Synthese*,(80), 121-140.

APPENDIX A
Hypothesis Testing: A Course Description

This is a course in applied statistics for biologists planning to conduct manipulative experiments. The intent of the course is to provide biologists with training in experimental design and quantitative analyses in cases where specific experimental hypotheses are to be tested. Material for the course will consist of lecture, readings from the textbook, and practical experience in computing statistical procedures by hand and with the aid of a computer. Theory and concepts will be covered in lectures and reading assignments. Practical experience is provided through assigned problems requiring both hand calculations and computer-based computation.

APPENDIX B
Experimentation and Investigation Techniques in Science:
A Course Description

This course deals with analytical thinking and experimental design in science -- processes that are vital to the success of the scientific enterprise. Together, these processes are often called the scientific method, but, in reality, there exists not one method, but a series of variations on the general investigative theme that can be logically applied to selected cases. In this course you will apply the skills of analytical thinking to scientific investigations.

This course is designed primarily to meet the needs of future science teachers, who must understand scientific processes in order to teach science effectively, but any student of science should understand these principles as part of their training. This course requires considerable independence and assumes the student is motivated to pursue the methodology of science. In effect, this course has more of the format of an independent research project than it does the format of conventional science courses. Furthermore, this course deals with the skill of scientific thinking, rather than with any specific knowledge of scientific content. Thus, evaluation of student performance in this course is based on the student's competence and productivity in the skills of scientific design, experimentation, and written communication.

Skills are learned by practice in a laboratory experience with students performing most of the design and analysis required. The instructor is a coach and not a lecturer and beginning in the second week of the semester, students will be the "thinkers in charge" of projects. Consequently, students must assume a high degree of independence and responsibility for your own learning.

Because scientific investigation requires the collection and data analysis, the experimental protocols must be exact and allow quantification when possible. To simplify the need for complex equipment while simultaneously providing experience with software that can be used in science teaching, students will use instruments that interface with a computer. The software will display and record the data so you will not be burdened with these details initially.

The course begins by having students gain familiarity with microcomputer-based laboratory probes followed by the generation of research questions that can be subjected to classic hypothesis testing procedures. Each student then performs the specific experiment to test the prediction. Students are evaluated based on their experimental design, analysis (including the use of basic statistical tests) and written presentation.

YVONNE MEICHTRY

14. ELEMENTARY SCIENCE TEACHING METHODS: DEVELOPING AND MEASURING STUDENT VIEWS ABOUT THE NATURE OF SCIENCE

This chapter provides an overview of the nature of science components within a 3-credit hour elementary science methods course. These components include both scientific inquiry and the character of scientific knowledge. Scientific inquiry is described as those processes used to generate and test scientific knowledge. The nature of scientific knowledge, as defined by Rubba and Anderson (1978), is portrayed as developmental (tentative), testable (capable of empirical test), creative (partially a product of human creativity), and unified (the specialized sciences contribute to an interrelated network of laws, theories, and concepts). Although a distinction is made between scientific inquiry and the character of scientific knowledge, the interdependent nature of these two areas is made evident. Defining the nature of science in this way has proven useful with this student population not only because their understanding of these aspects of the nature of science is very limited, but because these definitions are applicable to the teaching of elementary grade students.

DESCRIPTION OF THE COURSE

During the course, strategies are used to reveal, change, and assess student views about the nature of science prior to, during, and upon completion of the course. Various teaching approaches and activities are integrated within the course to help students develop more adequate views about the nature of science. In addition to strategies designed to develop understandings about the nature of science, an attempt is made to help these students relate what they learned in their methods class about the nature of science to the teaching of elementary science.

Instructional strategies applied throughout the semester enhance the learning of preservice teachers in a number of ways. By modeling effective teaching strategies for use with elementary students, providing "real" science experiences for preservice teachers to construct their own knowledge of the nature of science, allowing time for reflection on new understandings, and making an explicit connection between the learning of preservice and elementary students, preservice teachers not only develop their own understandings while learning effective teaching strategies; they also gain insights about the ways that their future students experience learning.

The study of the nature of science throughout the semester is addressed through the broader context of scientific literacy. The definition of scientific literacy used in this science methods course is characterized by four general themes. These themes are: 1) basic understandings of the knowledge base of science, 2) the understanding of and ability to use scientific processes, 3) the use of higher order thinking skills, and 4) scientific attitudes and values. Specific attitudes and values discussed include objectivity, openness, the value of trial and error, intellectual honesty, and tolerance for ambiguity.

DESCRIPTION OF TEACHING METHODS

During the one-semester elementary science methods course, students receive direct classroom instruction for a period of 12 weeks and complete a 4-week field experience in the schools. The class meets for a period of 75 minutes, twice weekly. Major course requirements and experiences include the following elements designed to integrate the nature of science:

1) daily hands-on, inquiry learning,
2) teaching a learning cycle lesson to peers and to elementary students, (including a reflective analysis about what was learned about science and science teaching),
3) conducting a controlled research experiment involving observations made over time, writing a research report, and sharing results with peers, as well as a follow-up reflective analysis of the long-range experiment, summarizing what was learned about the nature of scientific inquiry and how such an experiment might be structured in an elementary classroom,
4) field assignments involving science teaching, interviewing children for a variety of purposes, and describing science programs used in schools and classrooms,
5) development of a unit that integrates science with other subject areas,
6) quizzes and a final exam.

Students enrolled in this methods course are college seniors and graduate students seeking teacher certification. The vast majority of these students participate in the methods sequence the semester before student teaching. At minimum, students have taken a biology course taught by science faculty members and two physical science courses taught by science education faculty specifically for elementary education majors. Each of these three science courses has a laboratory component.

Learning Cycle Lessons

One strategy used to develop student understanding of the nature of science is the learning cycle approach. This approach is taught and modeled by the instructor during the beginning of the semester. After students have participated in three different lessons taught by the instructor using a learning cycle approach, the various phases of concept exploration, invention, and application are defined and discussed. The discussion of this approach focuses on the learning benefits in the context of the four areas of scientific literacy mentioned previously: the science content knowledge base is learned through an exploration and discovery process actually used by scientists in their work to generate scientific knowledge, students are engaged in higher-order thinking using the processes of science, and the attitudes and values of scientific inquiry are developed through a trial and error method and atmosphere of openness.

After experiencing the learning cycle in the role of an elementary student, these preservice teachers are then required to select a concept and teach this concept to their peers, using the learning cycle approach. As part of this requirement, students write a reflective analysis about what they learned about science and science teaching as a result of using the learning cycle approach to teach a peer lesson. Questions to which students respond are shown in Table I. Questions three and four, in particular, are designed to elicit changes in students' understandings and beliefs about scientific inquiry and the nature of scientific knowledge.

TABLE I
Reflective analysis of peer lesson

1) Describe what you believe went very well (strengths of the lesson). Explain your response.
2) Would you do anything differently next time you teach the same lesson? If so, what would you do differently and why? If not, why?
3) Describe what you have learned about science and science teaching as a result of teaching this lesson.
4) Explain any change(s) in attitudes or beliefs about science teaching that took place as a result of planning and teaching this lesson.
5) Describe any fears that existed about teaching this lesson to your peers. Were these fears realized or diminished as a result of teaching the lesson?

Students' peer lessons are scheduled throughout the semester. The length of peer lessons averages thirty minutes. An attempt is made to schedule all of the peer lessons prior to the field experience. The idea is to allow students to "practice" with their peers prior to teaching a learning cycle lesson with elementary students.

Design and Implementation of a Long-Range Research Experiment

Another teaching strategy resulting in improved understanding of the nature of science is to have students conduct a long-range experiment. Research questions have included how light affects the quality of plant life, which household cleaner is most effective at stopping bacterial growth, what factors affect the spoiling of a banana, and how will the addition of oil affect the evaporation rate of water?

Students, working alone or in small groups, are in complete control of this experience, from the development of a research question which interests them to the formulation of a hypothesis, the design of a controlled experiment, the collection and recording of data, the interpretation of results, and the drawing of conclusions based on their results. Students are required to submit a written report of their work and share their results with the "scientific community" of their peers. The guidelines for the written report are presented in Figure II.

TABLE II
Guidelines for the completion of the long-range experiment

1. Research question - state clearly the question you are investigating.

2. Hypothesis - state your prediction regarding the answer to the research question.

3. Materials - list all materials and amounts/number of materials used to conduct the experiment.

4. Procedure written clearly enough so that other "scientists" could easily replicate the experiment with a step by step list including how you set up the experiment, what are the controls and variables, and what methods were used to collect and record data.

5. Record of data collected - include all data collected during the experiment, in an easy to interpret form. This form could be a table,

graph, organized notes, photographs, or a combination of these. Be as specific as possible in your descriptions of data.

6. Results - write a narrative which describes the data collected throughout the experiment. Information presented in visual displays should be interpreted for the reader.

7. Conclusions - begin this part of your report by stating whether the hypothesis is accepted or rejected. State your reasons for this decision. Explain what the "research community" can learn from the results of your experiment. Include any research questions which you would recommend for follow-up or future research about the same topic or similar topics.

8. Bibliography - list references used to help conduct the experiment.

Students submit their written research report and make an oral presentation of their research study to classmates, and are then given a follow-up written assignment to be due the following class period. This follow-up assignment requires that students summarize the individual learning that occurred as a result of conducting the long-range experiment. Students are asked what they learned about the nature of scientific inquiry as a result of conducting this experiment and what ideas about structuring long-range experimentation and observation with elementary students they developed.

The questions and instructions provided for this summary are open-ended. The intent of the instructor, employing a constructivist approach, is to assess learning about the nature of science "constructed" by students as a result of actually using the scientific inquiry process. For that reason, the students are asked to respond to the two questions immediately upon completion of the experiment, written report and oral presentation, prior to any class discussion.

A constructivist approach is used during the learning cycle lessons and long-range experiment by providing students with a common, first-hand experience upon which they construct their own understandings about the nature of science. Topics related to the nature of science are not discussed by the instructor until students first have the opportunity to construct their own understandings. Instructional emphasis is placed on the inquiry process of generating knowledge, the developmental, testable and creative nature of scientific knowledge, and the scientific attitudes and values of objectivity, openness, importance of basing conclusions on scientific evidence, intelligent "failure," and the social context of science. Instructions for the reflective analyses following both experiences are consistent with the constructivist approach in that students are provided with open-ended questions, which allow them to construct and synthesize their own learning.

Quiz: The Nature of Science

Once the long-range experiment and reflective analysis are completed and discussed, students take one of four quizzes given during the semester. The quiz requires students to react to the appropriateness of any one of the following three statements below: The results of my experiment proved that mold grows best under moist and warm conditions; The hypothesis of my experiment was wrong, or The results of my experiment are wrong. Students are also presented with the following scenario and asked to explain what it tells them about the nature of scientific knowledge: Natural philosophers once held that the earth was the center of the solar system. They now have shifted their view and have put the sun in the center.

In addition to these two questions, students are presented with a hypothetical situation in which an elementary student has set up an experiment with two control variables. Students are to read the scenario and then determine whether the conclusions of the experiment were valid. They are not told that there are two control variables. This scenario is as follows:

> Clarietta Rutherford, a student at Mayfield Elementary School, developed the following research question, "How do the emotions of hate and love affect plant growth?" She plants Bermuda grass in two separate containers and shows hate toward one plant and love toward the other. After three days, the plant toward which she showed love is thriving, a healthy green color, and several inches taller than the plant toward which she showed hate, which is visibly dying. She concluded from this experiment that showing the emotion of love toward a plant positively affects its growth, while showing an emotion of hate toward a plant will have negative effects on the growth. When asked by her teacher if she had watered both plants equally. "No," she replied, "Why should I water I plant that I hate?" Are the conclusions drawn by Clarietta valid?

EVALUATING THIS STRATEGY:
MEASURES OF STUDENT VIEWS ABOUT THE NATURE OF SCIENCE

Several tools and techniques are used to measure student views on the first and last day of class, in addition to gauging their evolving views at several points during the semester. The intent is to determine the extent and nature of change in student views about the nature of science and science teaching as a result of different classroom experiences and as a result of the combined classroom work during the semester.

Learning Cycle Lesson

Student responses as part of the reflective analysis frequently indicate that they have developed understandings about the nature of science. These understandings include the objectivity and open-mindedness involved in drawing conclusions, the tentativeness of scientific results, the importance of "scientific evidence," the importance of trial and error, and the importance of replicating results. In addition to developing an understanding of the nature of science, students become aware of the importance of developing the same understandings of their future students.

Conducting Long-Range Experiments: Understandings Developed

Student responses to the question, "What did you learn about the nature of scientific inquiry as a result of conducting this experiment?" overwhelmingly reveal a greater depth of student understanding, and in many cases, the development of a more positive attitude toward the usefulness of scientific inquiry. Although a great number of students said explicitly that they developed a greater understanding of scientific inquiry as a result of conducting the long-range experiment, it is evident that all respondents acquire a greater understanding. Examples of student responses are presented in Table III.

TABLE III
Student responses to the long range experiment

Students have indicated that:

- The inquiry process generates more questions.
- Having an open mind about experimental results is important.
- The inquiry process is a never-ending search for knowledge.
- The usefulness of the steps of the scientific method is evident.
- The meaning of trial and error in science is more clear.
- Conclusions are often inconclusive and lead to other questions, hypotheses, and experiments.
- Results can change each time you do an experiment.

- When an experiment has been done many times, you should still say that the results "indicate" that . . .
- Repeating experiments is important to get valid results.
- Hypotheses are never wrong. They're just not supported by results.
- Experiments don't fail - you learn from results.
- The results of an experiment don't "prove" anything.
- Inquiry involves creative thinking.
- Science experiments can and should be simple.
- The simplest of experiments can yield much information.

An area of positive change that occurs as a result of students conducting the long-range experiment, not reflected in Table III, concerns an internal shift in student belief systems that is very important to understanding the nature of science. Many of the students describe a shift in feelings from initially being very disappointed by having to reject their hypothesis to feelings of "it's really okay if the hypothesis is rejected." One student even said she was glad her hypothesis was rejected because of what she learned!

Another interesting result is that many students state the experience of conducting a long-range experiment was their first ever. One student said it was his second experience, the first being a 7th grade science fair project.

Conducting Long-Range Experiments With Elementary Students

Although many responses to the question, "What are your ideas about structuring long-range experimentation with elementary students?" relate to areas of learning not directly related to the nature of science, such as structure of student groups and time issues, there are numerous and varied responses related to the nature of science. First, virtually every student notes the importance of doing long-range experimentation with elementary students. This in itself, is viewed as a remarkable and significant result of students conducting their own experiment in a methods course.

The most frequent response given is the importance of teaching children that they have not failed if their experimental results are not what they expected. Another frequent response is the importance of children being involved in the development and progress of the experiment from beginning to end. Several students mention the importance of having students develop their own questions to investigate. Other responses include the importance of teaching the processes of scientific methods, the relevance of experimentation to real life, development of clear research questions, hypothesizing, observation skills, the concept of controls and variables, methods of reliable data collection, and basing conclusions on experimental evidence.

The developmental, testable, and creative nature of scientific knowledge are represented in student responses to how long-range experimentation should be conducted with elementary students. The most significant growth in understanding, as reported by students, relates to the developmental nature of scientific knowledge.

Semester Results: Qualitative Analysis

As part of the final semester assessment, students are asked to respond to the questions, "What is science?," and "How do you think science should be taught at the elementary level?"

Regarding the question, "What is science?," on the first day of class, students heavily emphasize the knowledge element of science in their definition of science and description of how it should be taught. Their views about the processes of science are largely incomplete. Pre- and post-responses relating to science content include the environment, the branches of science, use of fact, theories, and laws, ways of describing the world, physical and abstract view of world, an answer to many questions, and core subjects dealing with both known and unknown factors. Most important, the number of student responses related to science as a process doubled when comparing the post responses to the pre-responses. Examples of process-oriented responses include a way of thinking about problems and curiosities, a method of discovery, an organized process in which ideas are tested, conducting an experiment to test a hypothesis, science is ever-changing and growing with new information, systematic approach to obtain knowledge, involves repeated trials, science is an ever-changing experience, discovery-inquiry-exploration, and going through a process that involves thinking and may involve attitudes and values.

With respect to the question about how elementary science should be taught, there is a significant decrease in the number of content-oriented responses relating the teaching of elementary science when comparing the post- and pre-responses. Examples of content-oriented responses include how science affects students, teaching basic facts, understanding versus memory, teaching of concepts sequentially, and teaching about animals, habitats, substances, etc. The number of responses which relate directly to teaching about the nature of science more than double from pre- to post-response. Examples of responses which relate to the nature of science include the following: teach basic methods of discovery, teach and apply scientific methods, hands-on learning for the purpose of teaching process, as a process to gain knowledge, finding answers to self-questions, design their own means of solving problems, and allow students to test theories.

Semester Results: Quantitative Analysis

The Modified Nature of Scientific Knowledge Scale (MNSKS), developed by Meichtry (1992), is also used to measure and compare student views about four dimensions of the nature of scientific knowledge at the beginning and end of the semester. The four dimensions of scientific knowledge measured are the developmental, testable, creative, and unified nature of science. The MNSKS is a modified version of the Nature of Scientific Knowledge Scale (NSKS), developed by Rubba and Anderson (1978). Original modifications of the NSKS, based on readability, reliability issues, and changing views of the nature of scientific knowledge, were made to create a more suitable instrument for use with middle school students.

The MNSKS is administered as a pretest on the first day of class after students responded to the questions, "What is science" and "How should science be taught at the elementary level?" The MNSKS is administered after students have responded to the same two questions as a pretest. In addition to an overall score on the MNSKS, mean scores for each of the four subscales (developmental, testable, creative, and unified) are calculated.

To determine whether the post-test mean scores of each of the four subscales of the MNSKS and the overall instrument differ significantly from pre-test mean scores, a paired comparison t-test analysis is conducted. To determine statistical significance at the .05 level, t statistics are calculated.

TABLE IV
Paired-comparison t-test for the MNSKS pretest and posttest

Variable	Mean Difference	STD Error	t PROB
Developmental	1.31	0.38	0.0009
Testable	2.03	0.53	0.0003
Creative	2.30	0.56	0.0001
Unified	1.39	0.39	0.0007
MNSKS Overall	6.49	1.19	0.0001

N=67

The results of the paired t-test analysis, presented in Table IV, indicate that the increase in student understanding of the nature of scientific knowledge during the semester is statistically significant at the 0.001 level. The results presented here, based on three classes of students during the same semester, are typical.

CONCLUSIONS

Given the time constraints on the one-semester elementary science methods course, instructors must be selective about what they teach to best prepare students to teach science. Developing student understandings of the nature of science is one of many areas to be addressed. The results reported here indicate that students begin their methods course with understandings of the nature of science which are largely incomplete. The results also indicate the potential of developing more complete understandings about the nature of science as a result of integrating teaching strategies throughout the semester. The students enrolled in this elementary science methods course develop significantly greater understanding about the nature of science throughout the semester, providing a sound foundation upon which students can continue to develop their understandings. These newly-acquired understandings increase the potential that these prospective teachers will assist their own students in developing authentic understandings and beliefs related to the nature of science.

Northern Kentucky University, Highland Heights, Kentucky, USA

REFERENCES

Meichtry, Y. (1992). 'Influencing student understanding of the nature of science: Data from a case of curriculum development' *Journal of Research in Science Teaching*, (29), 389-407.

Rubba, P. & Anderson, H. (1978). 'Development of an instrument to assess secondary school students' understanding of the nature of scientific knowledge', *Science Education*, (62) 449-458.

15. NATURE OF SCIENCE: IMPLICATIONS FOR EDUCATION AN UNDERGRADUATE COURSE FOR PROSPECTIVE TEACHERS

The *Nature of Science: Implications for Education* is an introductory science education course offered as part of a Bachelor of Education degree program by the Faculty of Education at the University of New Brunswick in Fredericton, New Brunswick, Canada. The course was developed and is taught by a science educator and a historian of science, who share concerns about the limited understanding of science and the nature of science that most prospective teachers demonstrate.

After teaching undergraduate science methods courses for six years, the science educator believes that the majority of preservice teachers interested in teaching secondary level science represent the minority who found high school science a successful experience; they represent those who are interested and succeed in university science programs. Further, the science educator believes that the greatest barrier to most beginning teachers' becoming effective teachers of science is their own science experience. Those who had predominantly positive science experiences see no need to change either their views of science or the understanding of science we should expect of high school graduates. Those who had negative and less successful science experiences have little self-confidence in either discussing what science is or how it should be taught. Regardless of background, the science educator has yet to find a preservice teacher who challenges the notion of science as defined by topics alone. Prospective elementary teachers generally talk about science as the study of such topics as rocks, butterflies, trees, clouds, or magnets; these students portray science as "owning parts of the world." Prospective secondary teachers often describe science as disciplines, such as biology, chemistry, physics, or sometimes more specifically, astronomy, biochemistry, or ecology; these students parcel science as knowledge into clearly bounded units and subunits. The textbooks, curricula, and science classes these students experienced, which also generally portray science as a series of topics or disciplines to be covered, are the context within which their notions of science were constructed.

The historian of science was aware that students, including science students, lack understanding of the radically revised conceptions of science that have come out of work in the philosophy, history, and sociology of science in the past four decades. In this period, a new and historically-informed philosophy of science began to challenge traditional models of scientific change, especially falsificationist models.

More recent trends, mainly rooted in sociology, have emphasized the social-constructivist aspect of science and the "negotiated" nature of scientific knowledge-claims. More recently still, science has been subjected to more explicitly political criticism for its alleged masculine and Western ethnocentric assumptions. In different ways, each of these trains of thought envisions science as an enterprise much more firmly embedded in, and more deeply conditioned by, society and culture than was believed to be the case thirty years ago or than is widely assumed today. Yet most science teachers still enter upon their careers with only a fleeting awareness of these revised concepts of science.

Together we wanted to find a way to encourage preservice teachers to clarify and articulate their own tacit notions of science and teaching science by challenging them with alternate conceptions of science shown within the history of science, and held by the science education and science communities. We not only wanted to introduce prospective teachers to the debates concerning science and its nature, but we wanted to involve them as participants in these debates. Further, we wanted prospective science teachers to construct more complete understandings of science, particularly for making pedagogical decisions.

After sixteen years of schooling, students are adept at being intellectually engaged in learning the content of their courses without becoming personally involved. However, we believe being personally involved is critical if preservice teachers' beliefs about science and its teaching are to be influenced. Consequently, we tried to create a course that would be a context for involving prospective teachers in conversations concerning what is science, what is the nature of science, what counts as understanding science, and what our beliefs about these questions mean for the way we teach science. The course allows those who were successful at and interested in science to hear the voices of those with science anxiety.

THE COURSE

This course is a modified seminar with the instructors identifying and introducing ideas and issues for discussion through readings, lectures, guests, and assignments. The course aims to help prospective teachers identify and articulate their current beliefs about science and about teaching science, exposes them to new arguments and perspectives, and shares examples of other educators who question traditional practices and approaches to science teaching.

Course Participants

The course currently draws prospective teachers from the elementary, middle, and high school-level education programs. Course participants include prospective teachers who have wide differences in their science backgrounds and with distressing to highly successful science experiences. The majority of prospective teachers who enroll are at the beginning of their education program, having already completed an undergraduate degree. A few participants are pursing two bachelor's degrees concurrently, and are taking this course as their first education course. Some participants are completing their four-year B.Ed. degree (a program currently being phased out); these students have substantial education background but little or no science background. We found the participants' varying backgrounds a worthwhile challenge and considered that diversity critical to the success of the course.

We try to limit the course to thirty participants, but due to the demand we have accepted more. Most of the preservice teachers in the course are specializing in science education. For them, the course is one of four required for a concentration. A few preservice teachers who are in the generalist program which requires only one science methods course use this course to fulfill that requirement.

Course Structure and Format

The prospective teachers in our course are responsible for participating in discussions, asking questions, and synthesizing ideas. Most prospective teachers in this introductory course are generally unfamiliar with the ideas, issues, and language associated with education in general, and with science education and the history of science, in particular. To feel comfortable in their class interactions, we found that participants need a level of background concerning the ideas and issues being introduced, a level of trust that conversations/discussions are safe contexts in which to speak, and a sense that such conversations lead to new understandings either about themselves, others' experiences, or science and the teaching/learning of science. The first half of the course is designed to provide the background, create such a context, and coax conversations that explore alternative perspectives, questions, and explanations. As the course progresses, the instructors withdraw more and more from the class conversations, leaving the structural mechanisms of the course, e.g., the weekly question, readings, lectures, and class assignments, to create the context for participant interaction.

To bring prospective teachers to a level of interaction where they feel comfortable participating in discussions, the course is structured in two ways. First, classes meet for three, one-hour sessions each week for thirteen weeks. On Mondays and Wednesdays we introduce ideas through invited lectures and guest participants, discussion of readings, and sharing of experiences. Fridays are reserved for small group discussion in which participants share their responses to the week's question followed by class discussion. The course syllabus is provided in Appendix A.

Second, the term is organized into three sections, introducing increasing complexity. The sections move from discussions about the understanding of science that guides current teaching practices, to introductory discussions of the nature of science, to more specific discussions about the relationship between one's beliefs about the nature of science and teaching science. We believe that prospective teachers usually have strong beliefs about teaching, especially about how they will teach, when they enter the education program. Given their often-naive notions about science as a method and enterprise, our first challenge is to make these beliefs about science and about teaching science problematic. In the first section of the course, for example, we present prospective teachers with data and articles indicating the limited number of high school students who pursue science studies beyond the time mandated and the negative attitudes that the majority of students and adults evince about science. Participants are asked to reflect on weekly questions such as "Why teach science in schools?" "What isn't science?" "Is there a difference between the public's views of science and those of scientists'? In the views of science as taught in schools and science as practiced?" and "Are there boundaries to science?" Examples of assigned readings are located in Appendix A.

The goal of the first section is to encourage participants to critically examine their notions of science, but another goal is to help participants learn to talk and listen to one another. Reflecting on the weekly questions in journals gives everyone something to share and encourages everyone to speak. To make the whole class discussion equally interactive is more challenging. The instructor must provide links between comments, pose the provocative question, and open discussion by drawing comments and experiences from participants. The instructor must also object to unacceptable behavior, discussion techniques, and arguments. Prospective teachers need to be reminded of their unusual position as both "traditional student" and "prospective teacher." As prospective teachers, they are looking for strategies that encourage participation and open discussions. They are asked to consider how they would respond if the others in their class were students in their own classes and they wanted to encourage discussion.

The next section gives participants some insight into the issues and debates raised in contemporary science studies about the nature of science and the relationship between science and society and science and technology. The historian of science presents a series of five lectures over a period of as many weeks. These lectures all carry the general title, "The Nature of Science," and deal with science sequentially as (a) a body of knowledge, (b) as a putative method for getting knowledge about nature and ensuring consensus about that knowledge, (c) as a reflection of human society and social values, (d) as technology, and (e) as "Western metaphysics." Under these themes, the lectures briefly introduce students to contemporary challenges to the adequacy of scientific method for the purpose of establishing and maintaining consensus, to the ideas of Kuhn and other revisionist philosophers of science, and to Strong Programme theorists and the notion of science as "construction" rather than "discovery." The lectures invite students to consider science as "unnatural" knowledge, the concepts of which frequently run counter to commonsense perceptions of the world. Using several historical examples, they ask students to question the common assumption that technological change has been a simple outgrowth of deeper scientific understanding of the world. By discussing the "mechanical philosophy," of the seventeenth century, the lectures introduce the reductionist and quasi-mechanistic commitments of mainstream science, and how these commitments can conflict with the religious and philosophical beliefs of some minority groups.

Following the Monday lecture presentation, the lecture and accompanying readings are discussed at Wednesday's class meeting. The discussion is also used to synthesize ideas across the weeks and monitor understanding. Examples of weekly questions include, "Is science 'unnatural knowledge' and if so, what are the implications for education?" "Is science really about social issues and driven by social interests?" "Is a Science, Technology and Society (STS) curriculum the way to save science in our schools?" "Is science created or discovered?" and "What is science's contribution to understanding natural phenomena?"

During this part of the course we urge participants to begin to identify those ideas about the nature of science with which they agree or disagree. For many of the participants, these ideas are so new, and in some cases extreme, so that their ability to understand, much less take a stand is very limited. We do not force participants to choose, and for many their response after describing the various perspectives is a fully-acceptable, "I can't decide."

In the third section we consider how our beliefs about science and the nature of science influence our beliefs about teaching and learning science. We discuss such issues as gender equity and the masculinity of science, the relationship between science and science curricula, the nature of technology and teaching/learning science, and the pedagogical implications for portraying science as a socially- and culturally-influenced endeavor. We focus on the consequences of our beliefs for the decisions we make as teachers. Participants are asked to consider what practices or what teaching strategies they would choose if they held one belief about science rather than another. For this section, we select readings that illustrate the struggle for congruence between theory and practice, between one's beliefs about understanding science or what should count as an understanding of science and strategies for teaching science. Examples of readings used in this section can be found in Table I.

In this section, the weekly questions shift from the nature of science to teachers as professionals and pedagogical decision-makers. For example, "Is science itself responsible for the exclusion of women from scientific practice or are science educators more responsible?" "To what extent are science and science classrooms gender- or culture-biased?" "If science is about theory and technology is about practice, what is the role of teaching practice versus theory in schools?" "Are learners' alternative constructions or models or explanations of the world 'primitive science' or should we think of them as something else?" "Who is entitled to label something as science?" "Is an STS curriculum the way to save science in our schools?" "Should we also consider our beliefs about the nature of schools when we consider how our beliefs about the nature of science influence teaching decisions?"

Course Assignments and Assessment

There are no formal examinations in this course, but four types of assignments are designed to provide insight into the kinds of understandings each participant is constructing.

1) Reflection journals are assigned to encourage reactions and provide a mechanism for sharing ideas. Because of their varied backgrounds, many participants are hesitant at first to talk or share their ideas during the whole class discussion. Sharing the journal entries each Friday has been a successful way to equip each person with a ready contribution and help them to gain confidence in their ideas. Our goal is to encourage class members to synthesize, react, and share without censure. We hope having to explain and defend their ideas to others will help clarify ideas and arguments. Participants are asked to write about assigned readings, class discussions, and their response to the weekly question. These journals are not graded, but are turned in at the end of the course. Completing the reflection journal accounts for 20% of the course grade.

2) A Research Paper exploring some issue or argument of interest in more depth than can be considered in class discussion is assigned. Participants work individually or in groups of two or three. Few of these preservice teachers seem familiar with writing papers that go beyond reporting bare summaries of facts and others' opinions. For this reason, in introducing the research paper, we suggest that participants consider the research and scholarly articles, individuals whom they interview, and other resources. Our prospective teachers chose a diverse range of research paper topics. These included "What Are the Implications of Integrating Physics and Technology for Education?" "Integration and Science"; "History and Science: How are They Related and How Does that Affect Student's Views of Them as they are Taught in the Public School System?" "Girls and Science Education" "Do Inherent Differences Really Exist Between the Scientist and the Humanist?"

"What Counts as Understanding Science in Elementary Schools" "Science Anxiety" "Outdoor Education and the Science Curriculum: Can We Learn Science through Hands-On Discoveries in the Natural World?" "Informal Science Education" "Science Literacy" and "Science and Technology." Participants are asked to share their topics with one another early in the course and to jointly discuss the major ideas they uncover. The instructor reads drafts of the papers and assists with literature searches when asked. The research paper composes 30% of the course grade.

3) Class Assignments consisting of informal interviews, provocative questions, and reaction papers provide participants with opportunities to introduce new ideas and/or perspectives and to share what they have learned. They also provide the instructors with insight into the participants' thinking. To provide prospective teachers with opportunities to compare their thinking with practicing classroom teachers or research scientists, we suggest participants ask others for their response and reactions to questions and ideas that we are discussing in class. Often, we pose critical questions when we introduce readings or suggest a topic for discussion at the next class. These assignments are not graded, but we try to recognize each contribution by referring back to these resources and ideas during that and subsequent discussions. Often participants bring in articles from the newspaper, magazines, or journals. We set up a folder on reserve in the library to provide general access to the readings. Participants write response papers in reaction to sets of readings, guest speakers and the five lectures delivered by the historian of science.

A typical example of these exercises, and one that we have found very successful, is the following challenge given to students:

The Atlantic Provinces of Canada are in the process of developing a common science curriculum. One challenge that faces those with responsibility for developing this new science program is, what understanding of science should guide the committee's decisions about choosing curriculum approaches and materials. What do we want graduates of our public schools to have as an understanding of science? Our students are assigned to provide the committee with a two-page analysis of the concerns people have about the current understanding of science used to guide curricular decisions.

In addition to assigned readings and class discussions, students are encouraged to use other resources, including discussions with other members of the education or science communities. They are allowed to attach a separate page of references. For this and for similar exercises, the members of the class are encouraged to meet with one another, to discuss their ideas and to share resources. Our only requirement is that each person write their own response paper. Unlike other class assignments, the response papers are graded and figure 20% of the course grade.

4) A "Position Paper" is the final assignment of the course. This paper is an opportunity for personal reflection, delivered in the context of the students' describing their position on some issue or some teaching strategy related to the course. The paper is assigned to help participants make sense of their experience and begin to identify useful and sensible ideas. We also use the paper as an opportunity for prospective teachers to "talk" about their teaching practices and strategies as outgrowths of their beliefs.

STUDENTS' REACTIONS TO THE COURSE

We were curious about the students' reactions to the course and their experience in it. In the term following their participation, we asked the preservice teachers to share their reactions by responding to a questionnaire or participating in focus group interviews. Fourteen of the thirty-four participants responded. In their responses the term "frustration" surfaced. Some felt frustrated that the course offered what they perceived as insufficient practical direction for actual classroom procedure; one commented that students were frustrated "because they were not being told how to teach science." Another elaborated that "people would comment on how the approach was quite different from anything they had been exposed to before. I believe that they were here to learn how to teach science and were angry that they weren't spoon fed. I was one of them."

More commonly, however, students expressed frustration about the challenges the course had offered to their traditional conceptions of science and its nature. One asserted that the class had been primarily composed of successful science students, "yet I think very few had an understanding of 'science.' The questions posed to us forced us to rethink our entire school careers. This alone caused great discomfort." Many students were frustrated by the critical focus of the course, in which many competing historical, philosophical, and political conceptions of science were raised one after the other and offered for discussion and analysis, yet in which no single conception was labeled as adequate or correct. One complained that "every time we thought we had an understanding of [the professor's] view of science, we would be frustrated again in class." Another noted that "for a period of time students were disoriented, became confused about how to proceed because the professor acted only as a facilitator and responded to questions with even more questions." One noted that "I think you were frustrating us all the time. And so we would go talk it out and then we would start to understand and then we think we would have what you wanted us to think. And then you would go and frustrate us again." Several referred to intense discussions of the more controversial issues that had spilled over into time outside class. One noted that "the lunch hour conversations that took place on a regular basis allowed us to clarify what was said, why someone felt a certain way . . . The debate always continued until the next class." Students felt freer to express themselves outside class, and that in turn helped to facilitate freer discussions in class.

We asked if participants believed that the course had been a success in its long-term objectives: to problematize conceptions of science and prepare them to think about the deeper pedagogical problems of science teaching. Not all of our respondents were uniformly positive. One questioned the abstract approach of the course, including instructors' basic assumption that classroom methods would benefit from a more thoughtful and sophisticated understanding of the nature of science. "Ask yourself," he challenged us, "if you are doing justice by presenting this method of teaching in a preservice class. Survival is the name of the game for beginning teachers. Trust me -- they need substance -- otherwise we are perpetuating the textbook approach to teaching . . . "

Other respondents, however, were more positive. They agreed that for most individuals if not all, the course had produced changes in conceptions of science that would lead to their teaching differently than they had been taught. As well, they agreed that the open-ended nature of the course debates had promoted intellectual flexibility and a tolerance for disagreement. One person noted that "the biggest success of [the course] was in tearing down the building, not rebuilding . . . , using [it] as a tool to rethink, not necessarily to rebuild." "Yes," commented another, "the course was successful because it encouraged me to think about science teaching and education in general. It made me realize that teaching is about creating lessons instead of doing someone else's prepared work."

University of New Brunswick, Fredericton, New Brunswick, Canada

APPENDIX A
The Course Syllabus and Related Readings

Section I: **Questing Beliefs about Science**

Week 1 - Should We Be Teaching Science in Public Schools?
Week 2 - What Is This Thing Called Science?
Week 3 - What Do We Mean When We Say Someone Understands Science?

Examples of assigned readings (or portions of books as indicated by the sets of page numbers) to support the goals of section I:

Campbell, D. N. (1989). 'All talk: Why our students don't learn', *Educational Horizons,* (68), 3- 4.
Glasersfeld, E. von (1987). *The logic of scientific fallibility,* paper presented at the Eighth Biennial Conference of the Mental Research Institute, San Francisco, CA.
Hazen, R. M. (1991). 'Why my kids hate science', *Newsweek,* February 15.
Stinner, A. (1989). 'Science, humanities, and society -- The Snow-Leavis controversy', *Interchange,* (20), 16-23.

Jacobs, E. (1993). 'Bridging the gap between the two cultures', in Stephen Tchudi (ed), *The astonishing curriculum: Integrating science and humanities through language*, Urbana, IL, National Council of Teachers of English.

Winchester, I. (1990). 'Thought experiments and conceptual revision', *Studies in Philosophy and Education* (10) 73-80.

Wynstra, S. & Cummings, C. (1993). 'High school science anxiety', *The Science Teacher*, (60), 19-20.

Section II: Exploring Issues Concerning the Nature of Science

Weeks 4 thru 8 - What are ways in which scientists and people interested in science conceptualize (e.g. through theories and/or philosophies) and challenge the nature of science? A series of lectures by the historian of science.

Weeks 4 thru 8 (related discussions)

- Examining issues related to the nature of science
- Science literacy
- Who should determine what counts as knowing science in public schools?
- So what is the nature of science?

Examples of assigned readings (or portions of books as indicated by the sets of page numbers) to support the goals for section II:

Chalmers, A. F. (1976). *What is this thing called science?*, St. Lucia, University of Queensland Press, 63-97.

Collins, H. & Pinch, T. (1993). *The Golum: What everyone should know about science*, Cambridge, Cambridge University Press, 1-3, 57-78, 109-151.

Driver, R., Asoko, H., Leach, J., Mortimer, E., & Scott, P. (1994). 'Constructing scientific knowledge in the classroom', *Educational Researcher,* (23), 5-12.

Gross, P. R. & Levitt, N. (1994). *Higher superstition: The academic left and its quarrels with science*, Baltimore, John's Hopkins University Press, 1-15.

Raymond, J. (1982). 'Rhetoric: The methodology of the humanities', *College English,* (44), 778- 783.

Toffler, A. (1986). 'The third wave', in A. H. Teich (ed.), *Technology and the future, 4th edition,* New York, St. Martin's Press, 59-71.

Weinberg, A.M. (1986). 'Can technology replace social engineering?', in A. H. Teich (ed.), *Technology and the future, 4th edition,* New York, St. Martin's Press, 21-30.

Section III: Exploring the Relationship Between One's Beliefs about Science and the Nature of Science and One's Beliefs about Teaching and Learning Science

Week 9 - Whose Science?

Week 10 - What is the relationship between the nature of science and the science curriculum?

Week 11 - Do our beliefs about the nature of science match our pedagogy?

Week 12 - In what ways does how we learned science influence our beliefs about how we should be teaching science? Do we ever change our own thinking? How do we rethink the way we will teach when we haven't been in the classroom?

Week 13 - Reflection: What should count as knowing science in public schools, any answers yet?

Examples of assigned readings (or portions of books as indicated by the sets of page numbers) to support the goals of Section III:

Aikenhead, G. (1992). 'The integration of STS into science education', *Theory Into Practice*, (31), 27-35.
Apple, M. W. (1988). 'Social crisis and curriculum accords', *Educational Theory*, (38), 191-201.
Burbules, N. C & Linn, M.C. (1991). 'Science education and philosophy of science: Congruence or contradiction?', *International Journal of Science Education*, (13), 227-241.
DeBoer, G. E. (1991). *A history of ideas in science education: Implications for practice*, New York Teachers College Press.
Eisner, E. W. (1991). 'What really counts in schools', *Educational Leadership*, 948), 10-11, 14- 17.
Gaskell, P. J. (1992). 'Authentic science and school science', *International Journal of Science Education*, (14), 265-272.
Harding, S. (1991). *Whose Science? Whose Knowledge?: Thinking from women's lives*, Ithaca, NY, Cornell University Press.
Heath, P. A. (1992). 'Organizing for STS teaching and learning: The doing of STS', *Theory Into Practice*, (31), 52-58.
Hills, G. L. C. (1989). 'Students' "untutored" beliefs about natural phenomena: Primitive science or commonsense?', *Science Education*, (73), 155-186.
Hodson, D. (1988). 'Experiments in science and science teaching', *Educational Philosophy and Theory*, (20), 53-66.
Hodson, D. (1992). 'In search of a meaningful relationship: An exploration of some issues relating to integration in science and science education', *International Journal of Science Education*, (14), 541-562.
Mathews, M. (1994). *Science teaching: The role of history and of science*, New York, Routledge.
Mura, R. (1991). *Searching for subjectivity in the world of the sciences: Feminist Viewpoints*, Ottawa CRIAW/ICREF.
Pitt, J. C. (1990). 'The myth of science education', *Studies in Philosophy and Education*, (10), 7-17.
Selley, N. J. (1989). 'The philosophy of school science', *Interchange*, (20), 24-32.
Shamos, M. (1995). *The myth of scientific literacy*, New Brunswick, NJ, Rutgers University Press.
Stanley, W. B. & Brickhouse, N. W. (1994). 'Multiculturalism, universalism, and science education', *Science Education*, (78), 387-398.

DAVID BOERSEMA

16. THE USE OF REAL AND IMAGINARY CASES IN COMMUNICATING THE NATURE OF SCIENCE: A COURSE OUTLINE

This chapter features a discussion of a three-pronged course-based approach to the nature of science within an upper level philosophy of science course at a small liberal arts university. Students in this course are typically undergraduate science majors who have chosen it as an elective to fulfill a general education requirement in the humanities. The three elements that form the focus of instruction include an examination of science as doctrine (i.e., content), as process (i.e., methods), and as social institution (i.e., values). I suggest that this approach is more informative and productive for students than traditional approaches which focus on topics such as confirmation, explanation, realism, etc. In my course I have found it profitable and interesting to use case studies as the basis for discussion of science and its nature. One such case, the mass extinction debate, is particularly illustrative and will serve as a central example of my instructional strategy in this chapter.

ORGANIZATION OF THE COURSE

Science as doctrine means that one way in which the sciences are conceived and demarcated from non-science is in terms of what it investigates, or in terms of its content. Many students quickly accept this distinction. After all, what students study in biology classes is quite different from what they study in foreign language courses. Science as process characterizes the sciences not in terms of what they investigate — even poets and philosophers talk about evolution -- but in terms of *how* they study what they study. Most students enter the philosophy of science course proclaiming reliance on and a belief in "the scientific method." Science as a social institution looks at the sciences as enterprises conducted by scientists, real-live members of society, who reflect and shape social perspectives and values. Students are often quick to identity (and misidentify) where the sciences and society directly interface (e.g., nuclear power). Table I is an outline of the course, with the section on "epistemic concepts" focusing on doctrine, "change and progress" examining process, and "values and society" investigating social institutions.

TABLE I
Course Outline

Weeks 1-4	Epistemic concepts (e.g., observation, measurement, experimentation, models, theories, explanation, confirmation)
Weeks 5-7	Scientific change and progress (e.g., inductivism, falsificationism, historicism, rational reconstructionism, pragmatism)
Weeks 8-11	Values and society (e.g., science and values, technology and culture)
Weeks 12-15	Mass extinctions case study

What is Science?
An Analysis of an Imaginary Case Study

These three foci of doctrine, process, and institution are illustrated with an opening-day discussion based on an excerpt from an article entitled "Umbrellaology," by Somerville (1941). (See Appendix A for an abridged version). This article encourages students to offer their own assumptions, presuppositions, and intuitions about the nature of science. Not only has this piece generated immediate and profitable dissidence in class, but also, since students see it again on the last day of the term, it has served as a useful "pretest-posttest" tool for the students' own thinking about science relative to the course. Class discussions ensue after students spend a few minutes reading the excerpt, after which I call for a show of hands of students who believe that umbrellaology is a science and who believe it is not. Of course, students are asked to provide justification for their positions. Students typically offer reasons for and against umbrellaology that parallel the arguments given in the excerpt itself. After discussion, including the writing of arguments and responses on the board, we conclude by noting that many of the claims made during the discussion appear to be based on varying views of what science is. Some of those views reveal a notion of science as doctrine, some of science as process, and some of science as institution. This allows me to indicate to the class the sorts of topics which will be covered throughout the term.

While the point of this exercise is not to get to "the right answer" regarding the status of umbrellaology as a science, students often want to know the right answer.

A major course goal is to have students appreciate the complexities involved in identifying any endeavor as science, and I indicate that as the course proceeds fruitful perspectives will emerge on just what constitutes science. Nevertheless, there are features of umbrellaology that philosophers of science would see as scientific, though few philosophers of science would suggest that those features clearly constitute science. For example, while few current philosophers of science would claim that science can or does begin with interest-neutral observation, a focused gathering of data is necessary for science. Likewise, the extrapolation and subsequent testing of data, which occurs in umbrellaology, is necessary for science. Without question, there are features of science within umbrellaology, which is why it is a useful case to foster discussion and analysis among students. What, then, are arguments against umbrellaology's claim to be science? Besides the questionable assumption of disinterested observation as a starting point for investigation, umbrellaology seems to lack any theoretical explanatory power. In Lakatos's terms, there is no positive heuristic suggestive of a progressive research programme. In Laudan's terms there is no problem-solving focus, recognizing - and distinguishing - empirical and conceptual problems to be solved. The only fruitful aspect of umbrellaology seems to be that it provides more descriptive information. The course is set up, of course, to look at just these sorts of issues throughout the term.

Besides having students wrestle with their views on the nature of science, the discussion of this article allows us to pursue several pedagogical and philosophical goals. First, the types of arguments presented both pro and con on the scientific status of "umbrellaology" permit the analysis of science as doctrine, process, and institution. Second, it easily enables me to address (and investigate throughout the term) basic elements and activities of science such as modes of information gathering and organizing, of testing and accounting for phenomena, of enhancing upon and extrapolating data, etc. Third, it helps foster a conceptual understanding of science and the "doing" of philosophy of science (i.e., using philosophical methods of analysis on concepts within and about science). The enunciation and appreciation of these three goals are piqued by the umbrellaology discussion, and, as a result, they are more easily met later in the term by the incorporation of a sustained case study, namely, the presently ongoing debates concerning mass extinctions of life.

SCIENCE AS DOCTRINE, PROCESS AND SOCIAL INSTITUTION

Science As Doctrine

With respect to science as doctrine, the focus is not on specific scientific claims, but on conceptual issues related to scientific claims, or "epistemic concepts." Such concepts are connected to science as a body of knowledge. Included in this discussion are concepts and issues such as the nature and role of observation, measurement, experimentation, models, and theories. For example, in looking at the issue of observation, the article by Hanson (1972) argues for the interest- and theory-ladenness of observation, while the article by Scheffler (1982) challenges Hanson's view. Or as we consider the notion of experimentation in science, we look at the conceptual complexities of experimentation by analyzing Duhem's (1954) claim that no experiment is "crucial" in the sense of settling the status of hypotheses, since in the context of experimentation many auxiliary hypotheses are also being tacitly tested.

TABLE II

Readings to support the discussion of epistemic concepts

'The Lonely Bird' (August 17, 1991), *Science News*, (140), 104.

'Fossil jaw offers clue to human ancestry' (October 30, 1993), *Science News*, (144), 277.

'Earth's heart beats with a magnetic rhythm' (October 30, 1993), *Science News*, (144), 327.

'Babies' brains charge up to speech sounds' (July 30, 1994), *Science News*, (146), 71.

'Catching the flutter of a falling leaf' (September 17, 1994), *Science News*, (146), 183.

'Genetic sleuths explain insects' resistance' (September 17, 1994), *Science News* (148), 247.

Achinstein, P. (1968). 'Theories', in *Concepts of Science*, Baltimore, MD, Johns Hopkins University Press, 121-137.

Duhem, P. (1954). 'Physical theory and experiment', in *The aim and structure of physical theory*, Princeton, Princeton University Press, 180-218.

Ellis, B. (1966). 'Fundamental measurement', in *Basic concepts of measurement*, Cambridge, Cambridge University Press, 74-89.

Friedman, M. (1974). 'Explanation and scientific understanding', *The Journal of Philosophy* (71), 5-19.

Gooding, D. (1993). 'What is experimental about thought experiments?', D. Hull, M. Forbes, and K. Okruhlik (eds.), PSA 1992, Volume 2. East Lansing, Philosophy of Science Association, 280-290.

Hacking, I. (1983). 'Experimentation and scientific realism', *Philosophical Topics* (13), 71-87.

Hanson, N. R. (1972). 'Observation', in *Patterns of discovery*, Cambridge, Cambridge University Press, 4-30.

Hempel, C. (1965). 'Studies in the logic of confirmation', in *Aspects of Scientific Explanation*, New York, Free Press, 3-46.

Hempel, C. (1965). 'Studies in the logic of explanation', in *Aspects of Scientific Explanation*, New York, Free Press, 245-258.

Ravetz, R. (1971). 'Scientific inquiry: problem-solving on artificial objects', in *Scientific knowledge and its social problems*, Oxford, Oxford University press, 109-145.

Scheffler, I. (1982). 'Observation and objectivity', in *Science and subjectivity, 2nd edition*, Indianapolis, Hackett, 21-44.

Science as Process

With respect to science as process, the focus is on five perspectives on the nature of scientific change and progress including inductivism (such as espoused by Bacon or Mill), Popper's falsificationism, historicism (e.g., Kuhn), rational reconstructionism (e.g., Lakatos), and Laudan's version of pragmatism. After reviewing the philosophical views of these perspectives, we use historical case studies to apply these various perspectives to actual scientific practices. We explore these issues with reference to the sections in Table III. I have found the historical case of plate tectonics to be particularly useful. There are a number of accessible historical summaries available. (See Table IV). Students explain how the historical events do or do not match up with these philosophical perspectives on scientific change and progress.

TABLE III

Readings to support the discussion of scientific change and progress

Chalmers, A. F. (1976). 'Inductivism', in *What Is this thing called science?*, St. Lucia, University of Queensland Press, 1-11.

Kuhn, T. (1970). 'The nature of normal science and crisis and the emergence of scientific theories and the nature and necessity of scientific revolutions', in *The structure of scientific revolutions, 2nd edition*, Chicago, IL, University of Chicago Press, 23-34, 66-76 & 92-110.

Lakatos, I. (1981). 'History of science and its rational reconstructions', I. Hacking (ed.), *Scientific revolutions*, Oxford, Oxford University Press, 107-127.

Laudan, L. (1981). 'A problem-solving approach to scientific progress', I. Hacking (ed.), *Scientific revolutions*, Oxford, Oxford University Press, 144-155.
Popper, K. (1963). 'Science: Conjectures and refutations', in *Conjectures and refutations: The growth of scientific knowledge*, New York, Harper Torchbook, 33-65.

TABLE IV
Works on plate tectonics

Frankel, H. (1978). 'The non-Kuhnian nature of the recent discoveries in the earth sciences', P. Asquith and I. Hacking (eds.), *PSA 1978, Volume 2*, East Lansing, Philosophy of Science Association, 197-214.
Gohau, G. (1990). *A history of geology*, New Brunswick, Rutgers University Press, 187-200.
Hallam, A. (1989). *Great geological controversies*, Oxford, Oxford University Press, 135-183.
Hallam, A. (1973). *A revolution in the earth sciences*, Oxford, Clarendon.
LeGrand, H.E. (1988). *Drifting continents and shifting theories*, Cambridge, Cambridge University Press.

Science as Social Institution

With respect to science as a social institution, the focus is on three basic topics including science and values, science and technology, and science and culture. Under "science and values" we look at epistemic values, such as simplicity, experimental replicability, quantifiability (i.e., values relating to the reliability of science as an epistemic endeavor). We also look at ethical values, like whether, how, and to what extent ethical values enter into scientific practices. The reading by McMullin (1988) speaks to epistemic values directly while the reading by Rescher focuses on the persistence of ethical values throughout the workings of scientists, from the choice of research goals to the specification of standards of proof to the dissemination of research findings. "Science and technology" spotlights how and to what extent science is related to and distinct from technology. Feibleman (1972) and Jarvie (1972), for example, both deal with the epistemic viability and interrelations of pure science, applied science, and technology, while Mesthene (1990) and McDermott (1990) speak to the social role of technology. Finally, "science and culture" includes issues such as the politicization of science (Hiskes &

Hiskes, 1986), feminist critiques of science (Hubbard, 1989), and Snow's Two Cultures doctrine (Sellars, 1991 and Snow, 1959).

TABLE V
Readings to support the discussion of values and society

Feibleman, J. (1972). 'Pure science, applied science, and technology', in C. Mitchem and R. Mackey (eds.), *Philosophy and technology*, New York, Free Press, 33-41.

Hiskes, A. and Hiskes, R. (1986). 'Science and technology: public image and public policy', in *Science, technology, and policy decisions*, Boulder, CO, Westview Press, 5-33.

Hubbard, R. (1989). 'Science, facts, and feminism', in N. Tuana (ed.), Feminism and science, Bloomington, IN, Indiana University Press, 119-131.

Jarvie, I. C. (1972). 'Technology and the structure of knowledge', in C. Mitchem and R. Mackey (eds.), *Philosophy and technology*, New York, Free Press, 54-61.

McDermott, J. (1990). 'Technology: the opiate of the intellectuals', in A. Teich (ed.), *Technology and the future, 5th edition*, New York, St. Martin's Press, 89-107

McMullin, E. (1988). 'Values in science', E. D. Klemke, et al. (eds.), *Introductory readings in the philosophy of science, Revised Edition*, Buffalo, NY, Prometheus Books, 349-371.

Mesthene, E. (1990). 'The role of technology in society', in A. Teich (ed.), *Technology and the future, 5th edition*, New York, St. Martin's Press, 73-88.

Rescher, N. (1980). 'The ethical dimension of scientific research', in E. D. Klemke, et al. (eds.), *Introductory readings in the philosophy of science*, Buffalo, Prometheus Books, 238-253.

Sellars, W. (1991). 'Philosophy and the scientific image of man', in *Science, perception and reality*, Atascadero, Ridgeview Press, 1-40.

Snow, C. P. (1959). 'Two cultures: and a second look', in *Two cultures: And a second look*, New York: New American Library, 45-57.

WHAT IS SCIENCE?
AN ANALYSIS OF AN ACTUAL CASE STUDY: MASS EXTINCTION

Having addressed science as doctrine, process, and social institution, and having acquainted students with well-known and not-so-well-known readings related to these topics, we spend several weeks using a case study both to apply the philosophical issues and analyses and to test the philosophical claims made. A particularly effective case study is that of the debates about mass extinctions of life on the earth. This case works well for several reasons. First, because the controversy is ongoing, the scientific and philosophical issues are current and students are more interested in wrestling with those issues than with historical cases that are resolved. Second, the underlying science concepts can be easily accessed by students regardless of their disciplines and training. Third, there is a wealth of material to draw from, ranging from an insider's perspective (Raup, 1986) to edited volumes (Glen, 1994) to up-to-date secondary reports and original research articles. Finally, students of all ages are captivated by dinosaurs and are particularly interested in why they died. See Table V for a list of books and articles to help students investigate this controversy.

The Extinction Debate: An Overview

Approximately 65 million years ago at the end of the Cretaceous period the dinosaurs and all other large creatures on both land and in the seas became extinct in a relatively short time geologically. By the time the Tertiary period began, the world was a very different place. With the dominant dinosaurs removed from the scene, the small mammals formally existing in the margins, rapidly evolved to fill many of the niches vacated by the dinosaurs.

Many suggestions have been put forward to explain the cause for the Cretaceous extinction, but both consensus and evidence were lacking. In the early 1980's geologist Walter Alvarez, son of Nobel prize-winning physicist Luis Alvarez, discovered that a layer of clay deposited at the boundary of the Cretaceous and Tertiary periods outside the town of Gubbio, Italy contained a strange concentration of the element iridium. Extraterrestrial rocks such as meteorites were known to contain high concentrations of iridium. In addition, a variety of quartz called shocked quartz or schistovite, which forms under extreme pressure, was also found in the same rock layers. This discovery led to the suggestion by Luis Alvarez, that the earth was hit by a huge meteorite (Alvarez & Asaro, 1990). Such an impact would have caused a dust cloud that blocked out sunlight for years, stopping plant growth, destroying the food chain, and killing all large species of animal life including the dinosaurs. Others are not so sure that the extraterrestrial impact idea is the only viable explanation. Courtillot (1990), for example, has suggested that a series of volcanic eruptions could have caused the same effect while throwing iridium from deep within the earth into the atmosphere.

Using the Extinction Case Study: A Strategy

We first spend several weeks on the historical and scientific details of the case, and I make a point of presenting the events chronologically so that students can better observe the processes of working scientists. We then spend two weeks covering relevant nature of science issues. For example, with respect to epistemic concepts we look at the nature and role of measurement. Questions arise such as what exactly are the "things" that are being measured, what factors influence those measurements, how and to what extent are those measurements theory-dependent? With respect to models of scientific change and progress, we look at questions such as whether or not the details of this case fit a Kuhnian model; if so, what is the paradigm under which the researchers are working, is what is going on normal or extraordinary science? If events don't fit the Kuhnian model, is a Lakatosian model of competing research programmes more appropriate? If so, what exactly constitutes the positive and negative heuristics in this case? Looking at science as a social institution, we consider such questions as what epistemic values seem most salient to researchers, how, if at all, has the extra-scientific community influenced actual scientific practices.

Certainly, one can use other case studies to ask the sorts of questions just presented and pursue the sorts of goals mentioned at the beginning of this paper. In addition to the plate tectonics case mentioned above, one could use the 19th century debate between those holding Darwin's evolutionary views and those supporting Lamarck's inheritance of acquired characteristics or the debate over spontaneous generation. Another recent controversy is that of cold fusion. Nevertheless, because this specific case represents an ongoing and unresolved issue, and because it involves scientists across disciplines, and because it is accessible to students, it is particularly fruitful in terms of addressing issues in the nature of science. In addition, as noted earlier, by dealing with these issues under the rubric of science as doctrine, process, and institution, the various standard topics in the philosophy of science (such as the nature of explanation, the nature of theories, and the nature of values in science) make more sense to students. Consequently, students are more attuned to the importance of such topics and are more interested in pursuing them.

TABLE VI

Books and articles related to the mass extinction controversy

---

Allaby, M. and Lovelock, J. (1985). *The great extinction,* London, Paladin Books.
Alvarez, W. and Asaro, F. (October, 1990). 'An extraterrestrial impact', *Scientific American,* 78- 84.

Archibald, J. David. (1996). *Dinosaur extinction and the end of an era*, New York, Columbia University Press.
Carlisle, D. (1996). *Dinosaurs, diamonds, and things from outer space*, Stanford, Stanford University Press.
Courtillot, V. (October, 1990). 'A volcanic eruption', *Scientific American*, 85-92.
Glen, W. (1990). 'What killed the dinosaurs?', *American Scientist*, 78, 354-370.
Glen, W. (Ed.). (1994). *The Mass-extinction debates: How science works in a crisis*, Stanford, CA, Stanford University Press.
Hallam, A. (1989). *Great geological controversies*, Oxford, Oxford University Press.
Monastersky, R. (January, 1992). 'Closing in on the killer', *Science News*, (141), 56-58.
Monastersky, R. (February, 1992). 'Counting the dead', *Science News*, (141), 72-75.
Raup, D. (1986). *The Nemesis affair*, New York, W. W. Norton.
Raup, D. (1991) *Extinction: Bad genes or bad luck?*, New York, W. W. Norton.
Stanley, S. (1987). *Extinction*, New York: Scientific American Library.

Pacific University, Forest Grove, Oregon, USA

REFERENCES

Alvarez, W. and Asaro, F. (October, 1990). 'An extraterrestrial impact', *Scientific American*, 78-84.
Courtillot, V. (October, 1990). 'A volcanic eruption', *Scientific American*, 85-92.
Duhem, P. (1954). *Physical theory and experiment from the aim and structure of physical theory*, Princeton, MA, Princeton University Press, 180-218.
Fiebleman, J. (1972). 'Pure science, applied science, and technology' in C. Mitcham and R. Mackey (eds.), *Philosophy and technology*, New York, Free Press, 33-41.
Glen, W. (Ed.). (1994). *The mass-extinction debates: How science works in a crisis*, Stanford, CA, Stanford University Press.
Hanson, N.R. (1972). 'Observation' in *Patterns of discovery*, Cambridge, Cambridge University Press, 4-30.
Hiskes, A. and Hiskes, R. (1986). 'Science and technology: public image and public policy' in *Science, technology, and policy decisions*, Boulder, CO, Westview Press, 5-33.
Hubbard, R. (1989). 'Science, facts, and feminism' in N. Tuana (ed.), *Feminism and Science*, Bloomington, IN, Indiana University Press, 119-131.
Jarvie, I.C. (1972). 'Technology and the structure of knowledge' in C. Mitcham and R. Mackey (ed.), *Philosophy and technology*, New York, Free Press, 54-61.
McDermott, J. (1990). 'Technology: the opiate of the intellectuals' in A. Teich (ed.), *Technology and the future*, 5th edition, New York, St. Martin's Press, 89-107.
McMullin, E. (1988). 'Values in science' in E.D. Klemke, et al. (eds.), *Introductory readings in the philosophy of science*, revised edition, Buffalo, NY, Prometheus Books, 349-371.
Mesthene, E. (1990). 'The role of technology in Society' in A. Teich (ed), *Technology and the future*, 5th edition, New York, St. Martin's Press, 73-88.
Raup, D. (1986). *The Nemesis Affair*, New York, W.W. Norton.

Scheffler, I. (1982). 'Observation and objectivity' in *Science and subjectivity*, 2nd edition, Indianapolis, IN, Hackett, 21-44.
Sellars, W. (1991). 'Philosophy and the scientific image of man' in *Science, perception and reality*, Atascadero, CA, Ridgeview Press, 1-40.
Snow, C.P. (1959). 'Two cultures: and a second look' in *Two cultures: and a second look*, New York, New American Library, 45-57.
Somerville, J. (1941). 'Umbrellaology', *Philosophy of Science 8, 557-566*.

APPENDIX A
UMBRELLAOLOGY

Dear Sir:

I am taking the liberty of calling upon you to be a judge in a dispute between me and an acquaintance who is no longer a friend. The question at issue is this: Is my creation, umbrellaology, a science? Allow me to explain this situation. For the past eighteen years I have been collecting materials on a subject hitherto almost wholly neglected by scientists, the umbrella. The results of my investigation to date are embodied in the nine volumes which I am sending to you under separate cover. Pending their receipt, let me describe to you briefly the nature of their contents and the method I pursued in compiling them. I began on the Island of Manhattan. Proceeding block by block, house by house, family by family, and individual by individual, I ascertained (1) the number of umbrellas possessed, (2) their size, (3) their weight, (4) their color. Having covered Manhattan, I eventually extended the survey.

It was at this point that I approached my erstwhile friend. . . I felt I had the right to be recognized as the creator of a new science. He, on the other hand, claimed that umbrellaology was not a science at all. First, he said, it was silly to investigate umbrellas. Now this argument is false because science scorns not to deal with any object, however humble, even to the 'hind leg of a flea.' Then why not umbrellas? Next he said that umbrellaology could not be recognized as a science because it was of no use or benefit to mankind. But is not the truth the most precious thing in life? And are not my nine volumes filled with the truth about my subject? ...When he asked me what was the object of umbrellaology I was proud to say, "To seek and discover the truth is object enough for me." I am a pure scientists: I have no ulterior motives . . . Next, he said my truths were dated and that any one of my findings might cease to be true tomorrow. But this, I pointed out, is not an argument against umbrellaology, but rather an argument for keeping it up to date, which is exactly what I propose. . . His next contention was that umbrellaology had entertained no hypotheses and had developed no theories or laws. This is a great error. In the course of my investigations, I employed numerous hypotheses. Before entering each new block and each new section of the city, I entertained an hypothesis as regards the number and characteristics of the umbrellas that would be found there, which hypotheses were either verified or nullified by my subsequent observations, in accordance with proper scientific procedure, as explained in authoritative texts... As for theories and laws, my work presents an abundance of them. I will here mention only a few by way of illustration. There is the Law of Color Variation Relative to Ownership by Sex. (Umbrellas owned by women tend to a great variety of color, whereas those owned by men are almost all black.) ... There is also the Law of Tendency towards Acquisition of Umbrellas in Rainy Weather. To this law I have given experimental verification ... Thus I feel that my creation is in all respects a genuine science, and I appeal to you for substantiation of my opinion.

Excerpted from J. Somerville: 1941, "Umbrellaology" *Philosophy of Science 8*, 557-566.

BARBARA SPECTOR, PASCHAL STRONG AND THOMAS LA PORTA

17. TEACHING THE NATURE OF SCIENCE AS AN ELEMENT OF SCIENCE, TECHNOLOGY AND SOCIETY

We came here not knowing what to expect, but we are leaving expecting not to know, but wanting to learn.

Student in STS class

This chapter describes the teaching and learning opportunities in a unit used to introduce aspects of the nature of science to preservice teachers within a Science/Technology/Society (STS) course. The course teaches the nature of science while modeling processes used in science to generate new knowledge. We begin with an overview of the course, its students, and the way the learning opportunities have been arranged to encourage students to use processes of science to generate personal knowledge. Then we describe the unit that introduces the nature of science explicitly, including commercially available materials used and excerpts from a lecture series that describe science as an outgrowth of biological evolution. Finally, we present strategies for assessment and comments about the impact of the course on students.

The nature of science unit in our STS course has been adapted and tested with a variety of audiences, including preservice and inservice secondary and elementary teachers of science. Some may elect to adapt this unit for use in preservice methods classes or as a freestanding workshop for inservice teachers.

SCIENCE/TECHNOLOGY/SOCIETY INTERACTION COURSE OVERVIEW

This five-credit STS course containing the nature of science unit is a required part of the preservice science education program at the University of South Florida for students wishing to obtain middle school or secondary school certification to teach science. (Those preparing to teach science in elementary schools are not required to take this course.) The course is taught by a team including university faculty members, experienced secondary level STS teachers, and business/industry representatives. The classes meet in five-hour blocks each week on campus for nine weeks (45 contact hours) and for five six-hour blocks (30 contact hours) in an STS demonstration school. The science teachers in the demonstration school were part of the group that developed the course and are an integral part of the instructional team for the course. These teachers model constructivist teaching and learning approaches both with the preservice teachers and with their own middle school students. The teachers answer questions with questions that foster systematic inquiry, and use open-ended tasks, concept mapping, cooperative learning, authentic assessment, assessment embedded in instruction and other related techniques. The BSCS (1991) *Middle School Science and Technology* program is the curriculum upon which teachers base much of the instruction in their schools. It has both a constructivist epistemology with emphasis on the nature of science.

Assignments beyond the contact hours require students to view videotapes for twenty-two hours, and to read books and journal articles. (More is said about these assignments and how they are integrated in classroom activities later.) Student products include writing weekly reflective journals, developing an action plan, teaching an STS topic to their peers, and doing a final project of their own design that illustrates their personal understandings constructed from experiences in the course.

THE STUDENTS

The students are completing, or have completed, a major in either physics, chemistry, or biology, but have not yet had school field experiences. The science and usually the middle and high school science methods courses they have completed are taught traditionally. Students typically report that their prior educational experiences failed to provide much background in the nature of science. Students come to class expecting to learn through didactic lectures and expect deductive science to be presented. They assume learning is memorizing information as it is presented. Learning in a formal education setting by "doing science" is usually a new experience for them. In this course they gather data from seemingly disparate experiences, search data for patterns, generate hypotheses, tie them into theory, and test these emergent hypotheses and theories by sharing their interpretations with the community of co-learners in class. This course is their first encounter with the nature of science as a topic of study and science as a way to learn personally.

STRUCTURE OF LEARNING OPPORTUNITIES

The way students "do science" by participating in structured learning opportunities in the STS course can be most easily understood if the learning opportunities are considered as two groups that are interdependent strands woven together and running concurrently. The first strand provides a sense of immersion in a wide variety of STS interactions. Strand One experiences constitute the information about STS provided to students. These are the course experiences from which students are expected to gather initial data to begin constructing understandings of the nature of science. The iterative design of the course fosters continuous reflection on science and its nature beyond the explicit unit on the nature of science. Thus, students' concepts of the nature of science continue to be enhanced throughout the semester.

The second strand stimulates the social interactions necessary to construct meanings and assimilate the concepts of science, technology, society and the interactions among the three. Students discuss their reflections, analyses, and views of experiences in Strand One activities with peers and experts as a community of co-learners. The dialogue enables everyone to articulate and compare the patterns emerging for each one personally and the interpretations, hypotheses, and theories they generate from the patterns. The evidence and reasoning used to establish each person's perspective are discussed. Potential biases are explored. Students are directed to rely on the expertise exhibited within the group and not on the authority of the instructional team as "the teacher." Constructions are thus tested within the community of co-learners. Ultimately a group meaning, though not necessarily a single perspective, is socially constructed.

Most Strand One learning opportunities take place outside of class. Strand Two learning opportunities take place in class. In Strand One, students read, view video tapes, reflect on these inputs, and record their reflections in journals. For the journals, students are asked to "reflect on the way you are integrating data into your thinking about the nature of science and STS from the various course experiences, from related outside events, and from your prior experiences and knowledge." In Strand Two, students discuss their journal reflections. The flow of conversation in class is then directed by the students as they share and explore each other's reflections. As discussions progress, they notice they are generating more questions than answers and seek additional data. Thus, students are beginning to experience the expansive and iterative nature of science.

Together, these learning opportunities promote an environment in which students perceive it is safe to use a blend of logic and imagination to speculate, hypothesize, and explore novel ideas to explain and predict phenomena. Encouraging students to ask questions and formulate lines of inquiry, in contrast to rewording a set of predetermined answers, and emphasizing the importance of multiple perspectives in generating new knowledge contribute to the perceptions of intellectual and emotional safety. Science and learning are experienced as a dynamic ongoing process, rather than one with fixed end points, because the knowledge being constructed feeds back to students' needs and desires for further knowledge. This positive feedback system emulates the positive feedback system that characterizes the expansive nature of science in which scientific activity leads to further increases in activity.

Students acknowledge that they are engaging in a process similar to that used by scientists to socially construct knowledge. They note that they share their reactions to, and syntheses of, out of class experiences with the class; challenge each other's observations, interpretations, and biases; and assess the soundness and validity of each other's understandings. They perceive these interactions as modeling the processes scientists use when they interpret data, and submit papers at conferences and journal manuscripts to peer review, before their findings are accepted by the community of scientists as new information to be added to the knowledge base in a particular field.

The Nature of Science Unit

Although many elements of the nature of science are embedded throughout the course, there is a unit explicitly introducing the nature of science which occupies three consecutive class sessions of five hours each. It follows three or four sessions in which the paradigm shifts in society and in science education are discussed.

The nature of science unit begins formally with students constructing individual concept maps of their personal understandings of the nature of science as an out of class assignment. The maps serve as a pretest upon which future instruction is based. Students' maps typically reveal they perceive science to be an accrued body of facts segmented into specific disciplines. Some students indicate the facts have accumulated through research which adheres to one arbitrary formal approach to knowledge gathering. They do not refer to the underlying causal roots of science, to science as a socio-cultural endeavor, to the nature of science as thought provoking or relevant, or to science as a way of knowing.

In order to ease our students, who are lecture bound, into reflecting on input about the nature of science, analyzing what they encounter, and integrating the nature of science into their own cognitive frameworks through induction, we combine a lecture- discussion series titled, "On the Nature of Science: What are its Roots and Characteristics?" (Explicated later in Appendices A and B with a cooperative learning concept mapping process.) These class sessions serve as a framework for integrating many perspectives on the nature of science.

During the three weeks while the lecture-discussion-concept mapping sessions are occurring in class, students concurrently complete additional learning opportunities outside of class that were assigned during the first class meeting. These include viewing the twenty video tapes in *Connections* and the *Day the Universe Changed* series (Burke, 1980, 1985) and reading the chapter titled, "Nature of Science" in *Science and Technology as Human Enterprises* (Spector and Lederman, 1990), followed by reading Kuhn's *Structure of Scientific Revolutions* (1970). The Burke tapes give students mosaic views of the history of science within the context of cultural and technological developments that shape it. Spector and

Lederman's chapter provides students with a primer about the nature of science. Kuhn's book provides a philosophical and sociological lens through which to construct an understanding of the nature of science.

Students are asked to use each of the preceding learning opportunities as a data base from which to infer the nature of science and to record these inferences in their weekly journals. These inferences then become an integral part of the lecture-discussion classes through questions and comments students share. In class, students pause an hour into the lecture-discussion format to work in cooperative groups, concept mapping their understandings of the first portion of information. Team members move from group to group responding to students' questions, usually with questions to encourage discussions. For example, a student may ask, "Have we related concept x to concept y correctly?" An instructor might respond, "Why did your group think it linked that way?" Groups then circulate to view each other's maps and discuss differences and similarities. Students are encouraged to evaluate the varied perspectives encountered as they review each other's maps for potential enhancements to their own maps.

This process of breaking the lecture-discussion by concept mapping and discussions of what the various groups interpreted and constructed is repeated five times in class. Outside of class students rearrange, delete, or add new understandings to their personal nature of science concept maps. They build on insights developed from the in-class group mapping during the lecture-discussion-concept mapping sessions and the preceding outside-of-class activities.

ON THE NATURE OF SCIENCE:
WHAT ARE ITS ROOTS AND CHARACTERISTICS?

In contrast with other sources of information about the nature of science, the lecture-discussion-concept mapping series uses a lens of biological evolution to introduce science as an outgrowth of the human genetic characteristic called, "curiosity" and a culturally derived method for systematically and efficiently exercising curiosity. Pseudoscience as an outgrowth of the human characteristic called, "safety seeking" is also addressed. The presentation concludes with a comparison of characteristics of science with characteristics of pseudoscience. This lecture series by Paschal Strong, an expert in the nature of science and in motivation theory, would require about six hours if it were presented without any discussion.

Blurring Discipline Boundaries

The approach to content used in these lecture-discussion-concept mapping sessions integrates perspectives from biology, physics, biochemistry, psychology, sociology, and philosophy. Evolution, energetics, uric acid, curiosity, language, Epicurus, religion, and pseudoscience are some of the seemingly disparate topics from the preceding disciplines that are integrated in these lecture-discussion-concept mapping sessions to explain the biological basis of curiosity and, therefore, of science. Landmarks along the evolutionary trail presented by Strong are noted in Appendix A. Appendix B contains a comparison of characteristics of science to characteristics of pseudoscience that is summarized in the final lecture-discussion-concept mapping session in this series explicitly focusing on the nature of science.

The processes used to search the topics in the preceding paragraph for patterns and tie them together to construct an understanding of the nature of science illustrate the usefulness of breaking down artificial barriers among traditional disciplines. Having the experience of searching for patterns in this unit helps students construct a cognitive template they use to continue elaborating their concepts of the nature of science throughout the remainder of the course.

Integrating Learning Opportunity Strands

Here is an example of the way students use this lecture-discussion-concept mapping as a framework to integrate information from an outside-of-class experience. Students talk about Kuhn's analysis (1970) of the scientific enterprise as an illustration of the curiosity/safety seeking conflict occurring in the scientific enterprise. The safety- seeking scientist is talked about as one who looks to existing theories as truth and verities and is threatened by attempts to upset current scientific dogma. The highly curious, exploratory scientist is talked about as one who sees existing dogma as something to be attacked, confronted, changed, and even destroyed. The latter is identified as a paradigm pioneer.

STRATEGIES FOR ASSESSMENT

Assessment is embedded throughout the course. Students generate numerous concept maps, one action (lesson) plan, an STS lesson to teach to their peers, and an original project illustrating what they have constructed from experiences with the course. They write ongoing reflective journals and weekly exit memos. The instructional team observes, records, and transcribes class interactions for analysis. These items comprise the data base for assessment.

The action plan, the STS lesson, and the final project are made public within the community of co-learners. Students are encouraged to be creative in the way they approach the process of making their products public. Poetry, drawing, music, dance, wood carving, video, and computer-generated multimedia are some vehicles students have elected to use for their final products to express what STS means to them.

A sample of a non-threatening technique used to critique a publicly shared product follows: After each student presents an action plan, the course participants are given five minutes to write answers to this question: "What more do you want to know if you had to implement this action plan tomorrow?" Each person, including the instructional team, reads his or her questions and gives the paper to the presenter. Presenters do not respond to the questions in class. They each have the option of revising the action plan to include whatever revisions seem meaningful to them before turning in a final product.

Some indicators of success developed by the students and instructors which are used to analyze the data base include the degree to which students do the following:

a) reveal they understand and value the expansive nature of science;
b) see science as a way of knowing, a way of learning;
c) make public presentations that reflect the nature of science;
d) initiate participation in the collaborative process;
e) ask questions and challenge each other's constructions;
f) illustrate pertinent connections in successive concept maps;
g) increase risk taking by testing creative constructions; and
h) look to the group to establish acceptable knowledge rather than to the instructors to know if they are "right."

An Appraisal of Course Impact

Indicators of the impact of this course have emerged from three qualitative research studies in progress. Students perceived the course made a difference in their understanding of science and what it means to be a science teacher. The nature of science went from being "so what" to being an exciting adventure central to understanding science and teaching and learning. In the minds of students, the nature of science changed from static facts in books to relevant, dynamic, and socially constructed knowledge. They demonstrated sensitivity to the way paradigms influence the work of scientists. Their views of scientists changed from persons engaged in a solitary endeavor to people engaged in vital collaborative processes. Experiencing science as a way of knowing (learning) that is expansive, open-ended inquiry generated excitement about changing teaching. Memorizing and the 3T's

(Teacher, Talk, and Textbook) gave way to learning science as an expression of curiosity and exploration that crosses traditional discipline boundaries. Students acknowledged themselves as empowered learners with increased self efficacy regarding their future as science teachers.

University of South Florida, Tampa, Florida

REFERENCES

BSCS, (1991). *Middle school science and technology,* Dubuque, IA, Kendall/Hunt.
Burke, J., Jackson, M. & Kennard, D. (Producers), (1980), *Connections.* [video tape]. NY, Ambrose Video Publishing
Burke, J., Lynch, J. (Producer). (1985). *The day the universe changed.* [video tape] Los Angeles, CA, Churchill Films.
Glickman, S.E., & Schiff, B.B. (1967). 'A biological theory of reinforcement', *Psychological Review,* 74, 83-87.
Kuhn, T. (1970). *The structure of scientific revolutions,* Chicago, IL, The University of Chicago Press.
Spector, B.S. & Lederman, N. (1990). 'Science and technology as human enterprises', Dubuque, IA, Kendall/Hunt.

APPENDIX A
THE EVOLUTIONARY TRAIL

The lecture posits that the uniquely human enterprise called "science" has its roots in the biological evolution of humans. The energetics of the evolutionary trail are traced from prokaryotic cells to eukaryotic cells, to multicellular organisms, to cell specialization, to large complex terrestrial forms, to ectotherms and endotherms that are herbivorous and carnivorous, and ultimately to omnivorous humans. Changes in food ingestion, energy production, and energy use are noted. Omnivores are highlighted as opportunistic feeders who must build cognitive maps of their environments and have a sense of time to capitalize on the seasonal nature of their diets.

A ten-fold increase in brain size and the development of new brain structures occurred along with the increase in energy consumption. Brain development included neural-hormonal structures which generate the urge we label "curiosity" and was accompanied by the evolution of a counterbalancing urge labeled "safety seeking". Omnivores, including humans, have a high level of curiosity that helps them learn and exploit the environment.

A mutation occurred that led to chimpanzees and humans with an increased uric acid level. Uric acid acts as a nervous system stimulant resulting in chimpanzees and humans being awake for many hours more than are needed for ingestion and metabolic house keeping alone. Exploratory and manipulative interactions with the environment increased.

Eventually, language developed which enabled the symbolic representation of the world and the ability to explore the natural world both physically and mentally. Humans were then able to substitute mental exploration for physical exploration. They could deal with symbolic representations of events previously experienced or not experienced which may not even exist.

Prescientific humans developed some systematic observational skills which led to a descriptive form of astronomy. They also developed technologies through their exploratory and manipulative behaviors which led to domestication of animals, smelting and metallurgy, grain and fruit strains, etc. Technology, developed through trial and error procedures, often outpaced understanding of the underlying processes involved. Trial and error exploration later evolved into the systematic processes today we call "science." The conventions of science represent one of the many ways humans exercise their instinctual curiosity and exploratory behaviors.

The two opposing tendencies, curiosity and safety seeking, lead to two basic opposing emotional and mental life styles in humans. The curiosity-based life style is explorative and adventurous, while the safety-seeking-based life style is constrictive. Both have genetic and cultural/environmental roots. The insatiable curiosity of the human organism is put forth as the underlying motive force for science. Science has developed as a highly efficient, effective means of exercising curiosity which is capable of being taught and culturally transmitted. It is a synergistic, positive feedback system which is expansive in nature.

The tendencies of safety-seeking and harm-avoidance lead to an emotional approach to life that seeks certainty, assurance, and stasis. It is, therefore, a negative feedback system which is constrictive as opposed to expansive. The urge for safety seeking that evolved along with curiosity is essentially constrictive and inimical to the exercise of curiosity and questioning. It discourages mental confrontation and skepticism and rather looks for unquestioning acceptance. These tendencies are characteristic of pseudoscience, charismatic religions, and ideological dogma.

Development of Formal Science

Science as the expression of curiosity in western European societies is a culturally derived pattern of thought grounded in Greek philosophical tradition. Two basic Greek philosophies developed; one that said the physical world, its structures and laws, are knowable and controllable and subject to rational study and understanding. This approach was embodied in the philosophical writings of the Ionian school of philosophers such as Democritus, Anaxamander, Epicurus, and Hippocrates. The second stated that humans are subject to supernatural forces which are essentially beyond human understanding and control. One can learn techniques of propitiation and seeking of divine intervention, but direct control is not possible. This represents the pseudoscientific and charismatic religious approach to understanding and manipulation of nature. Science is a manifestation of the curiosity instinct. The pseudosciences and charismatic religions are a manifestation of the safety seeking instinct. Since these derive from opposite neurological / hormonal systems it is little wonder that in society, the two forces are often in conflict.

Science has been so successful, because it leads to a highly satisfactory method of exercising curiosity. Note that this specifically does not say "to satisfy curiosity." According to the work of Glickman and Schiff (1967), the questions and discovering methods to answer them are the primary reinforcing mechanisms, not the answers. Certainly, there is satisfaction in finding an answer and having a theory supported, but there is also a postpartum depression at the end of an experiment unless the results suggest further experiments. Scientists don't look for end points and certainty, but rather new questions and pathways.

APPENDIX B: THE CHARACTERISTICS OF SCIENCE

What is it about "scientific method" that characterizes it and delineates it from the pseudoscientific or charismatic religious approaches to knowledge? Since exercising curiosity has been shown to be reinforcing in and of itself, science is a positive feedback system. Scientific activity leads to new questions and lines of inquiry that further increase the activity, thus a positive feedback system exists. Science is synergistic. Various parts come together to generate new, emergent results and findings that add to the feedback system. Science leads to a mental life style which is growth oriented and constantly expansive. It seeks and encourages skepticism, questioning of established doctrines, and intellectual confrontation. Scientists are willing to expose their theories and findings to the heartless scrutiny of their colleagues.

Science has developed into an enterprise with self-correcting mechanisms to save scientists from the all too human propensity to feel they are right and infallible. These include peer review, publications, replication of experiments, various control mechanisms to guard against bias, etc. Although human hubris, ambition, and sometimes even greed occasionally rear their ugly heads, they are pitfalls that scientists are aware of and consciously strive to avoid.

The enterprise consciously strives to achieve an egalitarian society in which a person's status in the enterprise is determined by his/her performance and contributions in contrast to potential societal barriers such as race, gender, ethnicity and other characteristics of the individual. Science is as close to a truly cooperative international enterprise as exists. As science has become more complex, cooperation and collaboration have emerged as important features of the scientific enterprise.

Characteristics of Pseudosciences

The pseudosciences, (e.g., parapsychology, astrology, etc.) charismatic religions, and ideological dogma stand in marked contrast to the above. These enterprises reflect safety seeking, the biological counter part of curiosity. They can be considered to be constrictive, not expanding, negative feedback systems seeking stasis and stability.

The practitioners do not seek new knowledge but merely confirmation of fixed, "eternal truths". Confrontation and skepticism within the group are discouraged. Conformity is the watchword. Often these movements center around a charismatic leader. Statements of certainty, not the probabilistic statements made by scientists, are typical. "Truth" is often revealed, not discovered.

MICHAEL L. BENTLEY AND STEVE C. FLEURY

18. OF STARTING POINTS AND DESTINATIONS: TEACHER EDUCATION AND THE NATURE OF SCIENCE

STARTING POINTS: THE NATURE OF SCIENCE IN TEACHER EDUCATION

An old Chinese saying is that the fish is the last one to know that it lives in the water. The current educational reform movement involving constructivism and reconceptualizing the nature of knowledge calls upon classroom teachers to change both the curriculum and patterns of instruction. But Battista (1994) reminds us that teachers themselves are the products of an old curriculum and have developed beliefs incompatible with the spirit and substance of these innovations (p. 468). If teachers are going to teach children about the nature of science, most will need first to examine their own understandings.

Unfortunately, tertiary (college) students preparing to become elementary teachers enter teacher education programs with largely unexamined beliefs about the nature of science. Typically these beliefs remain unchallenged in teacher education and science content courses (Haggerty, 1992; O'Brien & Korth, 1991). The strategy outlined here is an approach to address this situation.

Given the importance placed on understanding the nature of science by Project 2061 (American Association for the Advancement of Science, 1989) and by the National Science Education Standards (NRC, 1996), our purpose here is to describe how a focus on the history and philosophy of science can be incorporated through a science-technology-society approach.

The ideal approach for increasing pre-service teachers' awareness of the nature of science would be for tertiary science educators and science teacher educators to work together to assess the portrayal of science in the whole institution's program. In the best of all possible worlds, we would recommend thoroughly reengineering the science coursework taken by pre-service teachers.

But we are pragmatists and our intent is to make realistic suggestions--steps that can be taken in the right direction. A feasible step, we believe, is to create a formal science-technology-society course that would be required in addition to the standard science methods course in an institution's teacher education program. Alternatively, additional hours could be added to the science methods course for an extended history-philosophy-sociology-of-science-in-science-teaching (HPSS/ST) unit that could address much the same content.

OF DESTINATIONS: EDUCATING TEACHERS ABOUT THE NATURE OF SCIENCE

Hollon, Roth, and Anderson (1991) note that science teachers need to develop the type of knowledge that will enable them to make both curricular and instructional decisions. If prospective teachers are to learn how to engage children in conceptual change instruction related to the nature of science, they will need to further develop their own understanding of the nature of science in such a way as to enable them to plan the curriculum and choose appropriate teaching strategies in their classrooms. They will need to learn ways to guide and support children in considering alternative views and constructing meanings for science.

For science teacher educators, the instructional problem is how to help their students--present and future teachers--understand post-positivist perspectives of science and what they mean for curriculum and instruction. No teacher educator can transfer his or her own knowledge about science to the student, but he or she can provide opportunities for conceptual change and model appropriate instruction.

THE POTENTIAL OF AN STS COURSE

Bringing about structural changes is a daunting task given the tenacity of most teacher education practices. Foltz and Roy (1991) indicate that tertiary level STS courses have existed since the 1960s and are now offered at over 2,000 colleges and universities across the United States. Nearly all of these courses are offered as electives in the general education program, rather than designed specifically for teachers, as we advocate here.

An effective STS course addresses two distinct, but related, educational goals:

1. Teaching students about the nature and culture of science and technology as experienced by practitioners and understood through the conceptual lenses of sociology, history, psychology, anthropology, and philosophy, and

2. Introducing students to personal, local, national, and/or global issues at the interface between science, technology and society . . . (which) entails personal decision making and informed, premeditated action. (Cheek, 1992, p.199)

The unique strength of an STS approach is that it involves an interactive set of concepts, content, and skills that demonstrate how science and technology are socially mediated and value-laden fields; that every technology creates unanticipated effects; that the system is the matrix for work in science and technology, and that every technological device or system requires tradeoffs.

The Social Studies of Science

The STS approach we emphasize is on the social study of science. Rather than merely examining the impact of scientific and technological developments on society, the social study of science is concerned with how historic and contemporary human values influence the meanings and workings of science. In this sense we refer to science as both manufactured knowledge and as a way of knowing. This particular emphasis has implications for teacher education. Clough (1994, pp. 4-5) believes that pre-service teachers should be able to understand and appropriately use the terminology of the social studies of science, i.e., terms like theory, law, etc. Additionally, they should be able to explain possible consequences of the use or misuse of terms on how students will perceive the nature of science, and on the implications for learning activities, curriculum, etc.

Table I suggests some of the instructional outcomes for an STS course designed specifically for teachers.

TABLE I
Sample student outcomes for an STS course for teacher preparation.

Participation in an STS course should enable students (pre-service teachers) to:

- identify interactions between science and technology and society;
- relate the significance of particular persons and events in the historical development of science and technology;
- understand processes and models of knowledge generation and confirmation in the natural sciences;
- evaluate assumptions, propositions, evidence, and arguments in science;
- demarcate science from other disciplines and from non-science and pseudo-science;
- identify local, regional, national and global social and environmental conditions and trends related to science and technology;
- evaluate curriculum materials related to STS education;
- identify STS-related curriculum goals and instructional outcomes;
- develop STS-related teaching strategies and skills.

In a survey of two decades of literature related to the STS approach, Cutcliffe (1990) found no studies of STS course effectiveness at the tertiary level. Cutcliffe concluded that the jury was still out as to whether or not STS courses achieve the kinds of understandings that are intended by their instructors. More recently, Bradford, Rubba, and Harkness (1995) compared outcomes related to understanding the nature of science for university students who had completed a general education STS course to students who took a general physics course. As a measure they used selected items from the Views on Science-Technology-Society (VOSTS) item pool. They concluded the STS course enhanced student understanding of science, while the physics course had almost no impact on students' views of STS interactions . . ." (p. 369). The particular STS course studied by Bradford, Rubba, and Harkness was essentially a series of lectures given by different university faculty regarding STS issues in their fields. Student outcomes were not as strong as the researchers would have expected and they concluded that the course could be improved by the use of conceptual change teaching methods.

To our knowledge, there have been no studies of the outcomes of tertiary level STS courses designed specifically for pre-service elementary and middle school teachers. That kind of course would address the need for teachers who are more knowledgeable about the nature of science. Such a course addresses the two goals identified by Cheek, as well as the related curriculum and instruction issues.

Some study of the nature of science can be provided for within the regular pre-service science education methods course. The methods instructor can develop an HPSS/ST or STS unit consisting of a series of lessons that engage students in talking about key ideas. Students can compare views of science discussed in the methods course and the depiction of science drawn from their experiences in high school and university science courses. The dissonance that is likely to result in the minds of students who are unfamiliar with post-positivist and constructivist perspectives may produce the kind of intellectual doubt Dewey claimed was the foundation for serious learning.

One way to begin is to ask students to sketch what first comes to mind when a scientist is imagined--essentially the Draw-A-Scientist Test. The pictures can be discussed and compared, for example, on the basis of scientists' personal characteristics such as gender, age, dress, etc. Students also can be asked to respond to selected items on the VOSTS item bank, or to react to particular statements about science, indicating agreement/disagreement about science. Table II is a set of such statements that have been used successfully by the authors in previous classes. Note that positivists and constructivists would likely have divergent responses to many of these statements when asked to mark a Likert scale from "strongly agree" to "strongly disagree."

The methods instructor's pedagogical approach is a critical variable in catalyzing what is learned by the pre-service teachers. The instructor is the students' present model for science educators, if not scientists. The pedagogy should be symmetrical with the constructivist view of the nature of science the instructor wants the students to understand. The constructivist view of the nature of science has implications for science teaching and thus for instruction in the methods classroom (Duschl, 1988,

1990; Duschl & Hamilton, 1992; Matthews, 1994; Garrison & Bentley, 1990).

TABLE II
Statements to elicit student views about the nature of science

- Science is the search for truth.
- Compared to knowledge in other disciplines, scientific knowledge is more objective and unbiased.
- In research, scientists use the scientific method to solve problems and verify findings.
- In science, if a theory is proven true, it becomes a law.
- Biological knowledge would be different today if the historic proportion of men and women biologists had been reversed.
- Most of the new ideas in science are produced by brilliant individuals.
- If experimental results fail to support a theory, scientists reject the theory and seek other alternatives.
- Disagreements among scientists about a theory are normally solved when new data are generated related to the problem.

For example, in working with students to interpret their data in an investigation, it is more consistent with constructivism that they realize the refutation and elimination of alternatives gives power to the predictions that survive the investigative process, rather than that their prediction has been confirmed just because it is still a possibility. To promote this realization, teachers might ask questions that emphasize refutation rather than confirmation. Garrison and Hoskisson (1989) provide an example of this subtle but critical reframing with the questions, "Do you have any evidence that your prediction is wrong?" or "Have you found your prediction to be false?"

It is consistent with a constructivist perspective for the methods instructor to challenge students to critique and make explicit their initial understandings about science. This might sometimes include the instructor playing the role of agent provocateur in dealings with students (Lemke, 1990). Other relevant roles the instructor might enact are mapped in Figure 1.

Consistent with a constructivist perspective, methods instructors can structure activities and investigations in the field and in the classroom more like the open-ended investigations of real scientific work and less like the cookbook type labs often found in traditional instructional materials. Instructional activities also can be designed to engage students in more student-to-student discussion and debate focused upon scientific ideas, and more collaborative group work (Lemke, 1990;

Larochelle & Desautels, 1991). This kind of classroom experience communicates to students that ideas are negotiated in the world of science.

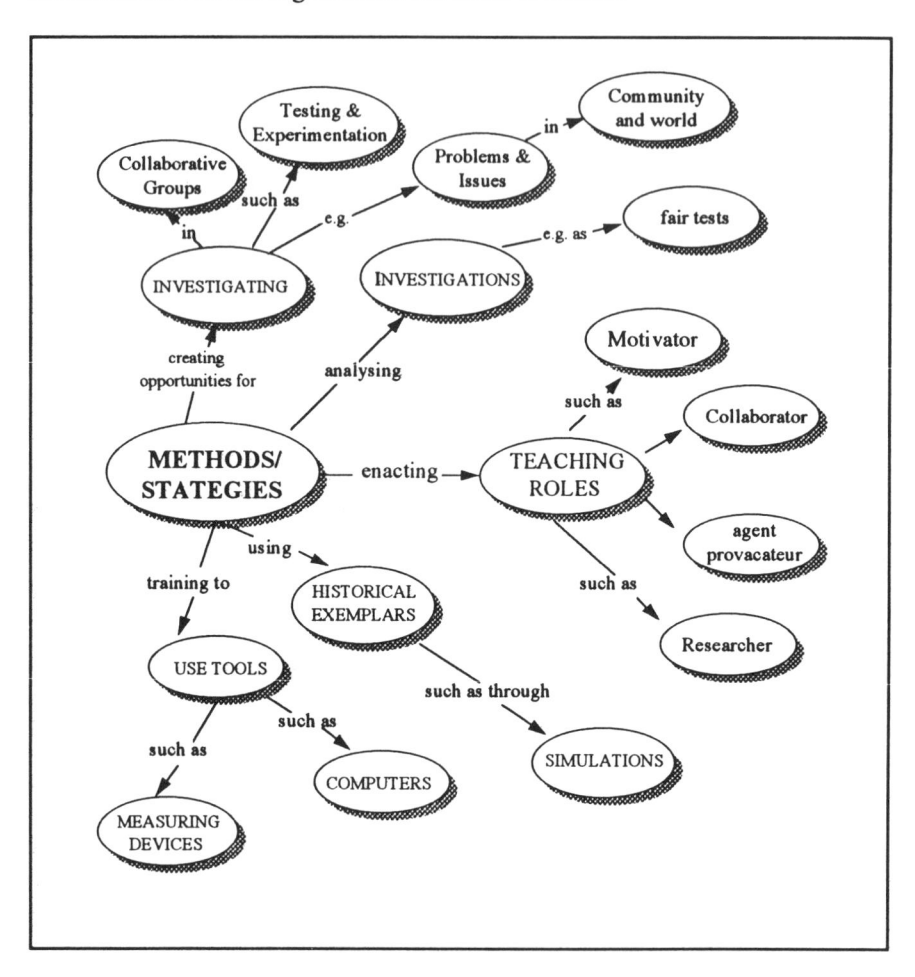

Figure 1. Methods related to teaching about the nature of science. Teaching about the nature of science, especially at the elementary school level, requires the flexible integration of a number of teaching strategies.

As a result of their HPSS/ST unit activities, students should examine actual science, technology, and society interactions in the world around them. Activities that facilitate this objective engage students in looking for STS connections in current events. Table III provides a list of STS topics students could investigate either individually or as group projects.

TABLE III
Potential STS topics for student investigation

- An emerging controversial technology in the physical sciences and its potential impact, such as superconduction, cryoscience, solar power, new synthetic materials, cold fusion.
- An emerging controversial technology in the life sciences and its potential impact, e.g., biotechnology, genetic engineering, human and non-human primate experimentation, invitro fertilization, embryo transfer, fetal tissue research, cloning, psychosurgery and behavior control, drugs and behavior control, organ transplantation, sex preselection and sex change.
- The history, purpose, and potential of the Human Genome Project.
- The history and future of nuclear power, including issue of nuclear waste disposal.
- Animal rights and the use of animals in research.
- Artificial intelligence and robotics.
- Case studies of the lives and work of scientists.

For STS education, it is important that students develop the skills and attitudes that enable them to understand different vantage points.

A problem becomes an "issue" when different people have different ideas of how to respond. This means that students need to define at least two different perspectives (sides) for each issue. To prevent polarized thinking in students, it is preferable to require the search for at least three perspectives about any particular issue.

"Players" refers to individuals, groups, organizations, and institutions having a vested role in how the problem will be resolved. The "positions" they hold will be powerfully influenced by the assumptions they hold about the particular issue, the nature of people, the nature of the world, and the nature of knowledge. The beliefs that different people bring to any social issue are guided by their values, which reflect the relative importance of beliefs in a given situation. Table IV describes different categories of values typically operating in STS issues.

SAMPLE COURSE OR UNIT ASSIGNMENTS AND ACTIVITIES

The final portion of this chapter has some practical classroom-tested suggestions for assignments and activities for an STS course or unit in a methods class. Student engagement in these activities likely will generate opportunities to introduce and work with many STS concepts.

TABLE IV
Types of values in STS issues
(Deep-seated preferred beliefs of how each should operate).

Political: The activities, functions, and policies of governments and their agents.
Economic: The distribution of resources and goods in society.
Religious: The use of belief systems based on faith or dogma.
Ecological: The maintenance of the integrity of natural systems.
Scientific: Concerning attributes associated with empirical studies.
Cultural: Pertaining to the continuation of societal knowledge, habits.
Educational: Concerning accumulation, use, communication of knowledge.
Aesthetic: The appreciation of form, composition, and color through senses.
Social: Pertaining to shared human empathy, feelings, and status.
Recreational: Pertaining to leisure activities.
Egocentric: Pertaining to a focus on individual self-satisfaction.
Health: The maintenance of positive human physiological conditions.
Ethical/moral: Pertaining to present/future responsibilities, rights and wrongs.

Sample #1 Assignment: The New Media

One sample assignment involving the analysis of news media addresses a number of our intended objectives of promoting student thinking skills about a particular topic as well as about the nature of knowledge itself.

Purpose: To increase student awareness of the range of STS issues covered by the news media and provide student practice in analyzing STS issues.

What to do: Collect and analyze three STS-related articles from newspapers or other media. The three articles should represent a range of issues, e.g., local to global, biological to engineering. Properly cite each article. Limit each analysis to two pages or fewer. Respond to the following for each case:

- What is (are) the main issue(s)?
- What are the different "sides" for each issue?
- Who are the "players" involved?
- What assumptions have not been examined for each position?
- What values are related to the decisions?
- What type of values are these?

- What kind of data would help inform decision-making?

Criteria for Grading Rubric:

> Complete Understanding 2 Points
> Partial Understanding   1 Point
> Little Understanding   0 Points

Points

1. Articles selected illustrate science and society issues. _____
2. Diversity of issues represented _____
3. Analysis identifies issue(s), positions, players, assumptions, values, needed data. _____
4. Analysis identifies and specifies values in conflict. _____
5. Mechanics at professional standard (organization, focus, clarity, references, format) _____

Classroom drama is a teaching strategy with particular potential in an HPSS/ST or STS unit or course. Joan Solomon (1989) has provided a model for this kind of activity in her "Retrial of Galileo" skit which students or teams of students play the roles of Galileo and the witnesses at his Inquisition trial in 1633. The drama illustrates the tenacity of an old theory and raises issues like rationality, reasons, evidence, the relationship of science and religion, and the use of authority.

Sample #2 Activity: Classroom Drama on Evolution & Creationism

Another example of a classroom drama activity involves the issue of evolution and creationism. Following the role play, the participants can form groups to discuss any of the questions and create webs linking concepts and issues. Relevant to this particular classroom drama activity would be the 1987 Louisiana Supreme Court ruling that no state can require the teaching of creationism in the public schools (U.S. vs. Louisiana).

The criteria, cited from testimony by Judge Overton, were that:

1) Science is guided by natural law;
2) Science has to be explanatory by reference to natural law;
3) Science is testable against the empirical world;
4) Science is tentative;
5) Science is falsifiable.

I. The Situation: Parkview Junior High School Science Department Meeting. The science department chairperson shares a memo from the Assistant Superintendent. She wants to hear from the science faculty and would like to be able to reach consensus on the place of evolution and creationism in the science program. She would like the group to draft a Science Department position statement on the issue.

II. The Memo:

Parkview School District
6543 Caramba Blvd.
Parkview, Illinois 60001

To: Chris Rhodes, Science Department Chair
From: Rhonda McElvoy, Assistant Superintendent
Date: July 10, 1997
Re: School Board request for information on science program

If you read last week's Pioneer Press coverage of the school board meeting, you know that a group of parents brought up the issue of the teaching of evolution in our science program. The Board has asked me for information about our program. I must rely on you in this regard since my background is not in science and I, too, need to be updated on what we are teaching in this area. The parents complained that their children are being taught the Big Bang and Darwinian evolution but not taught other theories of the origin of the universe, such as creationism.

Is this true? One parent claimed that many scientists are creationists--is this true? Can you briefly describe the content of our science program in grades 5 to 8 on these matters. If we do not also teach about creationism, do you feel we should change our curriculum to provide a more open forum about the alternative views? Please give me some reasons for your views on this, because we expect inquiries from the press. Also, how do you suggest we respond to this group of parents?

I'm sorry to drop this on you in the middle of your summer vacation, but the Board and my office were equally unprepared for this complaint. Please get something to me by Thursday, the 15th. You might want to check with other life science teachers too. Thank you for attending to this issue.

III. The Players

Chris - JHS Science Chairperson, life science teacher, good biology and chemistry background, single parent of an adolescent son, interested in environmental preservation, active member of mainstream church.

Mickey - Veteran physical science teacher from a family of engineers, parent of two college students, veteran of Viet Nam War, pro-nuclear power, non-church goer, considered "tough" by students but respected because also serves as basketball coach.

Sandy - Earth science and general science teacher, single/never married, former "1960s person," popular with students, leader of annual Earth Week activities, considers self spiritual but avoids organized religion.

Jerry - teaches sixth grade science, newest member of faculty, recently married, father a Methodist minister in a small town, knows a chemistry teacher from college who professed creationist beliefs.

Cam - Has taught fifth grade science for five years, married with a four-year-old daughter, non-practicing Jew, member of ACLU, active in the teacher's union, technology buff (surfs the Net in spare time), helped start local children's science museum.

Jo - teaches sections of life and physical science, in third year as teacher after leaving law career, divorced, no children, agnostic, sister an anthropologist.

IV. The Questions

1. What position will each player most likely take on the curriculum issue? Why?
2. What should middle school children be taught about human origins?
3. Does creationism merit consideration in the science program?
4. How should teachers respond to religious beliefs of students in conflict with the science content of the curriculum?
5. Are there real conflicts between science and religion? Is this one?
6. How should the public school system respond to patrons who question the curriculum?
7. What is the role in all of this of the Constitution's prohibition of the establishment of religion (i.e., the principle of separation of church and state)?
8. How should the conflict in the community be explained to the children?

Another activity which involves students in considering the role of evidence in science and in distinguishing what is science compared to non-science is borrowed from Giere (1984).

Sample #3 Activity: The Full Moon and Sexual Offenses

The following article appeared on the front page of the *Louisville Courier-Journal*, July 29, 1977.

Not only do people like to "spoon" by the light of the silvery moon, they are also more inclined to commit rape, sexual assault, and indecent exposure when the moon is full. The evidence of moon madness can be found in police statistics. A recent study by Dr. Ronald Holmes, a Jefferson Community College sociologist, found that reports of a range of sexual offenses increased significantly during a full moon.

Dr. Holmes analyzed the timing of the 1274 sexual crimes reported to the Louisville and Jefferson County Police Departments during 1974 and 1975. He found that 404 of those reported crimes (31.6% of the total) took place in the full moon cycle - 82 more cases (6.8% more) than the next most frequent of the moon's four phases.

Dr. Holmes cautioned, however, that the study does not point to the strength of moonbeams as the primary cause of a sex crime. "There's no one explanation for any kind of behavior," he said. "A person who is inclined to behave in a certain way may be more inclined to behave that way when his brain is under added pressure from certain acids normally produced by the body. When the moon is full, the bodily tides - levels of body water - are at their highest and more of those pressure-creating acids are shoved through the brain, he said.

Dr. Holmes rejected the tenets of astrology or lunar-based religions. It hasn't got anything to do with the position of the planets or any religion, he said. However, Nolaan Meyers, an astrologist with whom Holmes consulted on the project, said he thinks that the signs of the zodiac are important in analyzing the effect of the moon on sexual violence. The effect of those full moons during a person's sun sign will be especially powerful, he said.

Student activity: Based on the information given in the article, indicate your attitude toward the statements below by writing the letter a,b,c,d,or e in the space provided.

  a) Strongly inclined to agree
  b) Somewhat inclined to agree
  c) Cannot say. Insufficient information to form any opinion
  d) Somewhat inclined to disagree
  e) Strongly inclined to disagree

   A. In 1974-75, there were more reports of sexual crimes in the Louisville-Jefferson County area during full moons than during any of the other three phases of the moon.

B. In 1974-75, there were significantly more reports of sexual crimes in the Louisville-Jefferson County area during full moons than during any of the other three phases of the moon.
C. In 1974-75, there were more actual sexual crimes in the Louisville-Jefferson County area during the full moon than during the other three phases.
D. In general, there are more sexual crimes during the full moon than in other phases; that is, there will likely be more such crimes during full moons than during other phases this coming year in other cities.
E. The full moon causes some people to go off their beams with sexual offenses.
F. There are more sexual crimes during the full moon because of the stronger bodily tides that send pressure-creating acids to the brain.
G. The position of the planets has nothing to do with the higher percentages of sexual crimes during full moons.
H. The signs of the zodiac are important in analyzing the effect of the moon on sexual violence.
I. A person inclined to commit sexual crimes will be more inclined to do during full moons associated with his sun sign.
J. In 1974-75 one's chances of being the victim of a sexual crime in the Louisville-Jefferson county area were roughly 7% greater during full moons than during other times of the month.
K. Women should probably be a little more careful about going out alone during full moons than during other times of the month.
L. Women should avoid going out alone during full moons except in cases of extreme emergency.

History of Science Case Studies

Case studies in the history of science, like classroom drama, can be a useful strategy for those who learn from narrative. Researchers have documented the potential of narrative in teaching about the nature of science (Roach, 1992; Roach & Wandersee, 1995; Wastnage, 1994). Stories about scientists and historical events can be springboards for talking about any number of issues related to the nature of science. The discovery of Neptune, for example, illustrates how evidence which appears to refute a theory may actually lead to new knowledge. Astronomers in the 19th century noted that the perturbations of the orbit of Uranus seemed to refute Newtonian celestial mechanics. The French and English astronomers Leverrier and Adams independently proposed, ad hoc, the existence of a planet beyond Uranus to account for the observed discrepancy. Astronomers looked for, and found, Neptune. Selected resources for science stories are listed in the references. Numerous other resources are available, including the abundant resources of the Internet.

CONCLUSION

We began by arguing that the nature of science is often misrepresented in K-12 and tertiary (college) science education programs. As a result, most students preparing to become elementary and middle school teachers come into the teacher education program with unexamined beliefs about the nature of science. Scholarship in the history, philosophy, and sociology of science, over the past thirty years, has led to a post-positivist or constructivist understanding of science. This is a dramatic shift from the 19th century perspective. Because schooling has not kept pace with this change in the foundational disciplines, and also because what teachers believe about the nature of science influences their instructional planning and how they interact with children, teacher education programs should make instruction related to the nature of science a high priority.

In this chapter, we discussed how science teacher educators need to help teachers develop their understandings of the nature of science. They also need to help teachers come to terms with the curriculum and instruction implications of alternative perspectives of the nature of science. One approach is to add an STS course requirement in the teacher preparation programs. Another approach is to incorporate an HPSS/ST unit in existing science methods courses. We suggested a number of resources and instructional activities related to the nature of science which can be incorporated in either approach.

[1] *Virginia Polytechnic Institute and State University, Blacksburg, Virginia, USA*
[2] *State University of New York, Oswego, New York, USA*

REFERENCES

American Association for the Advancement of Science (1989). *Science for all Americans (Project 2061).* Washington, DC.

Battista, M.T. (1994, February). 'Teacher beliefs and the reform movement in mathematics education', *Phi Delta Kappan,* (75), 462-470.

Bradford, C. S., Rubba, P.A., & Harkness, W.L. (1995). 'Views about science-technology-society interactions held by college students in general education physics and STS courses', *Science Education,* (79), 355-373.

Cheek, D. (1992). 'Introduction', in Thirunarayanan, M.O. *Think and act - make and impact: Handbook of science, technology and society, Volumes I & II.,* Tempe, AZ, Arizona State University.

Clough, M.P. (1994, January). *Nature of science: Recommendations from the field,* paper presented at the annual meeting of the Association for the Education of Teachers of Science, El Paso, TX.

Cutcliffe, S.H. (1990). 'The STS curriculum: What have we learned in twenty years?', *Science, Technology, & Human Values,* (15), 360-372.

Duschl, R.A. (1988). 'Abandoning the scientific legacy of science education', *Science Education,* (72), 51-62.

Duschl, R.A. (1990). *Restructuring science education: The importance of theories and their development,* N.Y., Teachers College Press.

Duschl, R.A. & Hamilton, R.J. (Eds.). (1992). *Philosophy of science, cognitive psychology, and educational theory and practice,* Albany, New York, State University of New York Press.

Foltz, F. & Roy, R. (1991). 'Post secondary science for non scientists: STS, the new approach to the largest population' in. S.K. Majumdar, L.M. Rosenfeld, P.A. Rubba, E.W. Miller, & R.F. Schmalz (eds.), *Science education in the United States: Issues, crises and priorities,* Easton, PA, The Pennsylvania Academy of Science, 209-229.

Garrison, J. W. & Bentley, M.L. (1990). 'Teaching scientific method: The logic of confirmation and falsification', *School Science and Mathematics,* (90), 188-197.

Garrison, J., and Hoskisson, K. (1989, March). 'Confirmation bias in predictive reading', *The Reading Teacher,* 482-486.

Giere, R.N. (1984). *Understanding scientific reasoning, 2nd ed.* New York, Holt, Rinehart and Winston.

Haggerty, S.M. (1992). 'Student teachers' perceptions of science and science teaching' in S. Hills (ed.), *The history and philosophy of science in science education,* Kingston, Ontario, Queen's University, 483-94.

Hollon, R., Roth, K. J. & Anderson, C.W. (1991). 'Science teachers' conceptions of teaching and learning' in J. Brophy (ed.), Advances in research on teaching: *Teachers' knowledge of subject matter as it relates to their teaching practice,* Vol. 2. Greenwich, CN, Jai Press, Inc., 145-185.

Larochelle, M., & Desautels, J. (1991). 'Of course, it's just obvious: Adolescents' ideas of scientific knowledge', *International Journal of Science Education,* (13), 373-389.

Lemke, J. L. (1990). 'Talking science: Language', *Talking science: Language, learning, and values.* Norwood, NJ, Ablex Publishing Corporation.

Matthews, M. R. (1994). *Science teaching: The contribution of history and philosophy of science,* New York, Routledge.

National Research Council. (1996). *National science education standards,* Washington, D.C., National Academy Press.

O'Brien, G. E. & Korth, W. W. (1991). 'Teachers' self-examination of their understanding of the nature of science: A history and philosophy course responsive to science teachers' needs', *Journal of Science Teacher Education,* (2), 94-100.

Roach, L. E. (1992). *I have a story about that: Historical vignettes to enhance the teaching of the nature of science.* Natchitoches, LA, Roach Publishing Company.

Roach, L. E. & Wandersee, J. H. (1995, November). 'Putting people back into science: Using historical vignettes', *School Science and Mathematics,* (95), 365-370.

Solomon, J. (1989). 'The retrial of Galileo' in Herget, D.E. (ed.), *The History and Philosophy of Science in Science Teaching: Proceedings of the First International Conference,* Tallahassee, FL, Florida State University, 332-338.

Wastnage, R. (1994). *Storytelling in science: A note from Ron Wastnage.* MSTENews (Mathematics, Science and Technology Education Group, Queens, University, Canada, (3), 4-5.

MICK NOTT[1] AND JERRY WELLINGTON[2]

19. A PROGRAMME FOR DEVELOPING UNDERSTANDING OF THE NATURE OF SCIENCE IN TEACHER EDUCATION

This chapter outlines a teaching programme we have developed with both trainee and experienced teachers. The length of this chapter may look daunting but it includes comprehensively the things we say, the things we put on projectors and hand out to students. While this chapter is practical, it is not atheoretical. Our aim in this chapter is to explain the activities we did, why we did them and to discuss their value and purpose. But before we start the programme we begin by explaining why the NOS is important and where it lies in the curriculum

A NATURE OF SCIENCE PROGRAMME: AN INTRODUCTION

We describe three main strategies, a profiling activity which provides students and teachers with a 'profile' of their own view of the nature of science (NOS), a set of 'critical incidents' which explore participants' understandings of the NOS in a classroom context and exemplars of practical approaches which participants can use in their own teachings in schools. All these strategies have been used within a particular course - the one year post-graduate course for student teachers. They have also been used with experienced teachers participating in inservice courses. One of our expectations in running this programme is that science teachers will not be fully conversant with an accurate and up-to-date account of the history, sociology and philosophy of science, and nor should we expect them to be.

As an introduction we indicate to students that there are two domains of the nature of science in the science curriculum:

- Children's first hand experience; how and what are the children learning about science with practical experiences;

- Children's second hand experience; how and what are the children learning about science from stories about scientists and scientific ideas.

We tell participants that an important part for classroom practice is that pupils (i.e., school children) should reflect on their own experience. This could involve teachers asking questions like: "What do you think will happen and why?"; "What

W. F. McComas (ed.) The Nature of Science in Science Education, 293-313.
© 1998 *Kluwer Academic Publishers. Printed in the Netherlands.*

is in this set of results which convinces you that . . ."; "Can you explain why . . . ?" "Why do you think this experiment was a test of this theory?" so that their pupils consider questions concerning predictions, evidence, explanations and evaluations at different stages of practical work.

We also tell participants that pupils should develop from science teaching an appreciation and understanding that scientific ideas have developed in response to different needs both historically and in different cultures.

Pupils should move from first or second hand experience into constructing and perhaps altering their scientific models of the world back and forth (see Figure 1).

An example we share with teachers are the lessons that they may organize around the practical experience that copper turns black when heated in a flame. The initial experience would be organized by the teacher. After all, pupils are hardly going to be spontaneously putting copper foil into Bunsen flames. But the subsequent interpretations, predictions and practical experiences would be a matter of negotiation between the children themselves and between the children and the teacher. We offer teachers the following simple models of science (see Figures. 1 and 2).

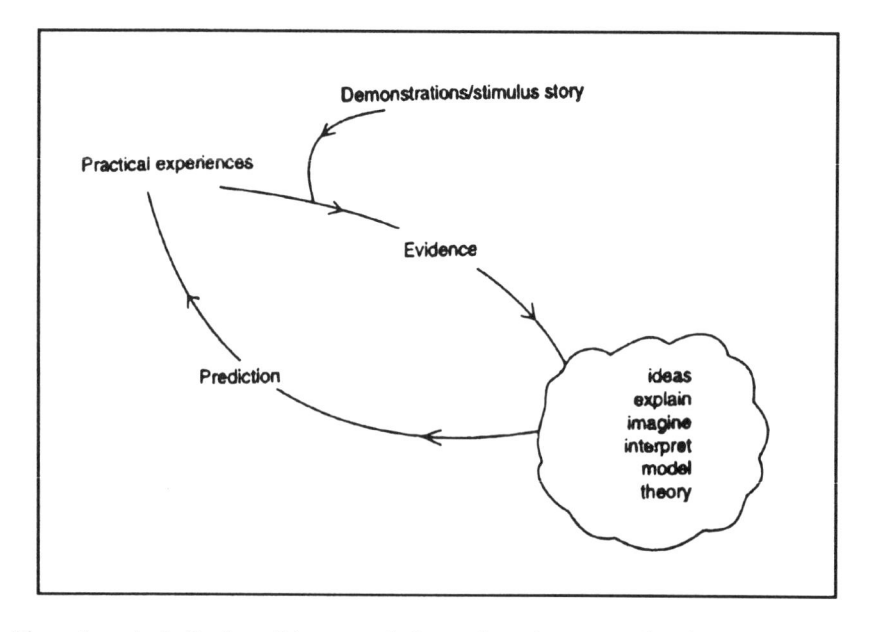

Figure 1. An indication of how practical experience is connected to theory.

The strand of second hand experience is the one that links the scientific ideas to culture and time as represented in Figure 2. This is linked with the model in Figure 1 because the "theories" on the right-hand side of Figure 1 are the ideas that scientists have which is central in Figure 2.

For example, we suggest that when teaching the germ theory, science teachers do demonstration work analogous to Pasteur's experiments with broth and swan-necked flasks. This would be an ideal time to highlight that the school demonstrations are frequently analogous replicas of previous experiments.

```
                    scientists as people
     technologies           ↓              past
                  \                       /
                   \                     /
       uses / needs —— scientific ideas —— present
                   /                     \
                  /                       \
        beliefs  /           ↓             \ future
          &
        values             grow
                            &
                          change
```

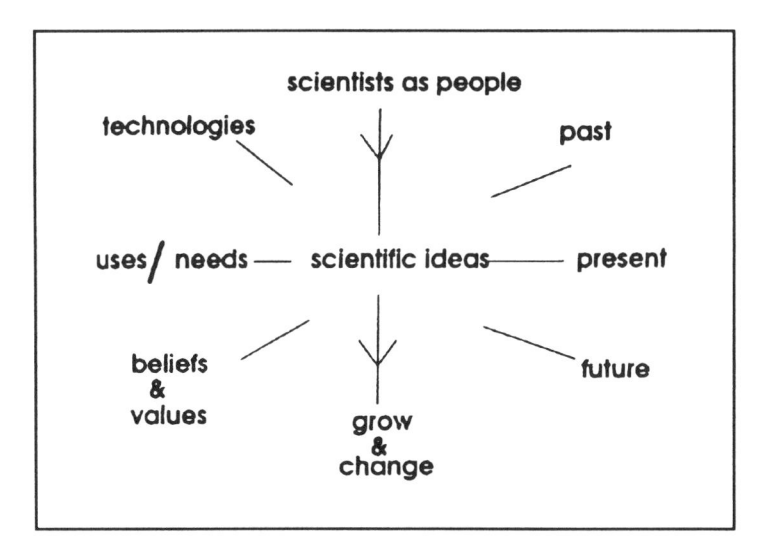

Figure 2. Our representation that scientific ideas grow and change, are rooted in people and vary across cultures and with time.

If time allows, we go on to mention further examples of how scientific ideas are created by people and will interact with the culture and times that the people live in. The kind of examples we outline are: how scientific ideas about the nature of matter, theories of disease etc. have changed in time; how science interacts with values and beliefs, e.g., changing from a geocentric picture to a heliocentric one changed people's conceptions of the position of mankind; how Malthus, market economics and Adam Smith influenced Darwin; how technologies can not only change the data collected but can actually provide new scientific ideas, e.g., Harvey explaining the heart as a pump whereas previous anatomists wouldn't be able to do this as they hadn't seen a pump; how computer architecture affects some explanations of the working of the brain.

Having given a brief exposition the programme moves onto a stage where the participants have to examine, and reflect upon, their own understanding of the NOS.

THE PROFILE ACTIVITY

The profile activity (requiring at least 40 minutes) is a way of eliciting teachers' views of the nature of science. This activity requires students to read through, and reflect upon, twenty-four statements about science. The views expressed in these statements come from a range of sources: science curriculum developers, philosophers and sociologists of science, and science teachers/educators. They reflect different perspectives on the nature of science, from both teachers and practicing scientists. We have tailored each one so that a student's response to it eventually indicates their position on five axes, all shown in Appendix A.

Students' responses, ranging from +5 (strongly agree) to -5 (strongly disagree) are added, using the scoring system in Appendix A. Participants then plot their own "profile of science" and check their profiles against our interpretation of each axis. These interpretations are shown in the third part of Appendix A. Participants then compare profiles with each other. This can be done in a rather tongue-in-cheek manner by asking each student to plot their own profile on a sticky label which can then be worn during the ensuing discussion! We have also found it valuable to collect all the profiles of the group and display them collectively either on a transparency or flip chart. It is then interesting to note any extreme positions (are there any raving relativists, pushy positivists, rampant realists or innocent instrumentalists?) What is the distribution of positions amongst the group? Are people happy with their position on the axes which the profiling activity has allocated? How do students' own backgrounds and life histories (e.g., as a school pupil, as a university student, perhaps as a trainee teacher, or as a researcher) determine their profile and their view of science as an activity?

We make it clear that the purpose of the profiling activity is not to determine their position on some objective scale, but simply to encourage them to reflect on and compare their individual views of science. The activity is an excellent way of breaking the ice and provoking discussion. However, it is no substitute for other "probes," such as critical incidents, which explore teachers' reactions to real classroom situations so that the participants articulate and add to their "pedagogical knowledge" (Shulman, 1987) or "knowledge-in-action" (Schön, 1983).

We think the profile is a good activity for getting people to talk about and reflect on the nature of science but we are not convinced that it either "measures" a Nature of Science or helps in the type of interactions that teachers have with pupils. Many of the participants, who had helped in the development of the profile, thought the results challenged their ideas about the nature of science and after the discussion work most recognized that having done the exercise once and discussed it they would not answer in the same way the next time. This confirmed our suspicions not to place too much faith in paper and pencil tests of teachers' understandings of the nature of science which is one of the reasons we developed the critical incidents.

USING CRITICAL INCIDENTS

We define a critical incident as an event which makes a teacher decide on a course of action which involves some kind of explanation of the scientific enterprise. It may be either an event like some practical work going wrong or an event which raises moral and ethical issues about scientific knowledge or the conduct of scientists. These events are often stimulated by pupils saying and doing things but they may also arise through the action of the teacher, particularly when a demonstration goes wrong.

Part of the incidents' criticality is that they evoke authentic responses from the teacher which provide an insight into the teacher's view of science as well as matters to do with teaching and learning. We have argued (Nott and Wellington, 1996) that teachers' knowledge of the nature of science will be grounded in their pedagogical content knowledge as well as subject content knowledge. By using critical incidents which are qualitative and rooted in classroom experience we believe that we get teachers to talk about the nature of science and then get them to reflect on that talk so that they have a wider repertoire of responses next time a similar incident occurs and reflect on their own understanding of the scientific enterprise.

We have used these incidents as an interactive small group discussion activity; alternatively, colleagues have used them with participants individually and then shared responses. A set of incidents is given to each group. They are then asked to work through them by each person choosing one, reading it to the others and then the entire group giving and discussing their responses in line with the guidance below. The groups discuss the incidents, the tutor circulates to join in and ensure that groups keep on task or that groups are discussing a range of incidents. We have found that the discussions stay focused, while frequently becoming intense and sometimes quite heated! After discussion we have selected practical incidents and non-practical incidents one at a time. Each time we have asked a group to give their responses and then asked other groups to comment. We can't stop them after that and it is very difficult to get them out of the room!

Participants are asked to respond in terms of what they *would* do (the convergent or reactive perspective), what they *could* do (divergent or proactive) and what they *should* do (ethical or moral). The first type of response involves a kind of off-the-cuff, pragmatic reaction to the incident. The second involves a more divergent approach, by asking what they could do if perhaps they had fewer constraints and more time to reflect. The third category of response leads perhaps from the ethical/conduct area to the area of what they really ought to do as science teachers (i.e., the moral area). Each response was thrown open to the larger group for discussion. Responses were noted from the lists and the discussion.

Examples of critical incidents

Four examples of critical incidents are provided below with paraphrases of the responses that teachers have made and with our interpretations of what they tell us about teachers' knowledge of the nature of science.

Incident A: A class of 14 year old students are heating magnesium ribbon in a crucible with a lid. The purpose of the lesson is to test a consequence of oxygen theory that materials gain in weight when burnt.

During the summary at the end of the lesson, four groups report a loss in weight, two groups report no difference and two groups report a gain in weight.

List the kinds of things you could say and do at this point.

Responses. Teachers talk about:

- the need for everybody to make sure they have followed the same procedure
- that experiments don't always "work" (i.e., fit accepted theory)
- that if experiments don't work then they need to be critically evaluated
- that results can be averaged and/or discounted

Understanding of NOS. Teachers demonstrate knowledge about the procedures that scientists use to check experimental results and that experiments are as much the result of the experimenters' skills as they are the mirror of nature.

Incident B: A class of 11 year olds are working with microscopes and you want them to observe and draw onion skin cells. They set up the slides and you check that they have focused the microscopes competently and then they start to look and draw. You find their drawings to be nothing like your image of onion skin cells.

List the kinds of things you could say and do at this point.

Responses. Teachers say and do things like:

- are you sure that you can't see something which looks like ...
- show the children drawings of what they are supposed to see
- arrange for prepared slides or images to be shown to the children

Understanding of NOS relative to incident B: Teachers demonstrate knowledge that the work of scientists is guided by other scientists' images and pictures of what is to be perceived - all observations are theory laden and neophytes have to be trained to see in the ways of more experienced scientists.

Incident C: You have set up a demonstration of the production of oxygen by photosynthesis with Canadian pond weed. Just before the lesson when the class are to look at the apparatus again, you notice that there is a small amount of gas in the test tube but not enough with which to do the oxygen test.

List the kinds of things you could say and do before or during the lesson.

Responses. Teachers say and do things like:

- explaining why it doesn't work, such as the lack of light intensity. Teachers will invoke auxiliary theories.
- "tweak" the experiment, for instance, increase light intensity, add sodium bicarbonate to the water
- Cheat, by adding oxygen from a tank prior to the lesson, for instance.

Understanding of NOS relative to incident C: Teachers will demonstrate either norms of criticality, e.g., use of auxiliary theories and/or evaluation of the experimental set up so that the discrepancy is explained or teachers will use counter norms which are fraudulent.

Incident D: The children are doing an investigation. They have all made predictions and are now well into their practical work. Some children's results conflict with their predictions. They go back and cross out their predictions and change them to agree with their results.

List the kinds of things you could say and do at this point.

Responses. Teachers say and do things like:

- they ought to understand why their predictions don't match their outcomes
- they ought to maintain an integrity in their written records
- require the children to amend their prediction to the original form.

Understanding of NOS relative to incident D: Teachers demonstrate knowledge that science is an activity which involves critical reflection and values such as integrity. It is the scientists' job to explain discrepancies and change their mind and theories.

All of these responses tell us something about the respondents' knowledge of the nature of science and their attitudes to the connections, and differences, between science and other ways of knowing the world. Following the presentation of each of these incidents, it will be useful to list the things that teachers could say or do in response.

Additional Critical Incidents

Incident E: 11 year old pupils are doing experiments with circuit boards. With two lamps in series, many find that one is lit brightly while the other appears to be unlit.

Incident F: You are demonstrating wave phenomena using a ripple tank. The children are unable to observe refraction clearly and frankly you find it hard to see with the apparatus available.

Incident G: You are conducting a lesson on the big bang theory of the origin of the universe. An 11 year old pupil at the front interrupts in the middle of your account of the big bang and says, "My family believe that the Earth was created by God in six days. This is what it says in Genesis and we believe the Bible to be true."

Incident H: You are six weeks into semester 1 with a group of eleven year olds. The unit you are doing is on Life and Living Processes. One of the pupils states impatiently at the start of a lesson, "When are we going to start cutting up rats then?"

Incident I: You are teaching a class of 14 year old students about contraception with a lesson about different methods of birth control. One of the pupils asks, "Do you believe it's right for the Catholic Church to say that only the rhythm method is acceptable? The rest are sinful."

Incident J: You have a particularly reluctant 15 year old learner in a chemistry class. The pupil is not aggressive but assertive that this work on chemistry is not something she or he likes doing. When you ask why, the pupil says, "Because if it hadn't been for chemists, we wouldn't have these chemicals ruining the Earth."

Incident K: Two episodes from lessons:

- A teacher is doing the starch test on leaves. For inexplicable reasons the tests are indecisive.

- A teacher is demonstrating the nonmagnetic properties of iron sulphide. However, the freshly made sample sticks to the magnet.

In both cases pupils say the following, how would you respond?:

"Oh well . . . science experiments never work"
"Anyway, we'll believe you, if you'll tell us the result."

Incident L: You have prepared a science lesson which involves discussion and debate, but no practical work. Several pupils ask as they enter the lab, "Are we getting the Bunsen burners out today?" What would you say or do?

During the course of a class doing their own investigations you experience the following incidents:

- A child, who has just finished planning an investigation, asks, "Can I do this investigation on my own?"

- A pupil writing in her exercise book looks up and asks, "Why do we have to write every part of the investigation up?"

- Twenty minutes before the end of the lesson a pupil comes up and says, "What shall we do now that we've finished our investigation?"

What could you say or do in response to these questions?

Incident M: You are teaching a Y10 group introductory theory about bonding and shell models of the atom when one of the class says, "We were taught before that atoms are hard and indivisible. What do you want us to believe?"

What could you say or do in response to this question?

Incident N: You are starting a module on light which will introduce the wave theory of light. One of the class puts their hand up - you stop, and ask, "Yes . . . what is it?". The pupil says: "But we've done light before."

What could you say and do at this point?

Ethical and Moral Dilemmas Associated with Critical Incidents

Participants recognize that they face ethical and moral dilemmas in dealing with the critical incidents. Initial responses are what they, as teachers, would do. We have found that participants need to be encouraged to be divergent and think of the range of things they could possibly do. These responses may well conflict with what they, as teachers, felt they should do.

The incidents where laboratory activities 'go wrong' appear to elicit three categories of response. These are 'talking your way out of it', 'rigging' and 'conjuring' (Nott and Smith, 1995). The majority of responses are in the first category of 'talking your way out of it' or as we would like to say more positively 'talking your way through it'. When science teachers talk their way through practicals that have gone wrong they often engage the children in a critical evaluation of practical work. 'Rigging' is the use of strategies that teachers have learned over the years to ensure that the apparatus or procedure works. The last category, 'conjuring', is where the teacher fraudulently produces the correct result for an experiment by sleight of hand. We have found that student teachers can start to conjure spontaneously or are inducted into it by science staff and technicians.

The answers to the practical incidents demonstrate knowledge of the experimental and social procedures of science. It is important to stress that talking your way through it is by far and away the majority response. However participants recognize the moral dilemma that what they *could* and *should* do, does not always coincide with what they *would* do. There are many other constraints operating on classroom life.

Responses to the critical incidents indicate that trainees and experienced teachers do have a depth of knowledge about the nature of science but the knowledge is expressed in the form of pedagogical content knowledge. We speculate that this pedagogical content knowledge may become subject content knowledge as teachers learn about the nature of science from having to teach science. Teachers frequently encounter classroom episodes where they have to make explicit how scientific knowledge is reasonably and ethically constructed and validated. Thus, much teaching about the Nature of Science takes place spontaneously in science classrooms as an everyday occurrence. We refer to this as covert teaching about NOS as opposed to overt teaching about NOS discussed in the next section on practical approaches.

PRACTICAL CLASSROOM APPROACHES TO TEACHING THE NATURE OF SCIENCE IN SCHOOLS

This next section (requiring at least one hour) illustrates practical approaches which can be used overtly to teach secondary school pupils about the nature of science. Three case studies are used to illustrate a range of classroom methods appropriate to teaching the NOS. Specific advice was written for teachers, in England and Wales, about teaching children the NOS (NCC, 1989). In order for children to learn to move between practical exploration and scientific ideas and to understand that scientific ideas have a history it was recommended that there should be:

- structured discussion amongst small groups of children;

- structured reading or listening or watching of items which involve stories about scientists or scientific ideas;

- drama and role play to develop qualities of sympathy and empathy with people in the past and different cultures;

- practical (hands-on or laboratory) work where children are asked both to explain and predict.

The case studies are drawn from curriculum materials published in Exploring the Nature of Science (Solomon, 1991) and in Practical Approaches to Teaching about the Nature of Science (Nott, 1994).

Using Case Studies to Communicate Aspects of the Nature of Science

If time is short then the case studies are talked through by a tutor. If more time is available, participants would be asked to work through the case studies themselves. The curriculum materials and their rationale are described and demonstrated to the participants and advice is offered on how to use them and what kind of learning may occur.

Case One. The first case study illustrates how the history of science could provide ideas to structure classroom activities. The materials are designed so that children will already have read about how Arabic scientists worked with a ray model. Then children go on to model rays of light using threads (an idea "borrowed" by the artists from Kepler). Paper cutouts of a pinhole camera and a carbon filament lamp (see Figure 3) can be placed on a wooden board and fixed with thumb tacks.

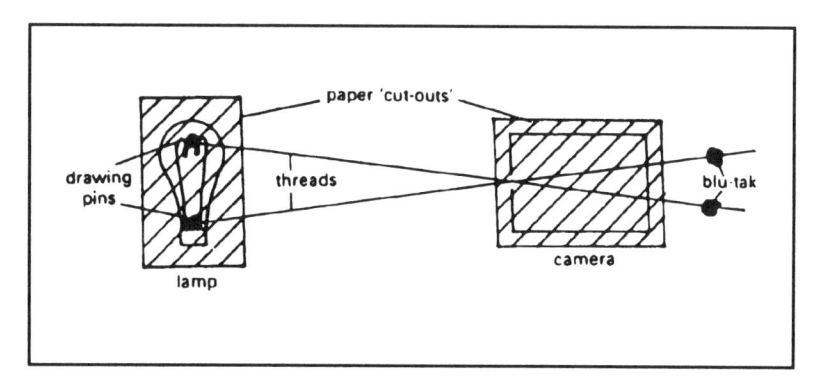

Figure 3. Using threads to model the passage of light rays through a pinhole camera.

A transparency prepared with threads is used to demonstrate these ideas that the pupils can use one thread "ray" from the top of the lamp and the other from the bottom of the lamp to make predictions about the image. The important part is that with careful organization the teacher should be able to encourage the children to move to and fro between practical experience and scientific model via evidence and prediction (Figure 1). If children experience these activities they are learning to use and apply standard scientific models. This example draws on the history of science but it isn't teaching children the history of science. Scientists' work from different times and cultures have been anachronistically juxtaposed to provide practical work to help the learning of a currently acceptable model. This doesn't stop teachers telling stories about past scientists or scientific ideas and this may help children understand that scientific ideas have been created and developed in different cultures and in different times (Figure 2).

However, it is stressed with participants that the prime motivation for creating the activity is for children to learn, use and understand the ray model of light and that they as teachers provide the model. This is not to say that time shouldn't be spent eliciting children's ideas on how shadows and images are formed but once they are elicited then they could test their ideas against the scientists' (teachers') model. We believe that children in school laboratories and with school apparatus will not rediscover or uncover many scientific theories. The teachers' role is to offer reasonable explanations which, in most circumstances, can be reasonably tested.

Case Two. The next case study concerns some curriculum materials called "Jabs for James Phipps." The specific topic is the story of Edward Jenner's work on vaccination. Participants are shown a short (five minutes) cartoon of the story of Edward Jenner and the famous vaccination of the small boy called James Phipps (Yorkshire Television, 1984). The experimental procedure is described in a clear chronological order. A rationale is also provided for Jenner's actions. The cartoon suggests with its imagery that James was not necessarily a willing volunteer to an "untested" procedure. We show participants the cartoon and then we ask them to do the following activities which are identical to what they can do with a class of pupils.

1) Telling a Story. This is an activity to process the information in the cartoon story in the video. The students are split into groups of approximately four in size. Each group is given an envelope with jumbled chunks of text which describe the story. The group task is to put the chunks of text in the right chronological sequence. All the curriculum materials are in Solomon (1991). The purpose of this activity is to provide an alternative way of processing the information in the cartoon, in other words to retell the story in a different manner. If done in mixed ability groups it provides an opportunity for all children to contribute to the telling of the story.

2) Evaluating an Experiment. The second part of the activity is to then open a second envelope with the statements "MAKING A HYPOTHESIS," "OBSERVING," "PREDICTION," "REACHING A CONCLUSION" and "DOING AN EXPERIMENT" each on a separate piece of paper with a brief explanation of its meaning. The task is then to match these words to the places where participants think they occur in the story of Jenner's experiment that they have just put in order. The purpose of this second activity is to help children to analyze the structure of the experiment.

We then tell participants that teachers who have used these materials agree that the evaluation of Jenner's experiment happens but it has also been reported that the analysis of the structure of the experiment has a transferability across to other scientific investigations (Nott, 1992a). It appears that the children, having analyzed (and criticized) the structure of experimental work as exemplified in the Jenner story, learned something about the structure of experiments. That learning had then been transferred to the planning of future experiments in terms of hypothesis, prediction, experiment, etc. The conclusion is that learning "processes" through stories is as important as learning them by doing experiments. The processes of experimental planning could be seen as the identical processes in the Jenner story.

3) Values and experiments. Participants can be invited to discuss whether they think Jenner's experiment was a "fair test." They can suggest improvements to Jenner's experimental design. They can be asked to discuss whether all improvements would be right and proper in terms of whether they would be allowed to do them, or whether they felt it would be right to do them. The purpose of this activity is to encourage the participants to evaluate the experiment and hence to consider whether there are any ethical and moral limits on the nature of the experiment and hence medical experiments in general. We then point out that they can do this when teaching pupils.

4) Role play. Watching the cartoon raises feelings. The cartoon implies that the procedure involved some risk and possible hurt to James. The materials contain a role play activity so that children can sympathize and empathize with the characters in the Jenner story. The characters are the obvious characters in the cartoon - Jenner, James Phipps and Sarah Nelmes - the milkmaid. Fictitious characters, whose views represent contemporary arguments for and against Jenner's work, are also added. A "role" card is available for each character and questions at the end stimulate discussion and help to build a character on the information presented. Participants are told how pupils can be split into small groups so that the roles can be built in groups and then one person from each group can act out the role - and even be prompted by her/his colleagues.

The scenario for the role play is a press conference where those not playing a role can act as journalists. This format has worked extremely well - perhaps because pupils are familiar with press conferences from the news. It is mentioned that teachers have been inventive and creative with it as well. One teacher reported creating the scenario for the role play as "the Phipp's family tea" - another familiar occurrence for children to use and there have also been children working the story up into the Jenner rap (Nott, 1992a) and even a ballet! These materials deal only with a very narrow case, i.e., the experiment on James Phipps by Edward Jenner. It is important to add that the textbook impressions that it was solely Jenner who invented a safe technique of immunization are wrong (see Smith, 1987). The story of immunization in England is one that involves other cultures and a determined woman, Lady Mary Wortley Montague (Alic, 1986). There is also available some excellent classroom material, "The long war against smallpox" (Science Education Group York, 1990) which adds to the historical dimension of the materials above and introduces the multicultural dimension.

Case three. The last case study we use is a case to help children understand that scientists have interpreted the same phenomena differently in order that children understand that different interpretations are possible and even desirable. The tutor talks the participants through the curriculum materials as in the account below. The particular case is studying Brownian Motion. Some details of this story can be found in a piece on the nature of science for school teachers (Nott 1992b). School texts never tell you why Brown was looking at pollen particles. He did have a clear research programme in mind but it was nothing to do with atomistic theories. The materials present pupils with one possible interpretation (Brown's original hypothesis) followed by our current interpretation and then the pupils have to suggest experiments and predicted results which will decide between the two interpretations.

The pupils work with Brown's ideas about pollen being active and then all organic material being active and then finally that all material, organic or inorganic, provided it was small enough would demonstrate activity. The materials lead the pupils through Brown's research programme in stages and get the pupils to discuss it and evaluate Brown's work from his perspective. This enables them to work with his interpretations. The next stage is the demonstration of the smoke cell and a three dimensional kinetic model of a gas. At this stage the teacher can present the contemporary interpretation of Brownian motion. In English high schools we have a piece of apparatus called a smoke cell. This is used in various years and pupils are expected to look at the smoke particles and explain their movement in terms of uneven bombardment from the air molecules in the cell. So the materials stress Figure 1 above, first hand experience, but do it by providing participation in a story and a practical demonstration. This also provides the opportunity to share Brown's motivation for looking at pollen through a microscope - hardly a chance occurrence!

The last element is to ask the pupils to imagine that they could write to or send a fax to Robert Brown to explain the modern interpretation and to suggest to Brown experiments he could do to persuade him of the modern point of view. The purpose being to work with both but to argue the case against Brown using our interpretations so that the contemporary scientists' model becomes more familiar and reasonable to the children. A full report on using the materials with pupils and the way that they responded is discussed in Nott (1994).

SUMMARY ACTIVITY WITHIN THE PROGRAMME: FINAL DISCUSSION WITH STUDENTS

We go on to say (in an activity taking about 30 minutes) to students and teachers participating in our nature of science courses that the above cases are based on stories of three scientists, but that it is not necessary to know every story in science to teach about the Nature of Science. In our view a teacher only needs a small range of stories to convey key ideas of the Nature of Science to children. Not every lesson is going to be, nor needs to be, based on historical information. An expectation that science teachers will be fully conversant with an accurate sociology, history and philosophy of science is unrealistic. The classroom resources cited and given out in further reading provide any teacher with an ample supply of ideas and stories to get going. We suggest to teachers that the point to recognize is that it is vital to have some stories and to recognize that stories are an important part of the culture of science. As Medawar has said, "Scientific theories are the stories that scientists tell each other."

In discussion, we also allay students' and teachers' anxieties that perhaps they are becoming history teachers. Similarly we argue that it would be a mistake to see the nature of science in the school curriculum as a course in the philosophy of science. Children have direct experiences of science and these can be used to present the NOS. There is little value in 14 year olds being told they are 'naive falsificationists' or discussing whether they are in a degenerative research programme or not!

REFLECTIONS ON AND DISCUSSION OF OUR NATURE OF SCIENCE PROGRAMME

Anecdotal evidence and course evaluations indicates that the session we do on the nature of science is very well received on teacher education programmes (pre-service and in-service). We also find that our approach of affirming what teachers do know rather than finding out what they don't know is welcomed and raises self-esteem.

The critical incidents have certainly proved to be productive as a springboard for discussion with practicing teachers, trainee teachers and even our own colleagues. They raise questions about the norms, counter-norms and values of science and of standard scientific education; about the purposes of practical activities in science lessons; about what is possible or legitimate in classrooms or laboratories, and about the knowledge and skills a teacher needs to realize those possibilities. Lastly, all participants find the activity on practical approaches useful and they perceive them as achievable and realizable in the classroom.

Underlying our teaching programme is the view that teachers do not necessarily have 'inadequate' views of the nature of science. They have teachers' views of the nature of science which are determined by their academic and professional histories. Teachers do not talk about science as science researchers or even science education researchers. Nor do they do talk about science as professional philosophers, historians or sociologists. They talk about science as teachers.

When teachers talk about school science as opposed to 'other' science then the 'other' science is science as done by science researchers - the people who create and discover scientific knowledge or apply scientific knowledge to particular technological problems. Teachers' knowledge about science is illustrated by examples from their own professional practice as much as scientific researchers' knowledge about the nature of science is illustrated by examples from their professional practice. Researcher-scientists know science through their professional practice and, we would argue, so do teachers.

However, we are not arguing that teachers' understandings of the nature of science are sufficient in themselves or immune from criticism. We believe that we will be left with extreme relativism if the nature of science is that 'anything goes' and that 'anything goes' for understanding the nature of science. There would be no point in having the nature of science in the curriculum! We recognize that teachers need to be aware of the nature of science as expressed by researchers, historians, philosophers and sociologists and those understandings should be reciprocated. We also believe there is a need to forge a broad consensus about the nature of science in school science - in other words what ought to be taught and how.

We would like to extend the time we have for this kind of work in teacher training and in-service work. We are currently extending our work on critical incidents to look at everyday incidents which at first sight don't seem critical but under analysis do become critical. However, we have never had the time to work with teachers on feedback and further coaching, all of which Joyce and Showers (1988) indicate are essential elements for professional development to effectively change teachers' practice.

Whether what we do actually changes the way that teachers' teach, we don't yet know. We do know that it is changing the way that we teach.

[1] Sheffield Hallam University, Sheffield, United Kingdom
[2] University of Sheffield, Sheffield, United Kingdom

REFERENCES

Alic, M (1986). *Hypatia's heritage*, London, Women's Press.
Joyce, B. and Showers, B. (1988). *Student achievement through staff development*, NY, Longmans.
National Curriculum Council (1989). *Non-statutory guidance for the national curriculum (science)*, York, England, NCC.
Nott, M. (1992a). 'History in the school science curriculum: Infection or immunity' in S. Hills (ed), Proceedings of the Second International Conference on the History and Philosophy of Science and Science Teaching, Kingston, Ontario, Queen's University, 215-228.
Nott M (1992b). 'The nature of science or Why teach Brownian motion?' in M. Atlay, et al. (eds.), *Open chemistry*, London, Hodder and Stoughton, 3-16.
Nott, M. (1994). 'Practical approaches to teaching about the nature of science', in J. Wellington *Secondary Science: Contemporary Issues and Practical Approaches*, London, Routledge, 258-283.
Nott, M. and Wellington, J. (1993). ' Your nature of science: An activity for science teachers', *School Science Review*, (75), 109-112
Nott, M. and Smith, R. (1995). 'Talking your way out of it', 'rigging' and 'conjuring': What science teachers do when practicals 'go wrong'', *International Journal of Science Education* (17), 399 - 410.
Nott, M. and Wellington, J. (1995). 'Critical incidents in the science classroom and the nature of science', *School Science Review*, (76), 41-46.
Nott, M. and Wellington, J. (1996) 'Probing teachers' views of the nature of science: How should we do it and where should we be looking?', in G. Welford, J. Osborne and P. Scott (eds.), *Science education research in Europe*, London, The Falmer Press, 283-294.
Science Education Group (1990). *The Salters' approach. Key Stage 4. Unit Guide: Keeping healthy*, London, Heinemann.
Schön, D. (1983). *The reflective practitioner.*, NY, Norton.
Shulman, L. (1987). 'Knowledge and teaching: Foundations of the new reforms', *Harvard Educational Review*, (57), 1-22.
Smith, J.R. (1987). *The speckled monster*, Chelmsford, Essex Record Office.
Solomon, J. (1991). *Exploring the nature of science*, Glasgow, Blackie.
Yorkshire Television (1984) Scientific Eye Leeds, YTV

APPENDIX A: 'YOUR OWN NATURE OF SCIENCE PROFILE'

The aim of the activity which follows (Nott and Wellington, 1993) is to encourage teachers to reflect upon their own view of the nature of science. It is intended to be a way of getting them to think, learn and reflect rather than a valid measurement of a position on some sort of objective scale. Teachers should not worry if, at the end of the activity, their profile is not as expected. The thing to do then is to consider why - this is an important part of the process.

PART ONE: Please read each of the twenty-four statements in Table 1 carefully.

Table I. Questionnaire for the nature of science profile.

This questionnaire is designed to give you some indication of your own philosophy of science. Score your response to each statement on a scale of +5 (strongly agree) to -5 (strongly disagree). A score of 0 will indicate a balanced view. (For the time being, ignore the initials in brackets.) Write down the score for each statement.

Questionnaire

1. Results that pupils get from their experiments are as valid as any others. (RP)
2. Science is essentially a masculine subject. (CD)
3. Science facts are what scientists agree they are. (CD, RP)
4. The object of scientific activity is to reveal reality. (IR)
5. Scientists have no idea of the outcome of an experiment before they do it. (ID)
6. Scientific research is economically and politically determined. (CD)
7. Science education should be more about the learning of scientific processes than the learning of scientific facts. (PC)
8. The processes of science are divorced from moral and ethical considerations. (CD)
9. The most valuable part of a scientific education is what remains after the facts have been forgotten. (PC)
10. Scientific theories are valid if they work. (IR)
11. Science proceeds by drawing generalizable conclusions (which later become theories) from available data. (ID)
12. There is no such thing as a true scientific theory. (RP, IR)
13. Human emotion plays no part in the creation of scientific knowledge. (CD)
14. Scientific theories describe a real external world which is independent of human perception. (RP, IR)
15. A good solid grounding in basic scientific facts and inherited scientific knowledge is essential before young scientists can go on to make discoveries of their own. (PC)
16. Scientific theories have changed over time simply because experimental techniques have improved. (RP, CD)
17. Scientific method is transferable from one scientific investigation to another. (PC)
18. In practice, choices between competing theories are made purely on the basis of experimental results. (CD, RP)
19. Scientific theories are as much a result of imagination and intuition as inference from experimental results. (ID)
20. Scientific knowledge is different from other kinds of knowledge in that it has higher status. (RP)
21. There are certain physical events in the universe which science can never explain. (RP, IR)
22. Scientific knowledge is morally neutral - only the application of the knowledge is ethically determined. (CD)
23. All scientific experiments and observations are determined by existing theories. (ID)
24. Science is essentially characterized by the methods and processes it uses. (PC)

TEACHING TEACHERS THE NATURE OF SCIENCE 311

PART TWO: Work out your profile by scoring each statement. Refer to Figure II.

Each statement has at least two letters in brackets after it e.g. PC - some have four e.g. RP, CD. Put your score for each question into the appropriate box or boxes in Figure Three i.e. some score once, some twice. NOTE CAREFULLY that some scores have to have their sign reversed (i.e. multiply by -1) before they can be entered into the box.

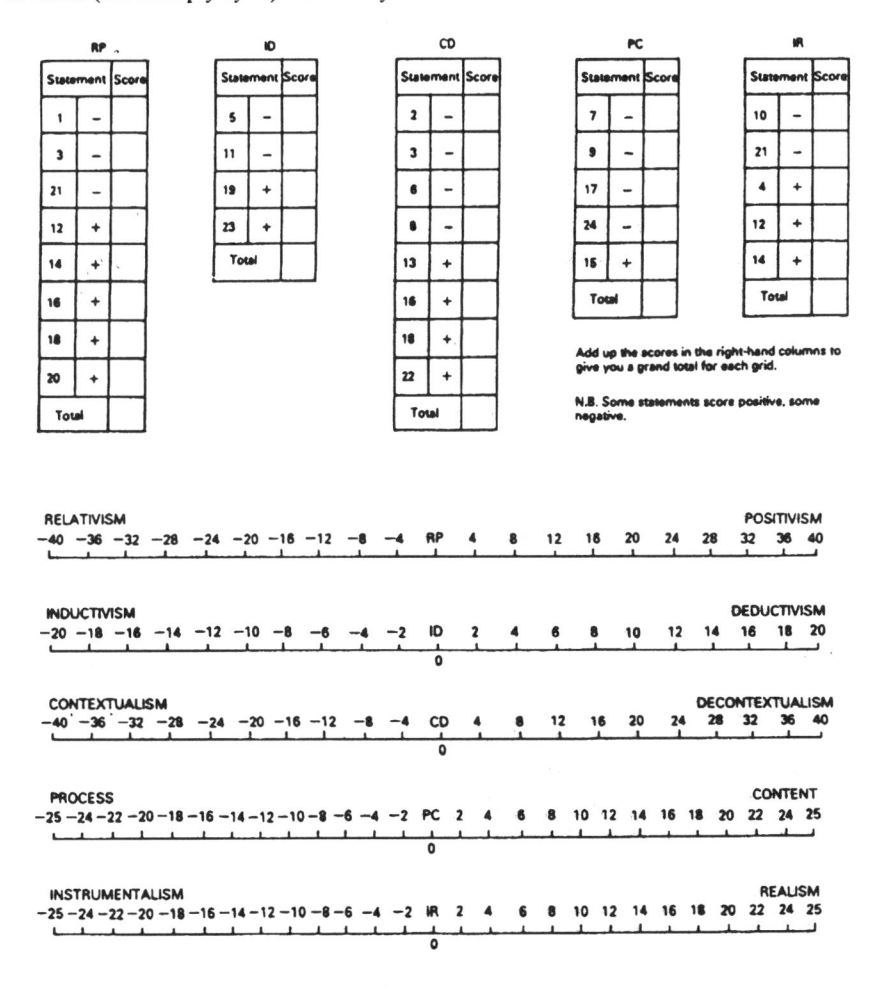

RP			ID			CD			PC			IR	
Statement	Score		Statement	Score		Statement	Score		Statement	Score		Statement	Score
1	−		5	−		2	−		7	−		10	−
3	−		11	−		3	−		9	−		21	−
21	−		19	+		6	−		17	−		4	+
12	+		23	+		8	−		24	−		12	+
14	+		Total			13	+		15	+		14	+
16	+					16	+		Total			Total	
18	+					18	+						
20	+					22	+						
Total						Total							

Add up the scores in the right-hand columns to give you a grand total for each grid.

N.B. Some statements score positive, some negative.

```
RELATIVISM                                                           POSITIVISM
-40  -36  -32  -28  -24  -20  -16  -12  -8  -4  RP  4  8  12  16  20  24  28  32  36  40


INDUCTIVISM                                                          DEDUCTIVISM
-20 -18 -16 -14 -12 -10 -8 -6 -4 -2  ID  2  4  6  8  10  12  14  16  18  20
                                     0

CONTEXTUALISM                                                    DECONTEXTUALISM
-40  -36  -32  -28  -24  -20  -16  -12  -8  -4  CD  4  8  12  16  20  24  28  32  36  40
                                                 0

PROCESS                                                                 CONTENT
-25 -24 -22 -20 -18 -16 -14 -12 -10 -8 -6 -4 -2  PC  2  4  6  8  10 12 14 16 18 20 22 24 25
                                                 0

INSTRUMENTALISM                                                         REALISM
-25 -24 -22 -20 -18 -16 -14 -12 -10 -8 -6 -4 -2  IR  2  4  6  8  10 12 14 16 18 20 22 24 25
                                                 0
```

After you have entered all your numbers into the boxes add up the totals, then transfer the marks from the columns to the correct position on each of the five relevant axes. Join up the five options to show your profile of science. Are you a raving relativist? A proud positivist? Or a coy contextualist? What do all these terms mean anyway?

PART THREE: Your Nature of Science

Many of the terms used may be unfamiliar. In fact, many of them are problematic and a matter of debate. Their meanings change and shift and can be seen as insults or praise depending on to whom you are talking.

Definitions for the meanings attached to the four continua above are offered below.

1. Relativism/Positivism Axis

   Relativist: You deny that things are true or false solely based on an independent reality. The 'truth' of a theory will depend on the norms and rationality of the social group considering it as well as the experimental techniques used to test it. Judgements as to the truth of scientific theories will vary from individual and from one culture to another i.e. truth is relative not absolute.

   Positivist: You believe strongly that scientific knowledge is more 'valid' than other forms of knowledge. The laws and theories generated by experiments are our descriptions of patterns we see in a real, external objective world. To the positivist, science is the primary source of truth. Positivism recognizes empirical facts and observable phenomena as the raw material of science. The scientist's job is to establish the objective relationships between the laws governing the facts and observables. Positivism rejects inquiry into underlying causes and ultimate origins.

2. Inductivism/Deductivism

   Inductivism: You believe that the scientist's job is the interrogation of Nature. By observing many particular instances, one is able to infer from the particular to the general and then determine the underlying laws and theories. According to inductivism, scientists generalize from a set of observations to a universal law 'inductively'. Scientific knowledge is built by induction from a secure set of observations.

   Deductivism: In our definition this means that you believe that scientists proceed by testing ideas produced by the logical consequences of current theories or of their bold imaginative ideas. According to deductivism (or hypothetico- deductivism) scientific reasoning consists of the forming of hypotheses which are not established by the empirical data but may be suggested by them. Science then proceeds by testing the observable consequences of these hypotheses, i.e. observations are directed or led by hypotheses - they are theory laden.

3. Contextualism/Decontextualism

   Contextualism: You hold the view that the truth of scientific knowledge and processes is interdependent with the culture in which the scientists live and in which it takes place.

   Decontextualism: You hold the view that scientific knowledge is independent of its cultural location and sociological structure.

4. Process/Content

   Process: You see science as a characteristic set of identifiable methods/processes. The learning of these is the essential part of science education.

   Content: You think that science is characterized by the facts and ideas it has and that the essential part of science education is the acquisition and mastery of this 'body of knowledge'.

5. Instrumentalism/Realism

   Instrumentalism: You believe that scientific theories and ideas are fine if they work, that is they allow correct predictions to be made. They are instruments which we can use but they say nothing about an independent reality or their own truth.

   Realism: You believe that scientific theories are statements about a world that exists in space and time independent of the scientists' perceptions. Correct theories describe things which are really there, independent of scientists e.g. atoms, electrons.

PART FOUR: Reflect on your profile by reading the working definitions and considering the points below:

- How do you feel about your profile? Has it really 'measured' your views about science?

- Do you feel confident that you understand it all?

- Do you think your views/opinions have been challenged or changed by this activity?

CRAIG E. NELSON[1], MARTIN K. NICKELS[2], AND JEAN BEARD[3]

20. THE NATURE OF SCIENCE AS A FOUNDATION FOR TEACHING SCIENCE: EVOLUTION AS A CASE STUDY

This chapter presents the approach applied in several inservice institutes for secondary school biology teachers with a focus on evolution and the nature of science. Here we present the rationale, review the pedagogical strategies and several of the lessons, and provide the evidence that the institutes made a major difference in how teachers taught both the science of evolution and the nature of science. In this chapter we will provide sufficient detail so that others can evaluate and adapt our approach in working with inservice and preservice high school teachers.

The combination of evolution and the nature of science is particularly effective in addressing many of the core issues within the nature of science. We feel that it is essential to present evolution in the context of the nature of science — specifically, to present it as the better choice of those so far proposed for explaining whole classes of facts in biology and anthropology. This allows teachers to emphasize that explanation is the central task of science and otherwise to explain the nature of science and the criteria used in science for selecting among alternative theories. It is important to help teachers and students understand that this exemplifies the nature of science generally. In addition, the teaching of evolution and related topics remains perhaps the single greatest content-related problem that high school science teachers must confront. Since evolution is the central organizing theory for all of biology, this problem is most severe in secondary school biology, but it also affects much of the rest of the science curriculum.

BACKGROUND OF THE PROBLEM

In working with college undergraduates, we each independently realized that two strategies made the teaching of evolutionary ideas much more effective. First, it seemed essential to present evolution in the context of the nature of science. Second, we found it fundamental to help students understand the nature of science generally. Otherwise, they often felt that our approach to teaching evolution was necessary because it was more dubious than the rest of science, which remained, in their minds, "truth." This is not an unreasonable attitude since science is so often taught more as stuff (the "right" answers) to be memorized than as comparisons of alternative explanations — the process.

Emphasizing the nature of science was also especially important since many of our students were not science majors. These students would be unlikely to learn about the processes of science while presented with a view of science as an unassailable body of facts. Furthermore, we encountered increasing numbers of students who had mistaken notions of the nature of scientific inquiry and knowledge. This was exacerbated by widely-disseminated "creationist" views that present a distorted view of the nature of science. Based on our experiences at the college level and our prior work with high school teachers in other contexts, we were convinced that our approach to evolution and the nature of science instruction would work at the high school level as well.

EVOLUTION AND NATURE OF SCIENCE INSTITUTES

In 1989, we began what became an externally funded six-year program for experienced high school biology teachers titled *Evolution and the Nature of Science Institutes (ENSI)*. The goal of this program was to improve the teaching of evolution in high school biology courses. We asked the teachers: 1) to begin their biology courses by considering the nature of science at length; 2) to introduce evolution early in the course as an illustration of the nature of science and modern scientific thinking; 3) to use evolution and the nature of science as central themes throughout the course; 4) to emphasize humans as a key example whenever possible; 5) to teach science in a manner compatible with how science is really done; and, 6) to monitor the effects of their revised courses.

Each ENSI began with a three-week, residential summer session followed by a two-day session in the fall and another in the spring. During six years (1989-1994) we worked directly with 180 biology teachers. In addition, from 1992 to 1997 we worked indirectly with nearly 650 additional teachers through forty-five "satellite" institutes conducted by 50 selected ENSI "graduates" whom we trained for a second summer to be "lead teachers."

The Summer ENSI program began with an extensive treatment of the nature of science. This was followed by the development of fundamental evolutionary processes and patterns, each presented within the nature of science framework. The latter part of each session included time devoted to the review, modification, and development of teaching activities on both the nature of science and evolution.

Also, each teacher was asked to design a teaching experiment (typically termed an action research project) for implementation during the following school year that monitored a designed change in their course content, their teaching practices, or both. These changes were intended to improve their teaching of some aspect of either the nature of science or evolution. The projects served both to foster improvements in their teaching and to illustrate for them and their students a practical application of investigative processes.

The focus of the fall follow-up (midway through the semester) was on their successes and problems in teaching their first units on the nature of science. The spring follow-up (midway through the semester) focused on their teaching of evolution. These pre-announced topics served to reinforce our first and second charges to the teachers.

A vital component of ENSI was the use of a variety of pedagogical activities that both illustrated and modeled different aspects of the scientific process. Open-ended brainstorming sessions on different topics (such as attributes of science) led to the teachers discovering one another's sometimes contradictory perceptions of what is most important to include. Concept mapping (Novak and Gowin, 1984) led to similar experiences. A traditional "Black Box" or Mystery Box activity (in which the teachers were asked to describe the barriers within a sealed box also containing a marble) was modified by not allowing them to open the boxes at the end of the session. This inability to get absolute verification of their descriptions underscored the uncertainty inherent in much of scientific knowledge--and generated much frustration as well! Hands-on labs (e.g., comparing ape, human and fossil skull casts) illustrated the inter-observer variation in scientific measurements. It also showed how preconceptions can differ as to what constitutes appropriate evidence for scientific decision-making (here, on phylogenetic relationships). Structured discussions (of both readings and lecture-posed questions) were used extensively to maximize the active processing of information and ideas, the sharing of alternative interpretations of the same material and a consideration of the consequences of adopting various alternatives.

Most of these activities were done in small groups, thus modeling the cooperative and collaborative nature of much scientific research. It is quite important that such groups be required to develop a consensus group-product generated as a social dynamic involving sharing, discussing and, usually, a compromising of diverse perspectives. Variation in the ethnic, socio-economic and gender composition of the groups contributed to this dynamic. These aspects of the small-group setting mirrored the socio-cultural context in which science is done. Taken together, these active, student-centered activities occupied well over two-thirds of the participants' time during the summer institute. Didactic, teacher-centered activities occupied less than one-third of the available time.

We believe that the most unique part of our approach was helping the teachers reconceptualize the nature of science in ways that made it easier for them to understand the strength of evolutionary theory and better able to deal with popular challenges to it. The rest of this chapter presents this conjunction in detail intended to be sufficient to allow those who like it to use it.

RECONCEPTUALIZING THE NATURE OF SCIENCE

We concentrated on four key aspects of the nature of science. The first examined the realm of science. The second focused on scientific knowledge as fundamentally uncertain. The third focused on science as a set of procedures for comparing alternative ideas and judging some to be "better," procedures that explain how science can be fundamentally uncertain yet quite useful and reliable. The fourth emphasized that scientific ideas are known with various degrees of confidence. It follows from this that one must consider the possible consequences of an idea in deciding when the evidence is good enough to accept the idea for a particular purpose.

We thus presented a constrained constructivist view of the nature of science. Here, the acceptability of alternative views is constrained by the extent to which they reflect and explain natural phenomena as well as by historically and socio-culturally influenced judgements of the value of their positive and negative consequences. To help develop these ideas, and to emphasize that they were not idiosyncratic to us, we made extensive use of selected readings from Strahler (1987), and more limited use of readings from Gould (1980a, 1980b, 1985), Kitcher (1982) (especially chapter 2), Ridley (1985), Strickberger (1996), and Glen (1994). Our overall approach to the nature of science and evolution was grounded in current understandings of intellectual development. Please refer to Nelson (1989, 1997) for further elaboration.

The Realm of Science

A delineation of the realm of scientific inquiry helps emphasize both the strengths and weaknesses of science as a way of knowing. For example, noting that scientists can only study the natural world helps alleviate any problems that might arise relating to issues involving the "supernatural." By "natural" scientists mean empirically and reliably detectable using the human senses, whether or not these are amplified technologically. Consequently, questions such as "How many angels can dance on the head of a pin?" or "Did God create the universe?" may be intriguing but are outside the province of science.

Within its realm, science can be seen as operating on two main levels. First, it provides summaries of empirical patterns. The summarization of patterns can be valid science even if the causes of the patterns are totally unknown. Galileo and Linneaus were great scientists even though they did not know the natural causes of the patterns they summarized. However, science is even better when the patterns can be explained. Thus, the second level of science is to provide natural explanations for the empirical patterns. Darwin's importance lay in providing explanations for patterns observed earlier by Linneaus and others.

The delineation of natural causes typically does not provide a scientific explanation for the origin of the systems studied. Newton and Einstein each did great science without explaining the origin of the systems with which they worked. Similarly, evolution can be great science if it delineates what happened after the origin of life and why it happened, even if our explanations for the origin of life are still quite tentative or even largely speculative at some points.

It is useful to contrast the two scientific levels with a third. One could say that God *makes* the planets go around the sun in ellipses (the pattern) due to the interaction of inertia and warped space (the natural causal explanation). Science provides no guidance as to whether one should attribute the existence of the system to a Creator. Similarly, religion is of no help in working out either the empirical patterns or their natural causes.

In the specific case of evolution, one further elaboration of these ideas is quite helpful. In current public discourse, "creationist" is essentially synonymous with the young-earth, flood-geology, no-new-kinds, perspective whose advocates call themselves "scientific creationists." Either terminology has two problems. First, it suggests that if one believes in a Creator, one must accept the tenets of scientific creation. Second, it suggests that the scientific creationist viewpoint is somehow more scientific than alternative theological viewpoints that combine at least some science with a belief in a Creator. Both of these ideas are patently false. It is quite helpful for both students and teachers to note that there is an array or spectrum of theological positions. These range from the "scientific" or (as we prefer) "quick" creation through the "progressive" creationist's idea of an old earth with the direct creation of new forms intermittently across hundreds of millions of years to "gradual" creation (often termed theistic evolution), the most scientific of the three. The gradual creation perspective states that God made organisms become more complex and diverse through time (the pattern) due primarily to the interaction of inheritance and natural selection (the causal natural explanation). Science is, in practice, non-theistic, meaning that it operates only at the levels of empirical patterns and natural causes. Once again, it provides no guidance as to whether or not we should attribute the existence of the biological systems to a creator.

Uncertainty in Science: Scope and Sources

There are two key questions about the nature of uncertainty in science. First, is the idea of scope. How much of science is really uncertain and subject to further major change? Second, after the answer to this first question is clear (i.e., most or all of science is really uncertain), the question of the sources of this uncertainty arises naturally. What prevents science from coming up with enduring truth?

Scope

The most powerful arguments that scientific knowledge is fundamentally uncertain come from the history of science. The history of virtually any scientific idea, such as evolution in biology or plate tectonics in geology, consists of any number of mistaken ideas and conclusions that may well once have been deemed the final word. For example: What kinds of things did physicists believe with absolute certainty in the last century that are now known to be quite wrong? Answers to this question include the notions that space is Euclidian, motion is Newtonian, and matter can neither be created nor destroyed. Complete agreement among scientists is not a guarantee of truth, even if that agreement persists for centuries (Newtonian motion) or even millennia (Euclidian rather than warped space)! And who is to say that our best ideas today will not be replaced eventually by better ones? In teaching that scientific knowledge if fundamentally uncertain it is important to invoke examples of both historical and current uncertainty from across all of science to counter the common impression that physics, in contrast to evolution and much of the rest of science, is largely free from doubt.

Sources

Once it appears that science is indeed broadly uncertain, a second question naturally arises. What prevents the processes of science from yielding certainty? The first task here is to understand that even direct observations can be misleading. Human senses have only a limited range of detection (for instance, we cannot "see" ultraviolet). Hence, we directly perceive only a limited part of the universe of potential stimuli. Compounding these sensory limitations, we actually perceive the world with our minds rather than our senses. For example we "see" depth only in the sense that our minds construct it. Optical illusions designed to mislead observers into thinking that two figures differ in size, when in reality they do not, illustrate this point well. Similarly, our minds are shaped by previous experiences and, thus, predisposed to recognize only some parts of reality. The famous "young woman or old hag" drawing illustrates that our minds unconsciously impose a biased interpretation on sensory data. (See Lederman and Abd-el-Khalick in this volume for a complete description of this activity.)

However, the more important sources of uncertainty in science are much deeper. It is useful to invert the question of uncertainty and ask: How might we get certainty? Logic and data are the two potential sources. Logic fails because nature often refuses to abide by it. Consider the question of whether 2 + 2 is always 4 in reality. An easy and surprising demonstration combines two liters of water and two of alcohol. They dissolve in each other, yielding visibly less than four liters. Thus, in reality 2 + 2 is only sometimes 4! Similarly, if logic yielded certain answers, space would have been Euclidian rather than warped. Logic tells us what would be true in an imaginary universe where the assumptions were absolutely true, but it can only provide approximations when dealing with the real universe.

Scientific data also fail to provide certainty. Two sets of assumptions are always problematic in any scientific study. One is that all of the uncontrolled variables are irrelevant--which they sometimes are not. The other is that one has at hand all the possible alternative interpretations. We could only claim that any explanation was absolutely correct if we knew that we had eliminated every possible other explanation. But how could we know that we had even conceived every possible alternative? Indeed, a scientist in the future may come up with a new idea that leads to a reinterpretation with no new data whatsoever (Kitcher, 1982).

More frequently in the history of science, new data elsewhere leads to a reinterpretation of other, older ideas. That is essentially what happened in geology in the 1960s when evidence suggesting that the sea floor was spreading led to the reinterpretation of many previously existing data sets on a wide array of geological phenomena (Glen, 1982).

Comparisons and Criteria: Science as the Set of Currently Better Models

Once general uncertainty in science is acknowledged, a key task becomes understanding how scientists pick better theories. Fundamentally, this task is one of comparing alternative patterns and explanations and using appropriate criteria to tell which are scientifically stronger. Both scientists and science teachers have often learned these criteria, to whatever extent they have learned them, more tacitly than explicitly. However, effective science teaching requires making both the comparisons and the criteria explicit.

Consider, for example, the fossil record. Darwin (1859) is sometimes credited with inventing a new form of scientific reasoning, one in which he explained not only the strengths of his theory but also the problems (Dewey, 1910). Among the chief problems was the lack of a clearly evolutionary sequence in the fossil record (the oldest fossiliferous rocks then known had quite diverse marine invertebrates). There was no prior reason to expect that the sequence of fossils in an adequately worked out record would have come to show the kinds of changes required by Darwin's theory. Indeed, several alternatives seemed much more likely to some of his contemporaries. Sedgewick, for example, thought he saw evidence for a series of independent creations. Lyell, in contrast, suggested that the fossil record would be found to be cyclical in nature, and even speculated that the giant reptiles might again come to rule the earth. Alternatively, the quick creationist idea that all organisms were created together in a single week would suggest that the oldest sets of fossiliferous rocks should contain fossils of all of the major kinds of organisms that have ever existed. We would like to have a fair test by which these various alternatives could be compared. An ideal fair test would have two key properties. It would be based on a line of evidence that was independent of alternative explanations. And, it would potentially allow any one of them to be supported as better than the others.

Fortunately, the methods used to put the fossil record in order are independent of any ideas of biological change or permanence and would have allowed it to support any of these alternatives. That is, the fossil sequence provides a fair test of evolution and of the various alternatives to it, including quick creation.

The initial criterion for deciding the temporal ordering of the rocks was the relative age as indicated by superposition. The sequence worked out on this basis has been confirmed, with some revisions, by radiometric or "absolute" dating (see Strahler, 1987) that does not depend on the position of the stratum. As is now well known, the fossil record starts, as Darwin explicitly predicted life must have started, with only a few kinds of simple organisms, all of them of the bacterial (prokaryotic) grade, and only slowly (not necessarily gradually) shows increases in the diversity and complexity of organisms.

Thus, scientists reject "quick" creation of the organisms on earth not because the original idea was based on religious texts. Rather, scientists reject it because it has failed a substantial fair test, a test that could have supported it but did not.

A number of additional criteria allow us to support or tentatively reject scientific theories. For example, radiometric dating is part of a larger body of nuclear science that has allowed the prediction and design of nuclear reactors and nuclear bombs. These successes increase our confidence in the power of nuclear theory, and, therefore, in the basic ideas underlying radiometric dating. Similarly, the successes of molecular genetics increase our confidence in the reality of the explanations we use for evolution.

We developed a number of such criteria in comparing evolution with "Lamarckian," quick creationist and other alternatives. By the end of ENSI, and by the ends of our courses, it is clear that evolution is supported by the results of a large series of fair tests, that it has immense predictive power, that its mechanisms are supported by all of modern molecular genetics, and that its acceptance illustrates a number of other criteria that are used in science to separate better from worse ideas. Further, because our approach has been comparative throughout, it has become clear that scientists reject quick creation, as science, because it has failed a substantial series of fair tests and is inferior to evolution on a number of other criteria, including the ad hoc or untestable nature of many of its assertions.

This same approach can and should be used repeatedly in any field of science. Ask what the alternatives are and what scientific criteria can be used to distinguish among them. Popular misconceptions and pseudoscience should be included in these comparisons wherever they can be made pertinent.

Additional Applications To Evolution

In dealing with each evolutionary pattern or process we laid out alternative hypotheses, asked how they could be compared, ideally by fair tests as defined above, and laid out the currently evident uncertainty for each conclusion. In looking at patterns, we took the approach, pioneered by Darwin (1859), that evolution uniquely explains "whole great classes of facts." We included an updated version of Darwin's argument that evolution both explains adaptation, via natural selection, and explains, via inheritance from variously remote ancestors, many aspects of development, structure and behavior that are not explicable in terms of current adaptation. We also developed a cladistically informed version of Darwin's insight that evolution explains the causes of the patterns of relationships first adequately captured in Linnaeus' eighteenth century, hierarchical classification as well as the causes of the dominant patterns of biogeography. Major sources of post-Darwinian support for evolution include the sequences in the fossil record (above), the abundance of intermediate fossils and the records of evolution that persist in molecular sequences. In each case, we emphasized examples from human evolution, both because of their intrinsic interest to students and because the evidence for human evolution provides one of the most comprehensive examples (Nickels, 1987).

Decisions in Context: Consequences and Their Value

So far we have summarized and provided explanations of how we asked teachers to understand and teach three key aspects of the nature of science. These ideas include the notion that science is limited to natural patterns and causes, that it is fundamentally uncertain, and that, nevertheless, by comparing alternative ideas and using a variety of criteria, scientists can distinguish better ideas from less satisfying ones. This process of comparison typically, but not inevitably, produces a series of increasingly more useful models of reality.

There remains one key property of the nature of science that we asked the teachers to understand and use. Scientists do not, and indeed, should not, always accept the idea which has the strongest support. Consider, the situation with traditional hypothesis testing. Conventionally, we reject hypotheses unless the chance that the most relevant null hypothesis adequately explains the data is less than five percent. That is, even if the new hypothesis is more probable we do not accept it unless it is substantially more probable. This means that with standard hypothesis testing we frequently reject the more probable hypothesis.

The nub of the idea here can be grasped better by asking: Why five percent, rather than ten percent, one percent or some other number? If we were to be more generous towards new hypotheses and accept them whenever the relevant null had less than, say, a ten percent chance of explaining the data, more of the ideas that were published would turn out to be wrong on further examination. Alternatively, if we were to be more demanding of new hypotheses and only accept them whenever the relevant null had less than, say, a two percent or one percent chance of explaining the data, then we would fail to publish many good ideas and research would become more time consuming and otherwise expensive.

Thus, the answer to "Why five percent?" is that it is a value-based tradeoff between the importance of avoiding the publication of too many wrong ideas and the slowing of scientific progress that being more rigorous would inevitably produce. The fundamental point is that, rather than being "objective," the tradeoff is "subjective" in the sense of being based on the relative value placed on these two different consequences.

A key to understanding most of the controversies at the interface of science and society is the recognition that scientific ideas often have substantial consequences that were not taken into account in choosing the standard five percent rule, consequences that vary from idea to idea and that may vary further with context. Consider for example, the safety of new medicines. The evidence we should demand for their effectiveness will vary with the kinds and severity of the side effects, with the severity of the condition being treated, and with the alternatives available. Thus individuals often choose to take drugs with substantial side effects when the condition itself is severe enough and there are no reasonable alternatives. Chemotherapy and radiation therapy for cancer provide clear examples of this phenomenon.

When the negative consequences are seen to be quite large and the benefits are seen to be small, we typically set aside large amounts of strong evidence in making decisions on what to do or on what ideas to accept. Some creationists likely see the negative consequences of mistakenly accepting the idea of evolution, were it actually false, as quite severe. In contrast, they see only small benefits, at best, from the idea of evolution, even if it were right. Would any rational person accept the idea of evolution if these were the relative benefits and risks envisioned?

Those of us who accept evolution do so not just because the evidence is strong but also because we see different benefits and risks than those emphasized by the quick creationists. Evolution explains substantial bodies of data and has led to predictions that are both scientifically important and economically beneficial. And we see the risks as different, different either because we believe that our risk of eternal damnation is essentially independent of our scientific ideas, or because, like Saint Augustine (1988), we believe that there are severe religious costs (specifically, increased risk of disbelief) in denying scientific ideas "the truth of which is written in the skies," or, even, in some cases, because we do not believe in damnation.

With the ENSI teachers, as with many of our own students, we have found that an approach which directly compares evolution with quick creation across a broad range of data and using a full array of scientific criteria, leads to the conclusion that the support for evolution has become so strong that its "truth" is essentially written in the skies. Overwhelmingly clear except to those who will not see. For many of the teachers and for many of our students, this led to a reassessment of both the scientific and religious risks and benefits involved in accepting evolution.

EVALUATION

Several instruments were used in assessing the effects of our efforts. Extensive informal assessment was also included, both for feedback, for modifying of our ongoing efforts and as a summative overview of the effects of the institute. A detailed presentation of the results is clearly beyond the scope of this chapter but is being prepared for separate publication. Only a few highlights can be included here.

Teachers in ENSI increased their coverage of the nature of science from an average of less than four class days prior to participation in ENSI to nearly fourteen days afterwards. By the end of the institute they projected a further increase of three days. This represented a quadrupling of the amount of time spent on a topic that cuts across all scientific disciplines and is basic to scientific literacy. The amount of time spent discussing evolution-related topics increased from an average of 19 days prior to teachers' exposure to the institute to almost 32 days during the ENSI year with a projection for the following year of an additional 6.5 days on evolution topics. This represented a doubling of the time devoted to evolution, the central organizing principle in all of biology. Our data also show a significant increase in the understanding by teachers of the ENSI concepts including both those on the nature of science and those on evolution. Teachers also indicated an increased teaching emphasis on the concepts we taught in ENSI. We conclude from this that after ENSI they were more comfortable with evolution concepts, dealt with evolution more completely in their classes, and linked it more clearly to the nature of science.

In addition, a shift occurred in the way the teachers thought about the nature of science and its relationship to teaching science. Almost all of the teachers came to the various versions of the institute because they wanted to learn more about evolution or wanted to learn how to teach it more safely in a system where it felt controversial. Generally, they felt that the nature of science was irrelevant to why they chose to come. Teachers felt that they already knew all about the scientific method and generally felt that in most of the areas they taught, science had reached a state not functionally different from the truth. By the end of the ENSI experience, the prevailing opinion was that their teaching had been transformed by the new approaches they had learned for thinking and teaching about the nature of science.

Perhaps even more important, was the overall pedagogical "paradigm shift" teachers experienced and brought into their teaching throughout their courses. They were more inclined to involve their own students in the process of scientific discovery and research by letting them design open-ended labs. They engaged them in comparative critical thinking as a method for determining better from worse proposed explanations (and not "absolutely true" ones from "absolutely false" ones). By understanding the social context in which every scientist works, the teachers were able to enrich their students' understanding of how science is conducted in the "real" world in contrast to the imaginary world of "the" scientific method. And these same active learning approaches were applied to evolution and to other units in biology.

CONCLUSIONS AND RECOMMENDATIONS

Our work in ENSI with experienced high school teachers of General Biology produced substantial changes in how the teachers taught both evolution and the nature of science. We attribute this success to the following conjunction of factors. We strongly recommend each of these individually to others who are working with either experienced or preservice teachers, and especially urge others to consider using several of them together.

1. A reconceptualizing by the teachers of the nature of science was clearly fundamental to both the changes in pedagogy and content. Central to this was the movement from a standard "steps of the scientific method" and "(almost) true" characterization of modern science to one emphasizing the uncertainty and comparative basis of scientific knowledge.

2. This reconceptualization allowed a reassessment of the strength both of the support for evolution and of its explanatory and predictive power. Strength is not nearly as evident until alternatives are directly compared.

3. Comparing evolution with alternatives including quick creation paradoxically makes it seem less challenging to fundamentalist religion, as it is no longer a confrontation of dogmas. This allowed the teachers to feel safer in teaching it. Other important pieces that contributed to this feeling of increased safety (besides point 2, above) were the idea of a spectrum of theological positions (quick to gradual creation), with its broaching of the dichotomy usually envisioned, and the provision to the teachers of copies of Strahler and group discussion of some of his analyses of quick creationist assertions. The teachers frequently commented that they felt newly prepared to deal with fundamentalist claims.

4. The focus on humans as a central case study whenever possible was also crucial. Using humans to illustrate various aspects of evolutionary processes and patterns throughout ENSI took advantage of perhaps the most interesting species that students can relate to. It also had the advantage of directly dealing with the species that most people have trouble accepting as

a product of natural processes. By learning about the abundant evidence for human evolution, the ENSI teachers were especially well-prepared to deal with concerns and objections to evolution.

These first four points represent in our minds the most innovative aspects of what we did. However, three more points were, we believe, critical to our success and thus also require emphasis here.

5. The extensive use of active social learning (concept mapping, small group work, structured small group discussions, open-ended laboratory projects, etc.) was clearly one of the strongest effects on the outcomes of the institute. With this was the switch from tests to more authentic forms of assessment.

6. It was clear that having the teachers find, present and, especially, directly participate in activities that were appropriate for use in their own classrooms was central to the transformations in both content and in pedagogy.

7. A final important lesson from ENSI is the immense value of providing continuing support for teachers who are attempting radical changes in their teaching and in their courses. In the case of ENSI, this consisted of requiring the teachers to attend a fall and a spring follow-up in addition to attending the residential summer session. The additional support that the ENSI faculty were able to provide may well have been secondary to that provided by the teachers themselves acting as a "support group" during the time of transition from the old teaching to the new. We strongly encourage the incorporation of such sequenced education opportunities for anyone working with in-service teachers. Such support opportunities can be emulated somewhat even in pre-service settings by having faculty coordinate and overlap their treatment of topics, such as the nature of science, that transcend any one scientific course or discipline.

[1] *Indiana University, Bloomington, Indiana, USA*
[2] *Illinois State University. Normal, Illinois, USA*
[3] *San Jose State University, San Jose, California, USA*

ACKNOWLEDGMENT

Funding for the Institutes came from the National Science Foundation (NSF/TPE: 88-555-60 and 90-555-85 to Indiana University and 91-552-59 to San Jose State University).

REFERENCES

Saint Augustine (1988). *The literal meaning of Genesis.*, translated and annotated by John Hammond Taylor, NY, Neuman Press.

Darwin, C. (1859). *On the origin of species,* London, John Murray.

Dewey, J. (1910). *The influence of Darwin and philosophy and other essays in contemporary thought,* New York, Holt.

Gould, S. J. (1980). *Ever since Darwin,* New York, W. W. Norton.

Gould, S. J. (1980). *The panda's thumb,* New York, W. W. Norton.

Gould, S. J. (1985). *The flamingo's smile,* New York, W. W. Norton.

Glen, W. (1982). *The road to Jaramillo: Critical years of the revolution in earth science,* Stanford, CA, Stanford University Press.

Glen, W. (Ed.) (1994). *The mass extinction debates: How science works in a crisis,* Stanford, CA, Stanford University Press.

Kitcher, P. (1982). *Abusing science: The case against creationism,* Cambridge, MA, MIT Press.

Nelson, C.E. (1997). 'Tools For Tampering With Teaching Taboos' in William E. Campbell and Karl A. Smith (eds.), *New paradigms for college teaching,* Edina, MN, Interaction Book Company, 51-77.

Nelson, C.E. (1986). 'Creation, Evolution, or Both? A Multiple Model Approach' in R. W. Hanson (ed.), *Science and Creation,* New York, Macmillian, 158-159.

Nickels, M. K. (1987). 'Human Evolution: A Challenge For Biology Teachers', *The American Biology Teacher,* (49),143-148.

Novak, J. D. and Gowin, R. (1984). *Learning How to Learn,* Cambridge University Press.

Ridley, M. (1985). *The Problems Of Evolution,* New York, Oxford University Press.

Strahler, A. (1987). *Science and Earth History: The Evolution/Creation Controversy,* Buffalo, NY, Prometheus.

Strickberger, M. W. (1996). *Evolution.* 2nd ed., Boston, MA, Jones and Bartlett.

SECTION IV

ASSESSING THE NATURE OF SCIENCE

NORMAN LEDERMAN, PHILIP WADE AND RANDY L. BELL

21. ASSESSING UNDERSTANDING OF THE NATURE OF SCIENCE: A HISTORICAL PERSPECTIVE

The nature of science typically refers to "the values and assumptions inherent to science, scientific knowledge, and/or the development of scientific knowledge" (Lederman, 1992). These values and assumptions include, but are not limited to, independence of thought, creativity, tentativeness, an empirical base, subjectivity, testability, and cultural and social embeddedness (Duschl, 1990; Lederman, 1992; Matthews, 1994). In brief, the nature of science is directly related to the epistemology of science as distinct from science process and content.

The development and assessment of students' and teachers' conceptions of the nature of science has been a concern of science educators for over 30 years. In spite of this focus on the nature of science as a goal for science instruction, a vast number of research investigations consistently indicate that students, as well as teachers, do not possess what are considered to be adequate conceptions of the nature of science (Aikenhead, 1973; Aikenhead, 1987; Cooley & Klopfer, 1963; Korth, 1969, Lederman & O'Malley, 1990; Lederman, 1992; Mackay, 1971, Rubba & Anderson, 1978; Wade & Lederman, 1995; Welch, 1981; among others). This finding makes the continued communication and assessment of aspects of the nature of science a vital part of the science curriculum as we enter the new century.

BACKGROUND ON NATURE OF SCIENCE ASSESSMENT INSTRUMENTS

The history of the assessment of the nature of science mirrors the evolution that has occurred in both psychometrics and in educational research design. The first formal assessments, beginning in the early 1960s, emphasized a quantitative approach to assessment, as was characteristic of the overwhelming majority of science education research. Prior to the mid-1980s, with few exceptions, researchers were content to develop instruments that allowed for easily "graded" and quantified measures of individuals' understandings. In some cases, standardized scores were derived. Within the context of the development of various instruments, some open-ended questioning was involved in construction and validation of items. However, little emphasis was placed on providing an expanded view of an individual's beliefs regarding the nature of science.

Research on the nature of science over the last three decades has provided at least four consistent findings, regardless of the instruments used in the investigations:

1) Science teachers appear to have inadequate conceptions of the nature of science.

2) Efforts to improve teachers' conceptions of the nature of science have achieved some success when either historical aspects of scientific knowledge or direct attention to the nature of science have been included.

3) Academic background variables have not been significantly related to teachers' conceptions of the nature of science.

4) The relationship between teachers' conceptions of the nature of science and classroom practice is not clear, and the relationship is mediated by a large array of instructional and situational concerns.

The assertion that there is a strong and continuing interest in the nature of science among science educators is supported by the wide number and variety of assessment instruments developed to gauge teachers' and students' understandings of the nature of science. Indeed, so much work has been done with regard to instrument development that it constitutes a distinct line of research. The purpose of this review is to describe and summarize the characteristics of the various instruments and techniques that have been used to ascertain a person's beliefs and knowledge about the nature of science. We hope that this review will better enable the reader to understand the meaning of research findings and claims related to the nature of science in general and to specific instruments in particular.

NATURE OF SCIENCE ASSESSMENT INSTRUMENTS: A CRITIQUE

Table I presents a comprehensive list of the more formal instruments constructed and validated to assess various aspects of the nature of science. Most of the instruments address only certain aspects of the nature of science and often inappropriately confuse the issue by addressing areas other than the nature of science, including science process skills and attitudes toward science. Instruments considered to have poor validity as nature of science assessments have the following characteristics:

1) largest proportion of the instrument concentrates on a student's ability and skill to engage in the process of science (e.g. to make a judgment and/or interpretation concerning data),

2) emphasis is on the affective domain (the realm of values and feelings) rather than knowledge (i.e., over 50% of items deal with a student's attitude toward or appreciation of science and scientists), and

3) primary emphasis is placed upon "science as an institution" with little or no emphasis placed upon the epistemological characteristics of the development of scientific knowledge.

TABLE I
Nature of science instruments

Date	Instrument	Author(s)
1954	Science Attitude Questionnaire	Wilson
1958	Facts About Science Test (FAST)	Stice
1959	Science Attitude Scale	Allen
1961	Test on Understanding Science (TOUS)	Cooley & Klopfer
1962	Processes of Science Test	BSCS
1966	Inventory of Science Attitudes, Interests, and Appreciations	Swan
1966	Science Process Inventory (SPI)	Welch
1967	Wisconsin Inventory of Science Processes (WISP)	Scientific Literacy Research Center
1968	Science Support Scale	Schwirian
1968	Nature of Science Scale (NOSS)	Kimball
1969	Test on the Social Aspects of Science (TSAS)	Korth
1970	Science Attitude Inventory (SAI)	Moore & Sutman
1974	Science Inventory (SI)	Hungerford & Walding
1975	Nature of Science Test (NOST)	Billeh & Hasan
1975	Views of Science Test (VOST)	Hillis
1976	Nature of Scientific Knowledge Scale (NSKS)	Rubba
1978	Test of Science-Related Attitudes (TOSRA)	Fraser
1980	Test of Enquiry Skills (TOES)	Fraser
1981	Conception of Scientific Theories Test (COST)	Cotham & Smith
1982	Language of Science (LOS)	Ogunniyi
1987	Views on Science-Technology-Society (VOSTS)	Aikenhead, Fleming & Ryan
1990	Nature of Science Survey	Lederman & O'Malley
1992	Modified Nature of Scientific Knowledge Scale (MNSKS)	Meichtry
1995	Critical Incidents	Nott & Wellington

The validity of many of these instruments is questionable because their primary focus is on areas beyond the scope of "nature of science." Consequently, to assess individuals' understanding of the nature of science, such instruments should be used in conjunction with other, more validly focused instruments. Those instruments with questionable validity (as measures of the nature of science) include the Science Attitude Questionnaire (Wilson, 1954), Facts About Science Test (Stice, 1958), Science Attitude Scale (Allen, 1959), Processes of Science Test (BSCS, 1962), Inventory Of Science Attitudes, Interests, and Appreciations (Swan, 1966), Science Support Scale (Schwirian, 1968), Test on the Social Aspects of Science (Korth, 1969), Science Attitude Inventory (Moore & Sutman, 1970), Science Inventory (Hungerford & Walding, 1974), Test of Science-Related Attitudes (Fraser, 1978), the Test of Enquiry Skills (Fraser, 1980), and the Language of Science, (Ogunniyi, 1982). For a comprehensive discussion and analysis of many of these instruments, the reader is referred to Munby's (1983), *An Investigation into the Measurement of Attitudes in Science Education*.

Nature of Science Instruments: A Look at the Best

The remaining instruments are considered for purposes of this review to be valid and reliable measures of the nature of science by virtue of their focus on one or more ideas that have been traditionally considered under the label of "nature of science," as well as their reported validity and reliability data. These instruments have been used in numerous studies and some continue to be used even though there is a significant movement away from such types of paper and pencil assessments. Although the validity of the assessment instruments described below has been severely criticized (and justifiably so) in the past few years, they are presented here as being the most valid (in focus) attempts to assess understandings of the nature of science using a written response format. Following is a brief discussion of each instrument.

Test on Understanding Science (TOUS) (Cooley & Klopfer, 1961). This instrument has been, by far, the most widely used assessment tool in "nature of science" research. It is a four-alternative, 60-item multiple choice test. In addition to an "overall" or "general" score, three subscale scores can be calculated: (I) understanding about the scientific enterprise; (II) the scientist; (III) the methods and aims of science. The topics of subscale (III) are specifically stated as follows:

1) Generalities about scientific methods
2) Tactics and strategy of sciencing
3) Theories and models
4) Aims of science
5) Accumulation and falsification
6) Controversies in science
7) Science and technology

8) Unity and interdependence of the sciences

It is this subscale (III) that most directly assesses characteristics relative to the nature of science. Subscales (I) and (II), however, are concerned with aspects of science that are quite distinct from the nature of science. A few sample items from these two subscales follow:

Item #2: Among the hundreds of scientific societies in various countries throughout the world, we find that

    a) Scientists voluntarily join the societies related to their specific field.
    b) National governments generally direct these societies.
    c) Membership is generally restricted to scientists of one nation.
    d) National governments are seldom interested in these societies.

Item #9: If we compare successful scientists with successful people in most other professions, we find that these

    a) Scientists tend to have higher incomes than other professionals.
    b) Scientists require more specialized training than other professionals.
    c) Scientists and other professionals have set rigid certification laws which keep out those who are not qualified.
    e) Scientists and other professionals tend to devote most of their energies to their work.

Item #27: The American Chemical Society (ACS) is one of the largest scientific societies in the United States. Which of the following functions would the ACS be least likely to carry on?

    a) Negotiate contracts with companies employing chemists.
    b) Assist its members in finding new jobs.
    c) Publish chemical journals and books.
    d) Establish standards of terminology in chemistry.

Although the three TOUS items cited here are concerned with science and the scientific enterprise, they are not related to a student's conception of scientific knowledge. Such questions are clearly more relevant to the institution of science and the profession of "scientist" than to one's understanding of the nature of science.

In addition to this criticism, others have stated objections to the TOUS. Welch (1969) suggested that the test could be improved through revision and stronger validity evidence. Wheeler (1968) stated that too many items embrace a negative viewpoint of science. He felt that items could be rewritten to minimize their reflection of current stereotypes of science and scientists and suggested the addition of more items to increase the test's comprehensiveness. In a factor analysis study, Hukins (1963) found that the TOUS loaded strongly on a verbal factor. He surmised that the complexity of some items obscured the meaning for his tenth grade students. Aikenhead (1973) suggested that some TOUS items evoke a response of attitude. He felt that students perceive the test as concerning their appreciation or lack of appreciation for science and scientists. Thus, he felt, some items are answered according to a scientist's "good guy" image.

Indeed, one of the developers, (Klopfer, personal communication, 1981), stated that he did not feel the TOUS to be a very good measure of the nature of science. In all fairness, one must consider the context and time within which the TOUS was developed. It was an excellent beginning for those interested in assessing understandings of the nature of science. Currently, however, the TOUS exam appears inappropriate as a sole assessment instrument for the study of an individual's understanding of the nature of science.

Wisconsin Inventory of Science Processes (WISP) (Scientific Literacy Research Center, 1967). The WISP consists of 93 statements that the respondent evaluates as "accurate," "inaccurate," or "not understood." However, in scoring the exam, "inaccurate" and "not understood" responses are combined to represent the opposite of "accurate." Thus, it is analogous to an "agree/disagree" response format. With the exception of the TOUS exam, this instrument has been used more than any other assessment instrument. The WISP was developed and validated for high school students. Subsequent to its development, WISP has been the preferred choice by researchers as a nature of science assessment instrument. Although this instrument has excellent validity and reliability data, a few concerns should be considered prior to its use. Of primary concern is its length. The 93-item test takes over an hour to administer, which precludes it from use in a single class period. However, with block schedules and laboratory periods, this concern may not continue to be a significant problem. In addition, this instrument does not possess discrete subscales which, unfortunately, means that only unitary scores can be calculated.

Science Process Inventory (SPI) (Welch, 1966). This instrument is a 135-item forced-choice inventory (agree/disagree) purporting to assess an understanding of the methods and processes by which scientific knowledge evolves. The content of the SPI is almost identical to that of WISP and TOUS subscale III. The validation of SPI was achieved in the usual manner for such instruments: literature review, devising a model, employing the judgment of "experts," getting feedback from pilot studies, and testing the instrument's ability to distinguish among different groups of respondents. Although the SPI appears to possess a multitude of the values and assumptions specific to one's understanding of the nature of science, caution must be applied. The length (135 items) is obviously too long for a single class period administration. Also of concern is its forced response nature. Students are unable to express "neutral" or uncertain answers. In general, we feel that Likert scale response formats are best for such assessment instruments as they afford respondents a maximum amount of freedom in expressing their views toward an instrument item. Finally, like the WISP, the SPI does not possess subscales. The importance of subscales has been recognized and there have been unsuccessful attempts to establish these for the SPI.

Aikenhead (1972) performed a factor analysis on the SPI and found that the factors did not correspond to Welch's predicted factors. Bates (1974) also factor analyzed the instrument but could not reveal any meaningful factors. Both studies emphasized the difficulty of interpreting scores obtained from instruments like the SPI that attempt to assess understanding of a wide variety of aspects of the nature of science.

Nature of Science Scale (NOSS) (Kimball, 1968). This instrument was developed to determine whether science teachers have the same view of science as scientists. It consists of 29 items to which the respondent may "agree," "disagree" or register a "neutral" response. Kimball's model of the nature of science is based upon the literature of the nature and philosophy of science and is consistent with the views of Bronowski (1956) and Conant (1951). The model is composed of the following eight assertions:

- The fundamental driving force in science is curiosity concerning the physical universe.
- Science is a dynamic, on-going activity, rather than a static accumulation of information.
- Science aims at ever-increasing comprehensiveness and simplifications using mathematics as a simple, precise method of stating relationships.
- There is no "one" scientific method, but as many methods as there are practitioners.
- The methods of science are better characterized by some value-type attributes than by techniques.
- A basic characteristic of science is faith in the susceptibility of the physical universe to human ordering and understanding.
- Science has a unique attribute of openness, both of mind and of the realm of investigation.
- Tentativeness and uncertainty are characteristics of all science.

The specific content of NOSS was validated by nine science educators who judged whether the items were related to the model. The development, validation, and reliability measures were carried out with college graduates. Thus, it lacks reliability and validity data with respect to high school populations. Another concern is that the instrument lacks subscales and is, therefore, subject to the same criticism as WISP or any other unitary measure of the nature of science.

Nature of Science Test (NOST) (Billeh & Hasan, 1975). This instrument consists of 60 multiple choice items addressing the following components of the nature of science:

- Assumptions of science (8 items)
- Products of science (22 items)
- Processes of science (25 items)
- Ethics of science (5 items)

The test consists of two types of items. The first type measures the individual's knowledge of the assumptions and processes of science, and the characteristics of scientific knowledge. The second type of question presents situations that require the individual to make judgments in view of his/her understanding of the nature of science. The major shortcoming of this instrument is not its content, but rather, that no subscales exist. Thus, only a global or unitary score can be calculated. Certainly, the nature of scientific knowledge is too complex to be measured in terms of such a unitary measure and, therefore, this assessment instrument should be used only in a broad context.

Views of Science Test (VOST) (Hillis, 1975). This instrument was developed specifically to measure understanding of the tentativeness of science. It consists of 40 statements that are judged to imply either that scientific knowledge is tentative or absolute. Respondents express their agreement with either view using a five-option Likert scale response format. For what it is measuring this instrument is quite effective. However, not only does it lack subscales but its measurement of the nature of science is restricted to a single attribute of scientific knowledge.

Nature of Scientific Knowledge Scale (NSKS) (Rubba, 1976). This instrument is a 48-item Likert scale response format consisting of five choices (strongly agree, agree, neutral, disagree, strongly disagree). The test is purported to be an objective measure of secondary students' understanding of the nature of science. The model of scientific knowledge used in the development of this instrument is quite basic and common to all delineations of the nature of science (Rubba & Andersen, 1978). The NSKS and its subscales are based upon the nine factors of the nature of science specified by Showalter (1974). Rubba (1977) listed these nine factors as tentative, public, replicable, probabilistic, humanistic, historic, unique, holistic, and empirical. He noted a certain amount of shared overlap between the factors and proceeded to collapse them into a six-factor or six-subscale model of the nature of science. Rubba and Andersen (1978) described the six subscales as follows:

1) Amoral: Scientific knowledge provides humans with many capabilities but does not instruct us on how to use them. Moral judgment can be passed only on the application of scientific knowledge, not on the knowledge itself;
2) Creative: Scientific knowledge is a product of the human intellect. Its invention requires as much creative imagination as does the work of an artist, a poet, or a composer. Scientific knowledge embodies the creative essence of the scientific inquiry process;
3) Developmental: Scientific knowledge is never "proven" in an absolute and final sense. It changes over time. The justification process limits scientific knowledge as probable. Beliefs that appear to be good ones at one time may be appraised differently when more evidence is at hand. Previously accepted beliefs should be judged in their historical context;
4) Parsimonious: Scientific knowledge tends towards simplicity, but not to the disdain of complexity. It is comprehensive as opposed to specific. There is a continuous effort in science to develop a minimum number of concepts to explain the greatest number of observations;
5) Testable: Scientific knowledge is capable of public empirical testing. Its validity is established through repeated testing against accepted observations. Consistency among test results is a necessary but not a sufficient condition for the validity of scientific knowledge;
6) Unified: Scientific knowledge is born out of an effort to understand the unity of nature. The knowledge produced by the various specialized sciences contributes to a network of laws, theories, and concepts. This systematized body of knowledge gives science its explanatory and predictive power.

These six subscales can be individually scored with validity and reliability established for each. The instrument was developed, validated and found to be reliable for high school level students. The five-option Likert scale response format affords maximum freedom of expression to the respondent. Cotham (1979) unleashed a rather strong attack aimed at invalidating many of the previously discussed assessment instruments; however, with respect to NSKS, he only remarked on its insensitivity to alternative viewpoints. However, despite Cotham's comparatively weak criticism of the NSKS, there is reason for concern about the instrument's validity. Many pairs of items within specific subscales are identical, except that one item is worded negatively. This redundancy could encourage respondents to refer back to their answers on previous, similarly-worded items. This cross-checking would result in inflated reliability estimates which could cause erroneous acceptance of the instrument's validity.

Conceptions of Scientific Theories Test (COST) (Cotham & Smith, 1981). The structure of this instrument was dictated by the developers' concern that previously existing instruments were based on single (supposedly enlightened) interpretations of the nature of science. Thus, the COST provides for non-judgmental acceptance of alternative conceptions of science. That is, no single point of view is considered to be the "correct" one. The instrument is an attitude inventory consisting of 40 Likert scale items (with four options) and four subscales, each corresponding to a particular aspect of scientific theories. These include (I) ontological implications of theories; (II) testing of theories; (III) generation of theories; and (IV) choice among competing theories.

The COST provides a theoretical context for four item-sets by prefacing each set with a brief description of a scientific theory and some episodes drawn from its history. The items following each theory description refer to that description. The four theoretical contexts are 1) Bohr's theory of the atom, 2) Darwin's theory of evolution, 3) Oparin's theory of abiogenesis, and 4) the theory of plate tectonics. A fifth context contains items that refer to general characteristics of scientific theories and is, therefore, not prefaced by a description.

Two concerns must be addressed prior to using COST as an instrument to assess high school students' understandings of the nature of science. The first of these is the cognitive level of the instrument. It was designed for teachers and validated with undergraduate college students. The four theory descriptions used to provide context for the items are presented at a level that may be above the capabilities of many high school students. Individual items may be difficult to decipher as well. Consider these two items that deal with the general characteristics of scientific theories:

Item #35: When a scientific theory is well supported by evidence, the objects postulated by the theory must be regarded as existing.

Item #37: Observation is not a basis for evaluating scientific theories because of the influence of theory on observation.

It is likely that the average high school student would find such items difficult to comprehend, which would negatively impact the reliability and validity of the instrument. Evidence for this problem exists in unpublished results (Bell & Lederman, 1996) where COST responses had very low reliability coefficients (Cronbach's alpha = 0.28 - 0.35) for even above-average high school students and a large sample of undergraduate college students.

A second concern with the COST instrument rests with the authors' claim that it, as opposed to all extant instruments, is sensitive to alternative conceptions of science. Unfortunately, close examination reveals this claim to be false. Cotham and Smith feel that it is extremely important for education to promote the view that science is tentative and revisionary. We certainly agree. In their zealous commitment to this concern, they actually specify which subscale viewpoints are consistent with the tentative and revisionary conception. Thus, although they claim to place no value judgments upon the various conceptions of science, Cotham and

Smith actually do just that by linking certain viewpoints to the "highly prized" tentative and revisionary conception of scientific knowledge. The problem which Cotham and Smith try to resolve (i.e., placing value judgments upon various conceptions of science, as if one view were the "correct one") can be eliminated through a slight variation in test scoring (discussed in the Some Thoughts About "Traditional" Assessment Instruments section of this review). Regardless, validity of the COST instrument remains suspect.

Views on Science-Technology-Society (VOSTS) (Aikenhead, Ryan & Fleming, 1989). The VOSTS was developed to assess students' understanding of the nature of science, technology, and their interactions with society. It consists of a "pool" of 114 multiple-choice items that address a number of science-technology-society (STS) issues. These issues include Science and Technology, Influence of Society on Science/Technology, Influence of Science/Technology on Society, Influence of School Science on Society, Characteristics of Scientists, Social Construction of Scientific Knowledge, Social Construction of Technology, and Nature of Scientific Knowledge.

The VOSTS was developed and validated for grade 11 and 12 students. A fundamental assumption underlying the development of this instrument was that students and researchers do not necessarily perceive the meanings of a particular concept in the same way. Aikenhead and Ryan (1992) recognized the importance of providing students with alternative viewpoints based upon student "self-generated" responses to avoid the "constructed" responses offered by most of the previous nature of science assessment instruments. The Science Process Inventory (SPI) provides a typical example of a "constructed" forced response. The SPI includes the statement "Scientific knowledge is tentative" (Welch, 1966) with which high school students are asked to agree or disagree. Aikenhead (1979) discovered that when offered the chance to respond "I do not understand," more than a quarter of grade 11 and 12 students did so. Therefore, whenever students responded "agree" or "disagree" to the SPI item "scientific knowledge is tentative," a number of those students may not have really understood the meaning of the statement. Unlike most other instruments, the VOSTS does not provide numerical scores; instead it provides a series of alternative "student position" statements. These statements were obtained from extensive open-ended student "argumentative" paragraphs in which students defended their stated position on a STS issue or topic. On the VOSTS, students choose from a number of choices (sometimes up to 10 different views) to respond to a "situation" based upon a STS issue or topic. Included as a choice in every question is the statement, "I don't understand," or "I don't know enough about this subject to make a choice." This response format eliminates the "forced" answer that many of the preceding multiple-choice response or Likert-type response instruments use.

It should be noted that several statements in this instrument refer to a specific nationality, such as Canadian scientist/issue or United States scientist/issue, so some caution or adjustments are suggested when administering this to students from other nationalities to avoid "nationalistic feelings" influencing responses. The extensive work on developing and validating the VOSTS instrument took approximately six years to complete (Aikenhead & Ryan, 1992). This effort included the enormous undertaking of surveying a stratified sample of approximately 10,800 Canadian students comprised principally of graduating high school students. Results of this undertaking are reported in a series of research articles published in a special edition of the journal *Science Education* (Volume 71, 1987).

Nature of Science Survey (Lederman & O'Malley, 1990). In an attempt to ameliorate some of the problems noted by Aikenhead, et. al (1987) during the development of the VOSTS and those noted in the use of the NSKS (Rubba, 1976) relative to the use of paper and pencil assessments, Lederman and O'Malley developed a rather open-ended survey consisting of seven items. This instrument was designed to be used in conjunction with follow-up interviews and each of the seven items focused on a different aspect of tentativeness in science. The seven items were as follows:

1) After scientists have developed a theory (e.g., atomic theory), does the theory ever change? If you believe that theories do change, explain why we bother to learn about theories. Defend your answer with examples.
2) What does an atom look like? How do scientists know that an atom looks like what you have described or drawn?
3) Is there a difference between a scientific theory and a scientific law? Give an example to illustrate your answer.
4) How are science and art similar? How are they different?
5) Scientists perform scientific experiments/investigations when trying to solve problems. Do scientists use their creativity and imagination when doing these experiments/investigations?
6) Is there a difference between scientific knowledge and opinion? Give an example to illustrate your answer.
7) Some astrophysicists believe that the universe is expanding while others believe that it is shrinking; still others believe that the universe is in a static state without any expansion or shrinkage. How are these different conclusions possible if all of these scientists are looking at the same experiments and data?

Several problems were noted in the wording of some of the questions. In particular, items 4, 5, 6 did not necessarily yield responses relative to students' views of the "tentativeness" aspect of the nature of science. For item #2, students tended to describe various experiments instead of discussing whether inferences had been made, as was intended. While these difficulties were alleviated by subsequent interviews, they served to reinforce the problems associated with attempting to interpret students' understandings solely from their written responses to researcher generated questions.

Modified Nature of Scientific Knowledge Scale (M-NSKS) (Meichtry, 1992). This instrument is a modified NSKS instrument with 32 statements from four of the NSKS subscales. These subscales are: (I) creative, (II) developmental, (III) testable, and (IV) unified. M-NSKS was developed, with reliability and validity reported, for use with 6th, 7th, and 8th graders.

Critical Incidents (Nott & Wellington, 1995). The use of "critical incidents" to assess teachers' conceptions of the nature of science is a significant departure from the usual paper and pencil assessment. In particular, Nott and Wellington are of the opinion that teachers do not effectively convey what they know about the nature of science in "direct response to abstract, context-free questions of the sort, 'What is science?' (Nott & Wellington, 1995). Instead, they have created a series of "critical incidents" that are descriptions/scenarios of actual classroom events. Teachers are expected to respond to the incidents by answering the following three questions: 1) What would you do?, 2) What could you do?, and 3) What should you do? So, for example, the teacher may be confronted with a situation in which a demonstration or laboratory activity does not yield the desired data. How the teacher responds to the aforementioned questions is believed to communicate what the teacher believes about the nature of science. The Critical Incidents instrument is very much a project under development in that additional incidents are continually being added to the total repertoire. Although the use of critical incidents appears to be an excellent instructional tool to generate meaningful discussions in preservice and inservice courses (indeed, Nott and Wellington have used the incidents in this manner), whether the teachers' responses are related to their views about the nature of science is still questionable. In short, the approach is based on the assumption that teachers' views of the nature of science automatically and necessarily influence classroom practice, an assumption that is simply not supported by the existing literature.

SOME THOUGHTS ABOUT "TRADITIONAL" ASSESSMENT INSTRUMENTS

The validity of many instruments purporting to assess the nature of science has long been criticized on the grounds that each instrument assumes its interpretation of science to be an enlightened view (Cotham & Smith, 1981). This assumption has been criticized because many aspects of the nature of science do not enjoy consensus within the community of scholars interested in the nature of science (Lederman,

1992; Lucas, 1975; Martin, 1972). Perhaps Cotham and Smith's (1981) example will best illustrate the point. An item from the Science Process Inventory (Welch, 1966) states, "Science is a series of successively closer approximations to the truth." The scoring key for the instrument indicates an "agree" response for this item to be correct. However, scholars such as Thomas Kuhn (1970) disallow any claims concerning the ontological implications of scientific developments. Lucas (1975) discussed the scoring of responses to the instrument item, "It is important for a physicist to be able to throw away widely held ideas and think without restriction." He postulated that a Popperian, who views science as progressing by experimental tests that attempt to refute a hypothesis arrived at by conjecture, would agree with this statement, while a Kuhnian who views science as progressing by the consolidation and articulation of a guiding paradigm during normal science, would favor a "disagree" response.

Such problems of "hidden" biases implicit in assessment instruments can be avoided quite easily and do not require the development of a new instrument as advocated by Cotham and Smith (1981). The problem lies not within the test, but rather in the interpretations of those scoring the test. If one interprets test scores simply as a measure of an individual's adherence to a particular conception of science, then no implicit value judgments are made. In such a manner no individual is said to understand science "better" than someone else. Thus, a score of 40, for example, on Rubba's (1976) NSKS Amoral subscale would be a strong indication that this individual believes that scientific knowledge is amoral, while a score of 5 would indicate a belief that scientific knowledge is moral by nature. No statement as to who understands science "better" would need to be made. To make such a statement would assume that a "correct" view exists. Thus, the valid concerns of Cotham and Smith (1981), Lucas (1975), and Martin (1972) can be eliminated by using non-judgmental scoring of student and teacher conceptions.

Having viewed the various research emphases specifically related to the nature of science, we are left with at least two salient points: 1) assessment instruments are interpreted in a biased manner, and 2) some assessment instruments appear to be poorly constructed. These criticisms not withstanding, it is interesting to note that research conclusions based on these instruments have been unusually uniform. Thus, although the various instruments suffer from specific weaknesses, if these were significant, it would seem improbable that the research conclusions would be so consistent.

There is a more critical concern, however, about the "traditional" paper and pencil approach to the assessment of an individual's understanding of the nature of science. Although not a new insight, Lederman and O'Malley's (1990) investigation clearly highlighted the problem of paper and pencil assessments. They documented discrepancies between their own interpretations of students' written responses and the interpretations that surfaced from actual interviews of the same students. This unexpected finding (i.e., the purpose of the interviews was to help validate the paper and pencil survey that was used) was quite timely, as it occurred when educational researchers were making a serious shift toward more qualitative, open-ended approaches to assess individuals' understanding of any concept.

THE FUTURE: CONCERNS, CAUTIONS, AND POSSIBLE DIRECTIONS

The VOSTS is arguably the result of the last systematic attempt to develop a paper and pencil assessment of the nature of science. However, the VOSTS is not similar to its "traditional" predecessors. There was a clear attempt to value students' views and an attempt to probe students' reasons for the responses they recorded. Investigations of students' and teachers' understandings of the nature of science have continued since the appearance of the VOSTS, but these investigations have carefully combined a number of research methodologies including field observations of classrooms and teachers, quantitative assessment (using many of the previously discussed instruments) and qualitative techniques involving both structured and informal interviews (Brickhouse, 1990; Briscoe, 1991; Duschl & Wright, 1989; Gallagher, 1991; Hodson, 1993; Lantz & Kass, 1987; Lederman & O'Malley, 1990; Lederman & Zeidler, 1987; Meichtry, 1992; Smith & Anderson, 1984; among others).

What seems to be recognized is that while paper and pencil instruments can reveal something about students' views of the nature of science, they cannot tell us everything we would like to know. Very often, the significant question is not whether a person's view of the nature of science conforms to a particular espoused viewpoint, but rather, what are the limits of the person's understandings and how do these understandings affect instructional choices and behaviors? These questions require more detailed descriptions of a person's beliefs than paper and pencil instruments alone are able to provide.

However, less traditional assessment techniques, such as interviews, observations, and reviews of lesson plans and classroom documents, are not without their own problems. Too often, researchers (e.g. Brickhouse, 1990; Gallagher, 1991) discuss the results of interviews without documenting the questions that were used to elicit the responses. Failing to record interview questions prevents adequate assessment of the validity of the interview and precludes the possibility of replication in other settings.

Furthermore, researchers commonly mention the use of classroom documents such as lesson plans, class notes, handouts, and assignments in their assessments of teachers' knowledge and practice concerning the nature of science. However, they frequently fail to adequately describe how these materials were collected and analyzed.

Classroom observations have been inadequately described as well. The purpose, number, length, and recording method are all important considerations that can impact the results of field observations. These factors should be, but are not always, included in the description of research methods. The observer's background and knowledge of the sampled subjects can also influence what is "seen" and should be described and considered in the research results. For example, to avoid the introduction of bias in making field observations, researchers should not assess teachers' views prior to the initiation of the field observations. Failure to take this precaution can cause observers to "see" what they expect to see and to interpret classroom behavior into a framework predetermined by the assessment results. Indeed, this is a clear instance of the biasing influence of the unsupported assumption that a teacher's understanding of the nature of science influences classroom practice. This assumption has been explicit since Robinson (1969) and it continues to be held independent of the empirical data from research on the nature of science. Quite simply, a circular loop exists in which the assumption guides the interpretation of classroom observation data, which is then used to support the initial assumption (e.g., Nott & Wellington, 1995). The obvious remedy is to sequence assessments of the subject's views of the nature of science after field observations and document analysis have been completed, or to keep results of the assessments hidden if assessments and observations are completed by different individuals. This is analogous to the "blind" studies often used in medical studies to reduce the impact of the researcher's and subjects' bias on the results.

In addition to using combinations of research methodologies to assess understandings of the nature of science, more authentic assessment tools should be developed. For example, students could be asked to develop an instructional model that they could use to teach other students about the characteristics of a scientific theory. If a student can create such a model, it would seem to be a much more trustworthy assessment of the student's understanding than with either directed questions on paper or interviews.

The uniform consistency of the results of studies that have used the instruments reviewed here attest to their ability to assess relevant aspects of teachers' and students' knowledge of the nature of science. We have, however, taken paper and pencil assessments about as far as they can be expected to go. What we need to concentrate more on now are assessments that help us answer other critical questions. We know that students have many misconceptions about the nature of science. What is the source of these views, how are they learned, and how might more accepted views of science be learned efficiently? We know that teachers' knowledge of the nature of science is often inadequate despite heroic and costly efforts in teacher education in this domain. What is the impact of teachers' beliefs about the nature of science on their classroom practice and their students' beliefs?

Given the strong evidence that teachers' beliefs do not necessarily influence classroom practice, perhaps a more appropriate question is what can be done to facilitate the translation of teachers' views into classroom practice? Emerging research (e.g., Lederman, 1995) seems to indicate that teachers' intentions and priorities and the uncontrollable realities of classroom life are critical factors to be investigated. Recent research attempting to answer such questions (Lederman, Abd-El-Khalick, & Bell, 1997) has necessitated the compilation of extensive data from diverse sources. These data sources have included classroom observations, video-taped lessons, supervisor notes, all unit and lesson plans for an entire term, and classroom documents, as well as responses to questionnaires and interviews. Obviously, this type of assessment is both more complete and more laborious than past efforts but has yielded important results. Although the teachers in this investigation possessed an adequate understanding of the critical aspects of the nature of science as expressed in the reforms and were well versed in various instructional approaches, little attention to the nature of science was evident in their classroom practice. Interestingly, these same teachers cited lack of confidence in their knowledge of the nature of science and inability to teach the nature of science as primary causes for not including it in classroom instruction. These findings are quite similar to what has previously been found in research on elementary teachers' attention to science. Up to this point, attempts to answer the research approaches to the question of why teachers spend little time addressing the nature of science have lacked a theoretical basis. It would seem that theories of self-efficacy and related research have clear implications for future research into the still critical, and unanswered, question concerning teachers' explicit attention to the nature of science during instruction. In terms of assessment, it is time that we move on to questions of classroom practice and lay to rest the continued focus of descriptive assessments of teachers' and students' conceptions.

Oregon State University, Corvallis, Oregon, USA

REFERENCES

Aikenhead, G. (1972). The measurement of knowledge about science and scientists: An investigation into the development of instruments for formative evaluation, Unpublished doctoral dissertation, Harvard University.

Aikenhead, G. (1973). 'The measurement of high school students' knowledge about science and scientists', *Science Education*, (57), 539-549.

Aikenhead, G. (1979). 'Science: A way of knowing', *The Science Teacher*, (46), 23-25.

Aikenhead, G. (1987). 'High school graduates' beliefs about science-technology- society: Characteristics and Limitations of science knowledge', *Science Education*, (71), 459-487.

Aikenhead, G., Fleming, R.W. & Ryan, A.G. (1987). 'High-school graduates beliefs about science-technology-society: Methods and issues in monitoring student views', *Science Education*, (71), 145-161.

Aikenhead, G. & Ryan, A. (1992). 'The development of a new instrument: "Views on science- technology-society" (VOSTS)', *Science Education*, (76) 477-491.

Aikenhead, G., Ryan, A.G. & Fleming, R.W. (1987). 'Views on science-technology-society' (Form CDN.mc.5). Saskatoon, Canada, Department of Curriculum Studies, University of Saskatchewan.

Allen, H. Jr. (1959). *Attitudes of certain high school seniors toward science and scientific careers*. New York, Teachers College Press.

Bates, G. (1974). A search for subscales in the science process inventory, paper presented at the 47th Annual Meeting for the National Association for Research in Science Teaching.
Bell, R. L. & Lederman, N.G. (1996). [COST results for high school students enrolled in a summer science apprenticeship program and undergraduate students enrolled in a geology course for nonmajors]. Unpublished raw data.
Billeh, V. Y. & Hasan, O. (1975). Factors affecting teachers' gain in understanding the nature of science, *Journal of Research in Science Teaching*, (12), 209-219.
Biological Sciences Curriculum Study (BSCS) (1962). *Processes of science test*, New York, The Psychological Corporation.
Brickhouse, N. W. (1990). Teachers' beliefs about the nature of science and their relationship to classroom practice, *Journal of Teacher Education*, (41), 53-62.
Briscoe, C. (1991). The dynamic interactions among beliefs, role metaphors, and teaching practices: A case study of teacher change, *Science Education*, (75), 185-199.
Bronowski, J. (1956). *Science and human values*, New York, Harper & Row.
Central Association of Science and Mathematics Teachers. (1907). 'A consideration of the principles that should determine the courses in biology in the secondary schools', *School Science and Mathematics*,(7), 241-247.
Conant, J. B. (1951). *Science and common sense*, New Haven, CT, Yale University Press.
Cooley, W. & Klopfer, L. (1961). Test on understanding science, Form W, Princeton, NJ, Educationa Testing Service.
Cooley, W. & Klopfer, L. (1963). 'The evaluation of specific education innovations', *Journal of Research in Science Teaching*, (1), 73-80.
Cotham, J. (1979). 'Development, validation, and application of the conceptions of scientific theories test, Unpublished doctoral dissertation, Michigan State University.
Cotham, J. & Smith, E. (1981). 'Development and validation of the conceptions of scientific theories test', *Journal of Research in Science Teaching*, (18), 387-396.
Duschl, R. (1990). *Restructuring science education: The importance of theories and their development*, New York, Teachers College Press.
Duschl, R. A. & Wright, E. (1989). 'A case study of high school teachers' decision making models for planning and teaching science', *Journal of Research in Science Teaching*, (26), 467-501.
Fraser, B. J. (1978). Development of a test of science-related attitudes, *Science Education*, (62), 509-515.
Fraser, B. J. (1980). 'Development and validation of a test of enquiry skills', *Journal of Research in Science Teaching*, (17), 7-16.
Gallagher, J.J. (1991). 'Prospective and practicing secondary school science teachers' knowledge and beliefs about the philosophy of science', *Science Education*, (75), 121-133.
Hillis, S. R. (1975). 'The development of an instrument to determine student views of the tentativeness of science', in Research and Curriculum Development in *Science Education: Science Teacher Behavior and Student Affective and Cognitive Learning (Vol. 3)*, Austin, TX, University of Texas Press.
Hodson, D.H. (1993). 'Philosophic stance of secondary school science teachers, curriculum experiences, and children's understanding of science: Some preliminary findings', *Interchange*, (24), 41-52.
Hukins, A.A. (1963) 'A factorial investigation of measures of achievement of objectives in science teaching, Unpublished doctoral dissertation, University of Alberta, Edmonton.
Hungerford, H. & Walding, H. (1974). *'The modification of elementary methods students' concepts concerning science and scientists'*, paper presented at the Annual Meeting of the National Science Teachers Association.
Kimball, M.E. (1967-1968). 'Understanding the nature of science: A comparison of scientists and science teachers', *Journal of Research in Science Teaching*, (5), 110-120.
Korth, W. (1969). *'Test every senior project: Understanding the social aspects of science'*, paper presented at the 42nd Annual Meeting of the National Association for Research in Science Teaching.
Kuhn, T.J. (1970). *The structure of scientific revolutions*, Chicago, IL, University of Chicago Press.
Lantz, O. & Kass, H. (1987). 'Chemistry teachers' functional paradigms', *Science Education*,(71), 117-134.
Lederman, N.G. (1992). 'Students' and teachers' conceptions of the nature of science: A review of the research',. *Journal of Research in Science Teaching*, (29), 331-359.

Lederman, N.G. (1995). 'The influence of teachers' conceptions of the nature of science on classroom practice: A story of five teachers', *Proceedings of the Third International History, Philosophy, and Science Teaching Conference*, 656-663.
Lederman, N.G., Abd-El-Khalick, F., & Bell, R. L. (1997). *Knowing and doing: The flight of the nature of science from the classroom*, paper presented at the Annual Meeting of the American Educational Research Association, Chicago, IL.
Lederman, N. & O'Malley, M. (1990). 'Students' perception of tentativeness in science: Development, use, and sources of change', *Science Education*, (74), 225-239.
Lederman, N. & Zeidler, D. (1987). 'Science teachers' conceptions of the nature of science: Do they really influence teacher behavior?', *Science Education*, (71), 721-734.
Lucas, A.M. (1975). 'Hidden assumptions in measures of 'Knowledge about science and scientists', *Science Education*, (59), 481-485.
Mackay, L. D. (1971). 'Development of understanding about the nature of science', *Journal of Research in Science Teaching*, (8), 57-66.
Martin, M. (1972). *Concepts of science education*, Glenview, Il., Scott Foresman & Co.
Matthews, M. R. (1994). *Science teaching: The role of history and philosophy of science*, New York, Routledge.
Meichtry, Y. J. (1992). 'Influencing student understanding of the nature of science: Data from a case of curriculum development', *Journal of Research in Science Teaching*, (29), 389-407.
Moore, R. & Sutman, F. (1970). 'The development, field test and validation of an inventory of scientific attitudes', *Journal of Research in Science Teaching*, (7), 85-94.
Munby, H. (1983). *An investigation into the measurement of attitudes in science education*, Columbus, OH, SMEAC Information Reference Center.
Nott, M. & Wellington, J. (1995). 'Probing teachers' views of the nature of science: How should we do it and where should we be looking?', *proceedings of the Third International History, Philosophy, and Science Teaching Conference*, 864-872.
Ogunniyi, M.B. (1982). 'An analysis of prospective science teachers' understanding of the nature of science, *Journal of Research in Science Teaching*, (19), 25-32.
Robinson, J. (1969). 'Philosophy of science: Implications for teacher education', *Journal of Research in Science Teaching* (6), 99-104.
Rubba, P. (1976). 'Nature of scientific knowledge scale,' School of Education, Indiana University, Bloomington, Indiana.
Rubba, P.A. (1977). 'The development, field testing, and validation of an instrument to assess secondary school students' understanding of the nature of scientific knowledge', Unpublished doctoral dissertation, Indiana University.
Rubba, P. & Anderson, H. (1978). 'Development of an instrument to assess secondary school students' understanding of the nature of scientific knowledge', *Science Education*, (62), 449-458.
Schwirian, P. (1968). 'On measuring attitudes toward science', *Science Education*, (52), 172-179.
Scientific Literacy Research Group. (1967). *Wisconsin inventory of science processes*, Madison, Wisconsin, The University of Wisconsin.
Sholalter, V.M. (1974). 'What is unified science education? Program objectives and scientific literacy', *Prism II*, 2, 3-4.
Smith, E. L., & Anderson, C. W. (1984). 'Plants as producers: A case study of elementary science teaching', *Jounal of Research in Science Teaching*, (21), 685-698.
Stice, G. (1958). *Facts about science test*, Princeton, NJ, Educational Testing Service.
Swan, M.D. (1966). 'Science achievement as it relates to science curricula and programs at the sixth grade level in Montana public schools', *Journal of Research in Science Teaching*, (4), 102-123.
Wade, P.D. & Lederman, N. (1995). 'College students' conceptions of science and science content', Proceedings of the Third International History, Philosophy, and Science Teaching Conference, 1271-1276.
Wheeler, S. (1968). *Critique and revision of an evaluation instrument to measure students' understanding of science and scientists*, Chicago, IL, University of Chicago.
Welch, W. (1966). *Science process inventory*, Cambridge, MA, Harvard University Press.
Welch, W. (1969). Curriculum evaluation, *Review of Educational Research*, (39), 429-443.

Welch, W. (1981). 'Inquiry in school science' in N.C. Harms & R. E. Yager (eds.), *What research says to the science teacher (Vol. 3)*, Washington DC, National Science Teachers Association.

Wilson, L. (1954). 'A study of opinions related to the nature of science and its purpose in society', *Science Education*, (38), 159-164.

NOTES ON CONTRIBUTORS

FOUAD ABD-EL-KHALICK

Fouad Abd-El-Khalick is a doctoral student in the Department of Science and Mathematics Education at Oregon State University. He is a former high school physical science instructor at the American School in Beirut. His research interests include teachers' and students' understandings of the nature of science and pedagogical content knowledge.

HIYA ALMAZROA

Hiya Almazroa received her Ph.D. in Curriculum and Instruction with a science education focus from the University of Southern California. She received her B.S. degree in Zoology from Girls College of Education at Riyadh, Saudi Arabia and her M.S. degree in curriculum and instruction from the University of Southern California. Her dissertation research focuses on Saudi science teachers conceptions of the nature of science and factors which influence them. Her research interests focus on science teacher education and the application of issues in the history and philosophy of science to science teaching and learning.

JEAN BEARD

Jean Beard is the Director of the Science Education Program in the College of Science at San Jose State University and a Professor in the Biological Science Department. She also has directed the NSF grant for the Evolution and the Nature of Science Institutes since 1992 and served on the National Research Council Committee on Biology Teacher In-service Programs. Outside of Science Education she most often teaches a general education course titled: Science and the Citizen. She holds a Ph.D. in Science Education from Oregon State University.

RANDY L. BELL

Randy L. Bell is a doctoral student in the Department of Science and Mathematics Education at Oregon State University. He is a former high school biology/physical science teacher and practicing scientist. Bell's research focuses on the relationship between teachers' beliefs and understandings of the nature of the science with actual classroom practice.

MICHAEL L. BENTLEY

Michael L. Bentley has worked for 27 years in the field of science education as a classroom teacher, science supervisor, museum educator, school principal, and researcher. Currently, he is an Associate Professor of Science Education at Virginia Polytechnic Institute and State University. His undergraduate degree in biology was earned at King's College, PA with graduate studies in science education at the University of Pennsylvania and the University of Virginia. His publications include journal articles, several chapters in books, and five books, the latest of which are, *Astronomy Smart, Jr.* (The Princeton Review, 1996) and *Science Timelines: A Multicultural Resource* (Rigby Education, 1996). Michael is interested in K-12 curriculum and teaching and learning in science, as well as in environmental education and informal science education. He lives in Salem, Virginia with his wife, the Rev. Susan E. Bentley, and three children, Sarah, Alexander, and Matthew.

DAVID BOERSEMA

David Boersema is the Douglas C. Strain professor of Natural Philosophy at Pacific University where he serves as a full professor in the philosophy department and chairs the Peace and Conflict Studies program. He received an undergraduate degree in philosophy and economics from Hope College in Holland, Michigan and the MA and Ph.D. degrees in philosophy from Michigan State University.

MICHAEL C. CLOUGH

Michael C. Clough taught high school biology and chemistry for seven years in Illinois and Wisconsin before accepting his current position as an assistant professor of science education at the University of Iowa. The nature of science has always been an integral part of his teaching, both at the high school and college level. Michael enjoys teaching, research, writing and making a difference in science education. His passion is to improve science teacher education and, in turn, enhance science teaching. Michael lives in Iowa City with his wife Sara, and their son Issac.

KAREN R. DAWKINS

Karen R. Dawkins now serves as interim Director of the Science and Mathematics Education Center at East Carolina University in Greenville, North Carolina following a twenty year science teaching career in public high schools in Kentucky, Mississippi and North Carolina. Karen specializes in providing professional development programs for K-12 science and mathematics teachers in the largely rural northeastern counties of the state.

STEPHEN C. FLEURY

Stephen C. Fleury is a social studies instructor and Associate Dean of the School of Education at the State University of New York at Oswego. His interest in science, technology, and society (STS) stems from his activity-oriented experiential approach to the study of current social issues, first developed while working as a middle-school social studies teacher. His research interest in social constructivism and the philosophy of science focuses on the relationship of science as a way of "human cooperative knowing" and democracy as a way of "human cooperative living." Publications include perspectives on the role of STS in the social studies curriculum, analyses of the politics of knowledge, and examinations of the impact of science on social knowledge as it is transmitted through public schooling.

ALLAN A. GALLTHORN

Allan A. Gallthorn is Distinguished Research Professor of Education in the School of Education, East Carolina University. He has written more than twenty books in the fields of curriculum and supervision. For many years, he was a high school teacher, supervisor, principal and faculty member at the University of Pennsylvania.

PENNY L. HAMMERICH

Penny L. Hammerich, is an Assistant Professor in Science Education at Temple University in Philadelphia, PA. In this capacity, she coordinates the elementary science education program and teaches both undergraduate and graduate courses. Dr. Hammrich is Principal Investigator of the National Science Foundation sponsored Model Project for Women and Girls in Science, Mathematics, and Engineering grant called "Sisters in Science," and the U.S. Department of Education sponsored programs called "A Head Start on Science." Dr. Hammrich is also a Senior Research Associate with the Laboratory for Student Success Mid-Atlantic

Regional Lab located at Temple University. She has presented numerous papers at international and national conferences and has published articles in the areas of gender equity, nature of science, conceptual development of science teachers, and world view theory.

FRED JANSEN

Fred Jansen is a Ph.D. candidate in biology education in the Department of Biological Education of the University of Utrecht. His dissertation research focuses on the identification of heuristics for biology education that allow students to participate in the process of theory development. The heuristic presented in this book has been developed for teaching and learning about immunology.

NAHUM KIPNIS

Nahum Kipnis is the science educator at the Bakken Library and Museum in Minneapolis. He was born and educated in the former U.S.S.R., where he received an MS in physics and mathematics. He taught high school and college physics and did research in experimental physics and the history of science. In 1979, he emigrated to the United States and in 1984, he received a Ph.D. in the history of science from the University of Minnesota. In addition to a number of articles on science education and the history of science, he is author of two books, *History of the Principle of Interference of Light* (Basel/Boston: Birkhäuser Verlad, 1991) and *Rediscovering Optics* (Minneapolis: BENA Press, 1992).

THOMAS LAPORTA

Thomas LaPorta is a social science teacher at Tarpon Springs High School in the Pinellas County School District of Florida and is a doctoral candidate at the University of South Florida (USF). His research area is secondary school curriculum and instruction emphasizing science, technology, and society (STS) issues. In 1994-96 he directed an STS research project at USF while he was on special assignment to the university and is currently a research associate for the SALISH project.

NORMAN LEDERMAN

Norman Lederman is a professor of science and mathematics education at Oregon State University. He is a past president of the Association for the Education of

Teachers in Science (AETS) and a former high school biology/general science/chemistry teacher in Illinois and New York. His research focuses primarily on students' and teachers' conceptions of the nature of science. He is currently National Science Teachers Association director of teacher education an editor of School Science and Mathematics.

CATHLEEN C. LOVING

Cathleen C. Loving is an assistant professor in Curriculum and Instruction at Texas A&M University where she recently completed a four year term as director of field experiences. She currently teaches undergraduate and graduate courses in science education, instructional supervision and middle school philosophy. Her research in the relationship between science education and the nature of science has been published in journals including Science Education, the Journal of Science Teacher Education and the Journal of Research in Science Teaching. Dr. Loving received her doctorate from the University of Texas at Austin following almost 20 years as a high school science teacher in New Jersey, Pennsylvania, California, North Carolina and Texas. She has led funded projects involving science teachers and professional development schools.

MICHAEL R. MATTHEWS

Michael R. Matthews is a senior lecturer in education at the University of New South Wales. He has degrees from the University of Sydney in science, philosophy, psychology, philosophy of science and education. He has taught in high school, Sydney Teachers' College and was the Foundation Professor of Science Education at the University of Auckland (1992-93). He has published in philosophy of education, philosophy of science and science education. His recent books include Science Teaching: The Role of History and Philosophy of Science (Routledge, 1994), Challenging New Zealand Science Education (Dunmore Press, 1995), Time for Science Education (Plenum, 1998). He has edited The Scientific Background to Modern Philosophy (Hackett, 1989), History, Philosophy and Science Teaching: Selected Readings (OISE Press/Teachers College Press, 1991) and Philosophy and Constructivism in Science Education (Kluwer, 1998). He is founding editor of the international quarterly Science & Education (Kluwer Academic Publishers)

JOHN O. MATSON

John O. Matson is Associate Professor in Biological Sciences and the Science Education Program at San Jose State University. He took his Ph.D. in Zoology from

Michigan State University. His biological research interests include mammalian biogeography, evolution, and systematics. The nature of science and school-university collaborations are his major interests in science education.

WILLIAM F. MCCOMAS

William F. McComas began his career as a physical science, biology and environmental science instructor in Pennsylvania where he taught for 13 years while earning masters degrees in biology and physical science. He received the Ph.D. in science education from the University of Iowa in 1991. He is a science education professor and founding director of the Center to Advance Science Education at the University of Southern California in Los Angeles teaching courses in the philosophy of science, issues in science education and advanced science teaching methods. His research interests include informal science learning, the relationship of the nature of science to science instruction, the problems associated with evolution education, and the role of the laboratory in science instruction. He recently received the *Outstanding Science Teacher Educator* award from AETS.

YVONNE MEICHTRY

Yvonne Meichtry has a B.S. degree in Elementary Education, a M.S. degree in Education and an Ed.D. in Curriculum and Instruction. Since 1988, she has taught courses in science methods and environmental science education as a member of the Teacher Education faculty. Previous to her work as a university faculty member, she taught middle school science for 10 years. Currently she is investigating the impact of curricula on student understanding, conceptual change and the varied definitions of the nature of science as a faulty member of the School of Education at Northern Kentucky University.

CRAIG E. NELSON

Craig E. Nelson is Professor of Biology and of Public and Environmental Affairs at Indiana University at Bloomington since 1966. He has received awards for distinguished teaching from IU, Vanderbilt and Northwestern and has been a Sigma Xi National Lecturer. He frequently presents invited workshops on fostering critical thinking and on diversity and teaching at national meetings, as Chautauqua Short Courses, and at individual colleges He is on the editorial boards of the *Journal for Excellence In College Teaching* and *Inquiry: Critical Thinking Across The Disciplines* and has been a consulting editor for *College Teaching*. His biological research explores the interface of evolution and ecology, resulting in the publication of over 70 papers.

MARTIN K. NICKELS

Martin K. Nickels is Professor of Physical Anthropology at Illinois State University in Normal. His research interests include the history of human evolutionary studies, hominid paleontology and prehistory, primate behavior, and the biological basis of human behavior. He received his Ph.D. from the University of Kansas in 1975. He was selected as the Outstanding University Teacher at Illinois State in 1991-92 and 1996-97. He was a Sigma Xi National Lecturer in 1995-96. He co-authored a college textbook titled *The Study of Physical Anthropology and Archaeology* and has authored or co-authored articles in *The Sciences*, *The American Biology Teacher* and *Creation/Evolution*. He is a member of several professional organizations.

MICK NOTT

Mick Nott is a senior lecturer in physics and science education at the Centre for Science Education, Sheffield Hallam University, Sheffield, UK. His curriculum development interests are in using the history and sociology of science to teach science. His current research interests are in teachers' understandings of the nature of science.

JOANNE K. OLSON

Joanne K. Olson is a Ph.D. candidate in Curriculum and Instruction with a science education focus at the University of Southern California. She received a Master's degree in Education from The Claremont Graduate School. She is currently teaching a multiage class in South Central Los Angeles, and instructs elementary science methods classes at USC. Her research interests include the nature of science and its instruction at the middle school level, as well as the intersection of neurobiology, cognitive psychology, and science learning.

SHARON PARSONS

Sharon Parsons is Assistant Professor of Science Education in the Teacher Education Science, Math, Technology Academy at San Jose State University. She holds a doctorate in science education from the University of British Columbia, Vancouver,

Canada. Other than nature of science her interests include multicultural and equity issues in science education, school-university collaborations, and international science education.

BARBARA SPECTOR

Barbara Spector is a professor of science education at the University of South Florida where she directs the Science/Technology/Society Center, an umbrella for multiple innovative programs in science teacher preparation implementing a vision of the schools of tomorrow and the Project 2061 Research and Development Center. She has been awarded 43 grants supporting the restructuring of science teacher education. She has authored more than 300 articles, chapters, books, and presentations and served as a consultant to state and local education agencies and professional associations. She has been Research Director for the National Science Teachers Association and the National Association for Science, Technology, and Society. She has served on the Board of Directors of the Association for the Education of Teachers in Science.

PASCHAL N. STRONG

Paschal N. Strong recently retired from the psychology department at the University of South Florida where he was a full professor for 30 years. His areas of specialities include neuropsychology, and comparative psychology. He has thirty publications, five book chapters, and several grants. He has worked in the areas of hippocampus functions, Alzheimer's disease, and comparative intelligence of animal species, including primates up to chimpanzees. He has also been a part of the laboratories involved in training chimpanzees for the National Aeronautics and Space Administration.

KAREN SULLENGER

Karen Sullenger is a science educator and member of the Curriculum and Instruction Department of the University of New Brunswick (Fredericton, N.B., Canada). Her professional and research interests include the natures of science, teacher/self-directed change, cooperative learning, and the role of writing in teaching and learning science. She holds a Ph.D. in science education from the University of Georgia.

STEVE TURNER

Steve Turner holds a Ph.D. in the history of science from Princeton University and has taught in the History Department of the University of New Brunswick (Fredericton, N.B., Canada) since 1971. His research and publications deal with science in the German university system of the nineteenth century, with the career of Hermann von Helmholtz, and with contemporary techno-scientific controversies over agricultural biotechnology. He is the author of *In the Eye's Mind: Vision and the Helmholtz-Hering Controversy* (Princeton University Press, 1994).

JERRY WELLINGTON

Jerry Wellington is reader in education and chair of the Educational Research Centre in the Division of Education, University of Sheffield, Sheffield, UK. He has published extensively in science education. His most recent book is *Secondary Science: Contemporary Issues and Practical Approaches* (Routledge, 1994)

PETER VOOGT

Peter Voogt is Emeritus Professor of Biology Education. Research within the Department focuses on biology education at the secondary level with special emphasis on the themes of stability and change in living systems.

PHILIP WADE

Philip Wade is interested in using an earth systems approach as a means to increase nonscience major understanding of the world and the scientific enterprise. His research work addresses nonscience majors' conceptions of science. He has developed new undergraduate baccalaureate core courses for the Geoscience Department at Oregon State University and recently completed his first book, Investigating the Earth. Currently he is completing a doctoral degree in Science Education from the Department of Science and Mathematics Education at Oregon State University.

Index

abduction 61
al-Haytham, Ibn 183f
attitudes 15, 22, 28, 192, 198, 222f, 232f, 235, 237, 246, 315, 331ff, 338
Aristotle 75, 184ff, 190f
beliefs, 14, 18, 21, 23, 34, 36, 38f, 79, 130, 140, 233, 247, 251ff, 277, 284, 287, 290, 347f, 352
biology instruction 13, 22, 34, 36f, 146ff, 155, 160, 163ff, 175, 207, 225, 229, 146ff, 243, 255, 325\6, 328, 348
biology 8, 13, 14, 34, 36, 126, 151, 155, 157, 159f, 162, 193, 225, 267, 271f, 274, 276, 281, 284, 319, 322, 348, 351, 354ff
black box 108ff, 148, 151, 217, 317
Brown, Robert 171, 306, 307
case studies xvii, xix, 34, 38, 77, 149, 165ff, 179, 193, 196, 202, 255, 259, 263, 283, 289, 302f
chemistry instruction xix, 7, 18f, 36, 149, 201, 203, 206, 229, 243, 268, 287, 300, 348
cognitive conflict 201
community 4, 11, 14f 26, 33, 51, 65, 68, 75, 100, 118, 127, 147, 164, 169, 170, 173f, 211f, 234f, 263, 268, 269f, 273, 287, 342
concept mapping 268, 270ff, 317, 327
Conant, J. xviii, 7, 178, 193, 221, 337
Conception of Scientific Theories Test 333
conceptions, teachers' 18ff, 23f, 39, 126, 136, 139, 148, 169, 207, 244, 291, 331f, 343, 347ff

conceptions, students' xx, 11, 23, 37, 39, 52f, 83, 108f, 126ff, 139, 148, 169, 201, 203f, 207, 243, 250f, 331, 344, 347ff
conceptual approach 225
conceptual change 14, 78, 135, 203, 205, 215, 216, 219, 224, 229, 278, 280, 356
confirmation xvii, 43, 55, 164, 169, 255f, 259, 276, 279, 281
constructivism xv, xxf, 34, 127, 149, 207, 224, 226, 228, 277, 281, 353, 355
cookbook science 22, 54, 55, 223, 281
cooperative controversy 7, 127ff, 131ff
cooperative learning 136, 268, 270, 358
cooperating teacher 197, 203ff
creativity 7, 23, 34, 47, 50, 53 58ff, 83ff, 91, 94ff, 109, 118, 135, 161, 172, 175, 192, 200, 207, 214, 219, 22, 231, 235, 238ff, 273, 306, 339, 342f
creation "science" 14, 61f
creationist 13, 35, 61, 286f, 316, 319, 321f, 324, 326
credential, teaching 223, 225, 227f,
critical incidents 29, 37, 297f, 300, 302, 308f, 343
critical thinking 127, 192, 198, 224, 228, 325
culture xi, xvi, xix, 7, 10f, 13, 46f, 51, 76ff, 81, 83f, 97, 137f, 142, 145ff, 192, 194, 222, 244, 247, 252, 256, 260f, 265, 275, 278, 284, 294f, 303f, 312f, 317f, 331, 352, 358f
curiosity 110, 187, 271f, 274f, 336
deduction 58f, 213, 219, 268

demonstration 18, 27, 29, 110ff, 180f, 184f, 295, 297, 306, 320, 343
discrepant 66, 110, 129
diversity 76ff, 222, 245, 285, 322, 356
Driver, R. xv, 11, 13, 126ff, 252
elementary teaching 17, 26, 74, 85, 91, 95, 114f, 118, 128, 133, 135, 223ff, 238ff, 243f, 249, 267, 277, 280, 282, 290, 347,
empirical 8, 10, 13f, 16f, 20f, 26, 38, 44, 49f, 54, 58, 76, 80, 83, 85, 91, 109, 113, 118, 154, 180, 206, 214, 216, 231, 257, 284f, 312, 318f, 331, 338f, 346
enquiry (see inquiry)
epistemic 255, 256, 257, 258, 260, 263
epistemology (epistemological) 10f, 13, 15, 16, 19, 21f, 34, 37f, 78, 140, 208, 214, 224, 268, 331
ethics 42, 338
ethnocentrism 244
Euclid xviii, 183, 184, 320, 321
evidence xvi, 3, 6, 7, 10, 11, 13, 19, 21, 23, 27, 41, 43ff, 49ff, 54, 56, 58ff, 76, 80, 85ff, 91, 96f, 100, 103, 108, 111ff, 118ff, 147, 164ff, 170ff, 202, 206, 216, 235, 237, 239, 262, 269, 279, 285, 288f, 294, 304, 317f, 321, 324, 327, 338
evolution 11ff, 34, 37, 38, 54, 97, 145, 147, 166, 169, 193ff, 208, 216f, 221, 227ff, 263, 267, 272, 274, 285f, 315ff, 319, 320, 323ff, 328, 340, 348, 351, 356f,
experiment xvi, 3, 6f, 11, 17f, 21f, 43, 50, 64f, 76, 80, 108f, 112, 118, 166, 170, 172, 177ff, 187ff, 196, 201f, 216, 221ff, 225, 227ff, 232, 234ff, 252f, 256, 258, 260, 264, 266, 275f, 281, 283, 295, 298ff, 304ff, 310, 312, 316, 342, 344
expectation 17, 49, 50, 63, 65, 66, 73, 79, 100, 101, 108, 293
explanation xvi, 8, 26, 54, 67, 78, 99f, 137ff, 145, 148, 170, 200, 245ff, 258f, 262, 295f, 304f, 315, 318, 322, 326
exploration 21, 64, 77, 135, 233, 239, 253, 274f, 302
extinction 99, 142, 255ff, 261ff, 328
extraordinary science 263
Facts About Science Test 333f, 349
falsification 61, 256, 259, 285
feminism xiv, xviii, xixf, 11, 97, 126, 253, 261, 264
Feyerabend, P. 138f, 149, 213
fossil 95ff, 126, 194, 196, 212, 258, 317, 321ff
gender 75, 143, 247f, 276, 280, 317, 354
Giere, R. 21, 138f, 147ff, 287, 291
Glymour, C. 138, 149
graduate students 27f, 30, 74f, 137, 147f, 232
goals for science instruction 68, 197
hands-on instruction (see inquiry)
Harding, S. 15, 20, 76, 79, 253
Hempel, C. 138, 149, 213, 229, 259
heuristic (design) 148, 151, 152, 154, 155, 157, 158, 159, 160, 257, 263, 354
history of science xvi, xxi, 7f,, 10, 14, 25, 34, 36ff, 43f, 49, 51, 63f, 75,77, 99, 168, 177, 179 191ff, 217, 244f, 259, 270, 289, 304, 320f, 354, 359
historical 7, 34, 177ff, 191, 193
historicism 256, 259
Holton, G. 138f, 149, 196, 214
hypothesis 9, 16, 18, 19, 46, 54ff, 69, 98, 111ff, 153f, 157, 174, 181f, 188ff, 222, 225ff, 230, 234ff, 238f, 265, 305f, 323, 344
imagination 35, 58, 83, 85, 91, 94f, 97, 109, 118, 149, 193, 196, 200, 216, 222, 269, 310, 339, 342
induction 35, 58ff, 63, 74f, 77f, 83, 85, 91, 94f, 97, 109, 118, 149, 193, 196, 200, 213, 216, 219, 222, 256, 259,

269f, 310, 312, 339, 342
inference 33, 46, 79, 84ff, 91, 95ff, 108ff, 118, 211, 213, 271, 310, 342
inquiry xiv, xviii, xxi, 5, 8, 11, 14, 34f, 37f, 62, 70, 80, 83, 134, 144, 150, 162, 168,179, 193, 208, 214, 222ff, 231ff, 237ff, 249, 259, 268f, 273, 276, 312, 316ff, 333f, 339, 348, 350
inservice 27, 29, 35, 84, 163, 223f, 227f, 230, 267, 293, 315, 343
interpretation 15, 22, 51, 60, 65, 96, 99f, 130f, 139, 180, 202, 214f, 222, 234, 268ff, 294, 296, 298, 306f, 317, 321, 332, 339, 343f, 346
instructional outcomes 15f, 279
instrumentalism 16, 21, 36, 215, 219, 313
Inventory of Science Attitudes, Interests and Appreciations 333f
investigation (see experiment)
Koertge, N. 15, 21
Kuhn, T. 14, 21, 26, 36, 63, 69, 126, 138, 149, 153, 162, 169, 193, 195, 208, 229, 259, 274, 348
laboratory instruction 18, 22, 27, 56, 60, 63, 177, 200, 202, 207, 230, 343
Lakatos, I. 60, 138, 149, 154, 162, 257, 259, 263
Laudan, L. 51, 138, 149, 257, 259f
law 6f, 10, 12, 20, 46, 51, 54ff,, 69f, 83, 85, 91, 109, 126, 130, 139, 164, 168f, 172f, 179f, 182f, 200, 208, 211ff, 217ff, 231, 239, 265f, 275, 279, 281, 285, 287, 312, 335, 339, 342
learning cycle 232ff, 237
learning game 73ff,
literacy; science 42, 48, 224
Locke, John 18, 19, 21
logic 17, 32, 45, 49f, 57, 67, 69, 80, 98, 135, 137, 146, 149, 164, 180, 208, 213, 216f, 251, 259, 269, 291, 312, 320f

logical positivist 224
measurement 3, 57, 180, 256, 258, 263, 309, 317, 334, 338, 347, 349
Medawar, P. 58, 69, 141, 202, 207f, 214, 307
metaphysics xvi, 62, 140, 246
methods instruction 27, 29f, 74, 84, 128, 223ff, 231f, 239, 241, 251, 267, 280ff, 302
Mill, J. S. 259
misconceptions xxi, 9, 11, 14, 38f, 53f, 68, 128, 144f, 149, 164, 167f, 197, 200, 205, 322, 346
Modified Nature of Scientific Knowledge Scale 240, 343
multicultural 11, 79, 228, 229, 306, 352, 358
nature of science defined 4
Nature of Science Scale 25, 333, 337
Nature of Science Survey 333, 342
Nature of Science Test 27, 333, 338
Nature of Scientific Knowledge Scale 18, 23, 25, 240, 333, 338, 343, 349
normal science 17, 26, 224, 259, 344
objective 13, 49, 62f, 98f, 164, 224, 232, 235, 237, 259, 265
observation xvi, 17f, 36, 43, 45ff, 50, 57f, 62ff, 69, 75, 79f, 82, 84ff, 91, 98ff, 108ff, 118f, 130, 141, 150, 167, 171ff, 182, 185, 190, 193, 202, 204, 215, 219, 222f, 232, 235, 238, 256ff, 264ff, 298, 310, 320, 339f, 345ff
paradigm 15, 26, 36, 47, 49, 63ff, 99f, 103, 149, 162, 166, 212, 215, 219, 221, 263, 270, 272f, 325, 328, 344, 348
patterns 23, 54, 109, 118ff, 128, 146, 259, 264, 268f, 272, 277, 312, 316, 318f, 321 323, 326
physics instruction xii, xix, 144, 147, 168, 177, 180, 193, 196, 243, 248, 268, 280, 290, 320

plate tectonics 218, 259, 250, 263, 320, 340
Popper, K. xiii, 49, 52, 60ff, 69, 138, 149f, 154, 162, 221f, 259f, 344
Portland Baseline Essay 14f, 21
positivism 227, 312
positivist 10, 14, 16, 18, 214, 224, 278, 280, 290, 296, 312
postmodern xiii, xv, 17, 19, 140
pragmatism 256, 259
preconceptions (see conceptions)
preservice xvi, xx, 10f, 18, 24, 27ff, 34f, 37, 39, 53, 73f, 78, 84, 197, 119ff, 206, 223ff, 227, 230f, 233, 243ff, 248, 250f, 267f, 315, 326, 343
prior knowledge 58, 63, 96, 101, 111, 156
problem solving 16, 154, 156 224
process science 33, 73, 224ff, 229, 332
Processes of Science Test 333f, 348
pseudoscience 14, 21, 139, 271f, 275f
rational reconstruction 256, 259
reductionism 13, 247
reflection (reflective thought) 16, 37, 129, 152, 168, 184, 202f, 205, 207f, 231ff, 237, 246, 248, 250, 252, 268f, 272, 299, 307, 309, 336
reflection-on-action 205f
relativism xvi, 308, 312
reliability 128, 169, 240, 260, 334, 336f, 339f, 343
research design 24, 223, 331
role playing 179
Science Attitude Inventory 333, 334
Science Attitude Questionnaire 333f
Science Attitude Scale 333, 334
science, characteristics of 6f, 271f
science, pure 67, 208, 260f, 264
science, applied 67, 200, 260f, 264
Science & Education 12, 16, 20ff, 35, 37, 39
Science Inventory 333f
Science Process Inventory 22, 333, 336, 341, 348, 349
Science Support Scale 333f
science, technology and society (STS) 33, 207, 277, 280, 290, 333, 341, 347
science wars xiii
scientific method(s) xviii, xxi, 6f, 10, 26, 53, 57f, 69, 81, 83, 164, 169, 178, 190f, 200, 202, 213f, 219, 230, 237ff, 246, 255, 276, 281, 291, 310, 325f, 334, 337
scientific theory profile 20, 37, 79, 137ff, 147ff
scientism 73, 76, 78, 81
secondary level teaching environment 243
Shapere, D. 138, 150
skepticism 6, 45, 50, 68, 168, 170, 202, 275f
social construction xvi, 53, 127f, 216, 269, 273, 341
sociological xv, xxf, 4f, 32, 37f, 41, 44, 46, 49f, 136, 142, 147, 149, 211ff, 243f, 271f, 278, 290, 293, 307, 313, 357
standards, science teaching xii, 6, 41ff, 166, 211, 216, 219, 227, 277
STS (see science, technology and society)
student teaching 28, 198f, 205f, 225, 228, 232
student teachers 28, 35, 197ff, 228, 291, 293, 302
subjective 23, 84, 97, 99, 100, 170, 253, 259, 265, 324, 331
Test of Enquiry Skills 333f, 348
Test of Science-Related Attitudes 333f
Test on the Social Aspects of Science, 333f
Test on Understanding Science 24, 38, 333f, 348
technology xix, xx, xxi, 3, 7, 10, 13, 33, 39, 47f, 51, 62, 67, 94, 126, 142, 160, 168, 171, 177, 193f, 200, 207,

228, 246ff, 252, 256, 260f, 264, 267ff, 274f, 277ff, 282f, 287, 290f, 333f, 341, 347
tentative 6, 12, 42f, 55f, 60, 83ff, 91, 97, 109, 112, 118, 137, 164f, 170, 174f, 188, 190, 224, 228, 231, 237, 285, 319, 322, 331, 337f, 340ff, 348f
testable 23, 45, 79f, 82, 231, 235, 239f, 285, 339, 343
textbooks 4, 9, 20ff, 35f, 53, 68f, 73, 85, 165, 178, 203, 205, 207, 212, 218, 243
theory xvii, xx, 3ff, 12, 14, 17f, 35, 37, 43, 46f, 51f, 54, 62f, 67, 69, 54ff,
theory, view of 151
theory development 151f, 154, 156ff
theory-laden observation 7, 47, 63, 219, 258, 298, 312, 340
Toulmin, S. 138, 150, 194, 195
trial and error 154, 232f, 237, 275
unified science 349
values xv, 9, 10, 14, 16, 20, 79f, 159, 164, 168, 222, 232f, 235, 239, 246, 255f, 260ff, 279, 283ff, 2990f, 299, 305, 308, 331f, 336, 348
variable 15, 18ff, 23ff, 37, 64, 181f, 187ff, 203, 234, 236, 238, 240, 280, 321 332
verification 16f, 22, 60, 172, 222f, 266, 317
von Glasersfeld, E. 224, 230
Wisconsin Inventory of Science Processes 336, 349

Science & Technology Education Library

Series editor: Ken Tobin, *University of Pennsylvania, Philadelphia, USA*

Publications
1. W.-M. Roth: *Authentic School Science.* Knowing and Learning in Open-Inquiry Science Laboratories. 1995   ISBN 0-7923-3088-9; Pb: 0-7923-3307-1
2. L.H. Parker, L.J. Rennie and B.J. Fraser (eds.): *Gender, Science and Mathematics.* Shortening the Shadow. 1996   ISBN 0-7923-3535-X; Pb: 0-7923-3582-1
3. W.-M. Roth: *Designing Communities.* 1997
     ISBN 0-7923-4703-X; Pb: 0-7923-4704-8
4. W.W. Cobern (ed.): *Socio-Cultural Perspectives on Science Education.* An International Dialogue. 1998   ISBN 0-7923-4987-3; Pb: 0-7923-4988-1
5. W.F. McComas (ed.): *The Nature of Science in Science Education.* Rationales and Strategies. 1998   ISBN 0-7923-5080-4

KLUWER ACADEMIC PUBLISHERS – DORDRECHT / BOSTON / LONDON